SMART URBAN REGENERATION

The role of real estate in our cities is crucial to building sustainable and resilient urban futures. *Smart Urban Regeneration* brings together institutional, planning and real estate insights into an innovative regeneration framework for academics, students and property professionals.

Starting by identifying key urban issues within the historical urban and planning backdrop, the book goes on to explore future visions, the role of institutions and key mechanisms for smart urban regeneration. Throughout the book, international case studies and discussion questions help to draw out global implications for urban stakeholders.

Real estate professionals face a real challenge to build visionary developments which resonate locally yet mitigate climate change and curb sprawl, and foster biodiversity. By avoiding the dangers of speculative excess on one side and complacency on the other, *Smart Urban Regeneration* shows how transformation aspirations can be achieved sustainably. Academics, students and professionals who are involved in real estate, urban planning, property investment, community development and sustainability will find this book an essential guide to smart urban regeneration investment.

Simon Huston is a Senior Lecturer at the Royal Agricultural University, UK.

SMART URBAN REGENERATION

Visions, Institutions and Mechanisms for Real Estate

Edited by Simon Huston

Routledge
Taylor & Francis Group
LONDON AND NEW YORK

First published 2018
by Routledge
2 Park Square, Milton Park, Abingdon, Oxon OX14 4RN

and by Routledge
711 Third Avenue, New York, NY 10017

Routledge is an imprint of the Taylor & Francis Group, an informa business

British Library Cataloguing-in-Publication Data
A catalogue record for this book is available from the British Library

Library of Congress Cataloging-in-Publication Data
Names: Huston, Simon, editor.
Title: Smart urban regeneration : visions, institutions and mechanisms for real estate / edited by Simon Huston.
Description: Abingdon, Oxon ; New York, NY : Routledge, 2017. | Includes bibliographical references and index.
Identifiers: LCCN 2017007414| ISBN 9781138935266 (hardback : alk. paper) | ISBN 9781138935280 (pbk. : alk. paper) | ISBN 9781315677521 (ebook : alk. paper)
Subjects: LCSH: City planning. | Sustainable urban development. | Urban renewal. | Real estate development.
Classification: LCC HT166 .S587779 2017 | DDC 307.1/216—dc23
LC record available at https://lccn.loc.gov/2017007414

ISBN: 978-1-138-93526-6 (hbk)
ISBN: 978-1-138-93528-0 (pbk)
ISBN: 978-1-315-67752-1 (ebk)

Typeset in Bembo and Stone Sans
by Florence Production Ltd, Stoodleigh, Devon

CONTENTS

CONTRIBUTORS

Matthew Axe BSc (Hons), PhD, MRAU is an agronomist and sustainability researcher. Matthew's chief interests lie in the sustainable management of land, predominantly agroforestry and linear land features such as hedgerows, roadside verges and canal banks. His research focus is enhancing carbon stocks of vegetation and soils to mitigate the rising levels of greenhouse gasses. Previously, Matthew worked in the public sector as an environmental and energy management adviser.

Richard Baines BSc (Hons), MSc, PhD is an independent consultant and senior academic at the Royal Agricultural University, UK. Research and consultancy interests embrace two converging themes: agricultural and rural development with special reference to emerging economies; and the management of food safety and quality along food chains in response to legislation, the food chain and consumers. Research and contract research carried out for governments, industry actors and NGOs has included: local–global studies of farm, food industry and retail stages in food supply; compliance of government policies with international (CODEX) guidance; environmental impacts along food chains; business models to link small farmers to value chains; and the development of private standards in relation to legislation, risk, trade, ethical issues and systems management.

Jim Berry BA, MSc, PhD, PGDPropInv, PGCPropLaw, FRICS MRTPI is Professor of Global Real Estate and Planning Strategy in the Built Environment Research Institute in the Faculty of Computing, Engineering and the Built Environment at the University of Ulster. Jim's work on international property markets, urban regeneration, project finance, infrastructure investment and public–private partnerships has attracted significant external research funding. He sits on the editorial boards of the *Journal of Property Research*, *Journal of European Real Estate Research*, *Journal of Property Investment and Finance* and *Journal of Place*

Management & Development. He holds the post of Executive Director of the European Real Estate Society; and is a fellow of the RICS and a member of the RTPI.

Jonas Hahn BA, MSc is a doctoral candidate at the International Real Estate Business School (IRE|BS) of the University of Regensburg, Germany. Jonas contributes to the EU-funded Horizon 2020 project 'RentalCal', which develops specific methodology and tools for evaluating the financial profitability of retrofit measures in rental housing. He holds an MSc in Real Estate from the University of Regensburg and a BA in Business Administration from University Erlangen-Nuremberg. He has ten years of professional experience in commercial functions, corporate finance and internal controls as well as real estate letting and transactions.

Simon Huston BSc (Econ), PGCE, MSc, PhD, ACMA CGMA is manager for the MSc Real Estate programme at the Royal Agricultural University in Cirencester, UK. Simon's research interests include land use contention, strategic development, urban design, project management and finance. During his 20-year academic career, Simon has taught a range of business, economics, and accounting and property modules in the Middle East, Australia and the UK. Previously, Simon worked as a military medic, auditor at Deloitte in the UK and commercial analyst for the Queensland Government. In 2009, Simon qualified as a CPA (Australia). In Commercial Advisory, Simon worked in due diligence and foreign investment advisory, and modelled the commercial viability and economic impact of major mining, housing and infrastructure projects.

Arvydas Jadevicius BSc, MSc, PhD is a real estate market analyst with extensive expertise in market research intelligence, data collection and investment analysis. In 2014, Arvydas gained a Doctorate for his work on UK commercial property forecasting accuracy and its improvement through combination forecasting. His research was awarded with the Construction Research and Innovation (CRI) prize (ARCOM, 2010), the Best Paper Award on Real Estate Market Analysis at the American Real Estate Society Annual conference (ARES, 2014) and Emerald Literati Network Awards for Excellence 2016. Prior to his transition to industry, Arvydas taught real estate at the Royal Agricultural University in the UK.

Anil Kashyap PhD, MRICS, MRTPI, FHEA is Head of the Department of Geography, Planning and Environmental Management at the University of the West of England. Anil has a wealth of national and international experience in academia, research and the public sector. He is Distinguished Professor at the School of Real Estate at the RICS School of Built Environment, India. He holds a bachelor degree in civil engineering, masters in urban planning and a PhD from the University of Ulster, UK. Dr Kashyap has strong research interests spanning energy efficiency in built form, healthy and smart cities, urban regeneration and infrastructure development and financing. He is a member of the International Land Measurement Standard Setting Committee (ILMS – SSC).

Marek Kozlowski MA Architecture, BRTP PhD is Senior Lecturer at the Department of Architecture, Faculty of Architecture and Design at the Universiti Putra Malaysia. Formerly he worked at the Gold Coast City Council designing the new metro system and at the University of Queensland in Brisbane, Australia. Marek has an interest in urban design, urban revitalisation and sustainable tourism. He has published many papers on the topic, including for the *International Journal on Sustainable Tropical Design Research and Practice*.

Ebraheim Ali Lahbash BA, MBA is a PhD candidate in the Centre for Real Estate at the Royal Agricultural University, UK. He manages a property development company in Fujairah, United Arab Emirates. Ebraheim developed strong relationships with major UAE developers and with the Dubai Land Department. Frustrated by significant valuation issues, he investigated possible solutions. The problem is now the focus of his thesis into the residential valuation system in the UAE. Information systems, professional standards and trust are central. Ebraheim has written several monographs on the topic and presented at several European Real Estate Society conferences.

Puying Li BA, MA is a PhD candidate in the Centre for Real Estate at the Royal Agricultural University, UK. In 2012, Puying finished her BA in Civil Engineering Management at Zhengzhou University of Aeronautics in China and joined the RAU to complete her MA in Business Management. She completed an internship at the Baise Construction Committee of Housing and Urban Planning in 2009 and at Jiantong Construction Ltd in 2012. From the knowledge distilled from these diverse academic and professional experiences, Puying identified a pressing need for further construction industry research. Her thesis investigates firm level competitiveness in the global construction market.

Will Manley BSc, M.Phil, C.Ecol. C.Env. FIEEM is a Principal Lecturer in the Centre for Rural Land Management at the Royal Agricultural University, UK. He is an environmental scientist with specialist research and project development centred on an applied background in agri-ecosystems, UK and European countryside policy, biodiversity and rural recreational activities. His work has placed particular emphasis on the integration of these areas with commercial land management and rural development. Projects for government and institutional agencies have included a range of studies on the environmental impacts of rural policies and the measures, and the economic implications of conservation measures.

Onur Mengi BSc, MSc, PhD is Assistant Professor in the Department of Industrial Design in Izmir University of Economics, Izmir, Turkey. He is also coordinator of the Design Management Masters Program in the Graduate School. His primary research areas are creative and cultural industries, strategic management in creative industries and ecology of industry clusters. Onur carries out research projects on the specific topics of urbanism, urban theories, and design projects including urban design, product design and design management.

Negin Minaei holds a PhD degree in Urbanism, an MSc degree in Architectural Engineering and an MSc in Environmental Psychology. She started lecturing architecture and urban courses as a full-time academic member at the IAUK in 2000 and has been involved with architectural supervision and consulting projects since 2002. She received a scholarship from Bauhaus in Germany in 2004 and attended the research programme of 'Transnational Spaces'. As a postdoctoral and an associate researcher, she worked on sustainable urbanism and on national projects in the UK. She has been lecturing in different universities across the world since 2012. She won the UK national Social Enterprise and Innovation Award in 2013 for her idea which led to establishing a social enterprise called 'EnDesign, Social Environmental Design'. She teaches 'Sustainable Smart Cities' at the Faculty of Engineering at the University of Windsor, Ontario, Canada.

Yusnani Mohd Yusof PhD is a Programme Leader and Lecturer in the Department of Geography and Development/Environmental Studies at the Universiti Brunei Darussalam in Brunei. She is a member of the Malaysian Institute of Planners. Formerly, Yusnani taught planning and development in the School of Geography, Planning and Environmental Management at the University of Queensland in Brisbane, Australia. Yusnani is interested in sustainable and culturally appropriate urban development and regeneration in Malaysia where important considerations include preserving face, a desire for harmony and compliance. Yusnani has written monographs on the role of urban planning for climate change and the accommodation of the elderly in Malaysia.

Andreas Oberheitmann, Diploma in Economics, PhD (Guanlixue Boshi, Tsinghua University) is an economist, Sinologist and Professor for Business Administration and International Management at the German-Sino School for Business and Technology of FOM University of Applied Science and Visiting Scholar to the RWI – Leibniz Institute for Economic Research. Since 2012, Andreas has been a member of the Editorial Board of the *American Journal of Climate Change* and was Expert Reviewer of the contribution of Working Group III (WG III) to the Fifth Assessment Report of the IPCC (AR5) until 2014. Andreas's interests include evidence-based climate change policy (energy efficiency, renewable energies, post-Kyoto issues, low carbon economy) and other fields of environmental policy in China, other East Asian countries as well as in Germany and the EU.

Ali Parsa M.Phil, PhD, FRICS is Professor and Head of the Centre for Real Estate at the Royal Agricultural University in Cirencester, UK. Formerly, Ali was Professor in Real Estate at the University of Salford and prior to that held the Rotary International Chair in Urban Planning and Business Development at the University of Ulster. Ali currently leads a number of research projects, including Economic Impact of RAU on Cirencester and Gloucestershire, Impact of Superfast Broadband on Rural Economy in Gloucestershire and Hertfordshire. Ali is a frequent speaker at national and international conferences and has published more than 150 research articles. His recent research has focused on international real estate markets, real estate investment, comparative planning and urban management.

Sophia Price MA (Cantab), MCIEEM is the Heritage and Design Manager for Cotswold District Council, based in Cirencester, UK. Sophia is responsible for landscape, biodiversity, historic environment, arboriculture and urban design. Sophia was one of the first ecologists to be appointed by a local authority and was a founding member of the Association of Local Authority Ecologists. Her main interests include the environmental aspects of Development Management and Planning Policy, and particularly the interrelationship between environmental, heritage and social issues.

Reyhaneh Rahimzad BA, MA is a doctoral candidate in the Centre for Real Estate at the Royal Agricultural University, UK. Reyhaneh has contributed to a number of EU-funded projects such as the MEGAPROJECT COST and the CLUDS. Reyhaneh holds an MA in Architecture from the University of Liverpool and a BA in Interior Design from Liverpool John Moores University. Her doctoral research aims to develop and validate a new screening tool (model) for the evaluation of mega urban regeneration projects. Reyhaneh has written several papers, including for *Cities: The International Journal of Urban Policy and Planning* and for the European Real Estate Society conferences.

Savaş Zafer Şahin PhD is Associate Professor at Atılım University in Ankara. Zafer worked as a civil servant in different tiers of the Turkish State, including municipalities, governorships and ministries. Working for the Turkish Chambers of Engineers, Zafir raised urban planning issues with architects and city planners. In 2009, he joined Atılım University where, in 2014, he became a professor. Zafer's interests span urban planning, political sociology, corruption, local governments and ethics in public administration. He has written papers and book chapters on Urban Planning and Turkish Public Administration. Using his awareness of important international and national planning cases, Zafer helped draft important Turkish national policy documents and development plans.

Nisa Semiz BSc, MSc, PhD is Assistant Professor in the Faculty of Architecture in Istanbul Kultur University. She studied architecture in Istanbul Technical University where she completed a master's degree on architectural conservation and restoration and proceeded to PhD study in the same field. Nisa's dissertation is on the preservation of Istanbul's Sea Walls. Nisa has participated in archaeological surveys and excavations in Hasankeyf, Sagalassos, Dara, Binbirkilise and Larisa in Turkey, and she participated in the restoration works at Zeyrek Mosque (the Pantokrator Church) and Çemberlitaş (the Column of Constantine) in Istanbul.

Peter Smith BSc (Hons), Dip Surv CEM, PGCHE, FHEA, MRICS is programme manager for the BSc (Hons) Real Estate degree at the Royal Agricultural

University, Cirencester, UK. Academic interests include landlord and tenant law, commercial property management and valuation for local taxation. He also runs the final year Integrating Project for the undergraduate programme. He has delivered short APC and CPD courses for surveyors, and delivered training courses and provided advice on real estate course design in a number of Eastern European countries. Prior to entering academia, Peter worked as a chartered surveyor in professional practice and with the Valuation Office Agency in both the residential and commercial property sectors.

Koray Velibeyoğlu PhD is Associate Professor at the Department of Urban and Regional Planning, Izmir Institute of Technology, Turkey. Koray is also director of the Center for Lifelong Learning. His research spans urban design, planning history, knowledge management, local asset-based development, urban ICT policy-making and knowledge-based development processes. Dr Velibeyoğlu is project manager of Izmir Peninsula Sustainable Local Development Project.

URBAN CONUNDRUMS

Simon Huston

Abstract

Smart urban regeneration is not a destination but a mind-set. It is participatory and responsive with economic, social, environmental and governance dimensions and invokes measured interventions in collaborative institutional networks. Its goal is a 'smart city' but this notion remains elusive with multiple, yet unsettled, blueprints. Regulated technological ones contrast with evolving multicultural melting pots. The gestation and propagation of human settlements invokes wider development conundrums. Over the past millennia, anthropogenic activities like farming, manufacturing and settlement construction have transformed Earth at an accelerating pace. Already, remorseless human pressures have significantly depleted biodiversity and, unless tempered, could destabilise the climate. In increasingly populous conurbations, the burden of maladministration and market failure manifests in housing stress, obesity, congestion and air pollution. Philanthropic governments, seeking to shape built environments, confront political wrangling, institutional challenges and resource constraints. Politically, neo and social liberals dispute the level, format and vehicles for intervention to tackle symptoms of urban malaise without compromising freedom or innovation. Smart development, without resolving the conundrum, blends astute processual strategy with evolution in regulated markets. Its realisation involves practical wisdom and partnerships between robust political, commercial and social institutions to select, administer and deliver substantial low-carbon investments or tackle multiple issues at different spatial scales in complex fragmented societies. To conserve or appropriately transform built environments needs ambition, expertise, deliberation, adaptable systems, trust and standards, nurtured by sound policy. As well as the vertical political system with its administrative tiers, the network encompasses horizontal commercial, non-government institutions and bottom-up feedback from local bodies or citizens, empowered by digital technology and land information.

Introduction

In the last hundred years, human manipulation of resources has transformed our planet beyond recognition. In the twentieth century, we used ten times more energy than our ancestors did in previous millennia (McNeill 2000). Whilst sustainable development grapples with humanity's multiple planetary impacts, Smart Urban Regeneration (SUR) confines itself to the management of its built environments. A treatise on it though confronts many complications, not least of which is ambiguity (Hollands 2008). The imprecise term has technological or progressive social connotations and vies with a host of similar multi-criteria city labels such as 'resilient', 'green', 'liveable', 'garden', 'inclusive', 'competitive', 'intelligent' or 'sustainable' – although Forman (2014: 29) consigns 'urban sustainability' to the oxymoron dustbin.

Second, philosophy and political contention explains the spectrum of technical or ecological 'smart' Eldorados on offer but political consensus remains elusive. Liberals concede some state intervention to counteract economic predation but spurn the utopian blueprints social engineers use to remodel society (Popper 1966 [1945]). Social planners as opposed to utopian ones tinker piecemeal or muddle through (*ibid.*; Lindblom 1959).

Third, a wide range of human settlement typologies exist (Benevolo 1980), in terms of size, function, morphology, demographics, social structure, history, culture, ecologies, institutions, administrative boundaries and geographical constraints. Institutional heterogeneity and complex dynamics limit smart urban development expositions to broad generalisations or confine them to specific geographical and temporal contexts.

Finally, a complex interplay of factors at various spatial scales drives or hinders urban development. Invariably, there are detractors to any development trajectory or questions concerning the distributive justice of its impacts. In the case of Dubai, for example, some laud its quixotic leadership, cosmopolitanism and regional status as a financial safe haven but Davis (2006: 53) condemns it as a 'monstrous caricature of futurism'.

Notwithstanding competing ideologies, smart interventions encompass environmental, social and economic concerns and span fiscal policy or regulatory instruments as well as low-carbon infrastructure, precinct regeneration or housing projects. In settled constitutions, market realism, governance and the rule of law curtail the excess, corruption or nepotism often tainting grandiose schemes.

Visions

Political philosophy

Notwithstanding debate since antiquity (Plato and Aristotle), the state's role in shaping a *polis* remains politically contentious. Prosperous cities emerge and propagate in propitious regional hinterlands, by administrative decree, religious impulse or, like Akrotiri, along trading routes. Strong, predatory states stifle regional expansion (Bosker *et al.* 2008). For Smith (1979), even in ostensibly laissez-faire regimes, the state surreptitiously facilitates suboptimal development to enrich sectorial interests. Philosophical contention aside, political flux and media distractions undermine attempts to articulate 'the public good' or an intrinsic settlement purpose. Presumably, all citizens in a *polis* should access basic nutritional, health, housing and educational needs so they can actually realise their human capabilities (Nussbaum 2000; Sen 2009). Beyond basic needs, *dirigisme*'s ultimate urban purpose and, therefore, instrumental direction, flounders. Candidates include hedonic or eudaimonic well-being, intelligence, progressive or entrepreneurial directions (Ryan and Deci 2001; Hollands 2008; Wadley 2010). Whatever its vision, a *dirigiste* smart development regime postulates a matrix of targets (teleology). Historical utopian precedents guide formulation but alert us to challenges (Campanella 2006 [1899; 1623]). Evolutionary, liberal narratives dismiss autocratically imposed social purpose and stress market freedom. For neo-liberals, Arcadian-inspired tweaking (notwithstanding deliberation, foresight, scrutiny, oversight or checks and balances) invariably backfires and, ultimately, leads to tyranny – presumably worse than the modern corporatist Panopticon. Neo-liberal sceptics question government plans and resent regulations to curtail, regulate or stimulate dwelling investment. Smith (1759: IV.1.6, 10) and Veblen (1994 [1899]) lamented the wasteful or conspicuous consumption factored into the price of freedom. For constitutional democrats, law and a strong civic culture circumscribe smart city political authority. Communities can freely dispute, contract or forgive in *agoras* or via other forums. Central is the adage, '*Fiat justitia, et pereat mundus*' (Let justice be done, though the world perish) (Arendt 2003: 52). In short, visions pit planning utopias against legal precedent, competitive capital markets and practical aspirations of local communities for self-determination.

Practicalities

Notwithstanding philosophical conundrum and political tensions, if the *polis* settles on *dirigisme*, practical evaluation (Lewis 1955) begins by cataloguing the instrumental features of a putative smart city. However, for Shelton *et al.* (2014: 1), smart city interventions are not 'constructed on *tabula rasa* according to the centralized plans of multinational technology corporations [but] are always the outcomes of, and awkwardly integrated into, existing social and spatial constellations of urban governance and the built environment'. Therefore, aside from its physical layout, a *dirigiste* smart city vision must articulate political and administrative frameworks as well as social goals and environmental thresholds.

Administration

Operationally, the smart city administrative system devolves to a network of private and public institutions at different spatial hierarchies. National policies and regulatory measures set parameters but local ones distribute environmental risks fairly. Whether at regional, district or local scale, the separation of powers between the executive, legislative and the judiciary keeps dictatorial impulses in check. Devolved systems block excessive plutocratic zeal or corporate malfeasance but facilitate appropriate (low-carbon, etc.) investments. Socially, settled constitutions and the rule of law at the national scale underpin local freedom and sound administrative praxis. Whilst excessive inequality is corrosive, fuels crime, undermines health and strangles talent (Giles-Corti *et al.* 2016; WHO 2010, 2014; GBD 2016; Stiglitz 2012; Piketty 2014; Wilkinson and Pickett 2010), punitive taxes or red tape to redress it can backfire and are equally pernicious. In tolerant and cosmopolitan smart cities, feedback and policy fine-tuning restores balance for shared prosperity. Stability and a wide tax base provide finance for transport networks and public realm enhancement.

Goals and constraints

In *dirigiste* development, technological, competitive, civic or ecological planning visions jostle for pre-eminence in diverse cultures and traditions (Pregill and Volkman 1999). Literature on Intelligent Urbanism or New Urbanism sketches alternate smart city visions (Doxiadis 1968; Benninger 2001; Jacobs 1993). Important contributors to the architectural and planning debate include Kevin Lynch, Lewis Mumford, Heinrich Tessenow, Léon Krier and Peter Calthorpe. Fractious future urban visions balance multiple and often contradictory considerations, including logistic infrastructure imperatives, social contracts and economic pressures. Experts, developers and locals often disagree on priorities. Important deliberations turn on built forms, local housing needs and local ecologies. In the smart city, aspirational or conservation tensions remain unresolved but, informed by science, planners restrict inappropriate growth or negotiate trade-offs. To internalise externalities and balance function with form, developments should consider aesthetics and reflect place character or local views. Multi-criteria evaluations stress anthropological, dwelling, community, logistic and ecological dimensions. Glitzy projects, gated communities, marketing hubris or technological wizardry are insufficient and can distract or divide. Fair, well-considered and legitimate plans help garner local support and facilitate project realisation. However, high taxes and burdensome conditions stifle housing supply and exacerbate housing stress. To implement research-informed policy involves integration of disparate regeneration practices (Leary and McCarthy 2013).

Urban form

Inspirational visions help shape markets and prevent 'default urbanism', characterised by nondescript, low-quality housing on infrastructure-depleted estates (Adams and Tiesdell 2013: 202). Instead of dreary banality, mean streets or sinister, 'pseudo-places' (Geddes 1915; Sorkin 1992), creative architects challenge conservative notions of beauty in an 'aesthetic–cultural battleground' (Arad and Jacob 2014: 1). The smart city is convivial and legible. It is aesthetically appealing, yet easy to navigate. It provides equitable access to work, housing and other essential utilities (water, power, education, health and social care) and leisure amenities (shopping, green space). As seemingly in Akrotiri, its free, tolerant and informed citizens dwell in a range of mostly affordable, yet energy-efficient, dwellings. Central are potent places with spiritual or civic meaning. In the genre are Pergamum's theatres, the baths of Caracalla, Mahabalipuram's Sea Temple or St Mark's Square in Venice. *Agoras*, piazzas or other quality public realms provide a convivial milieu where children can play safely and adults can reflect, argue or innovate. Well-designed public realms host a rich tapestry of ordinary social life but also communal rituals or spectacles (such as the *Palio* in Siena). Active streets, canals, green corridors, tramways, dedicated cycleways or pedestrian walkways connect squares to a broad mix of dwelling types and tenures. In the ideal city, children can conveniently access secure, clean and well-maintained open green spaces. Music, art and science flourish.

Evolution *or* dirigisme

Successful *dirigisme* is a directed strategy by state technocrats who promulgate indicative plans or deploy a suite of measures (e.g. credit policies, subsidies, technologies or regulation). Competition drives evolutionary adaptation or extinction, influenced by market, expert, public or legal pressures or geographical and environmental constraints. To adapt, the smart city mixes enlightened *dirigisme* and shapes markets. Judiciously selected (smart) interventions protect or enhance public realm, heritage or critical environmental assets or, where appropriate, catalyse or support private sector initiatives. Urban transformation aspirations include cohesion, green corridors, energy-efficient buildings, precinct connectivity, affordable and reasonably stable house prices, resilient utility networks and seamless public transport networks.

Institutions

Governance

Smart institutions encompass 'vertical' (socio-political), 'horizontal' (contracting) (Acemoglu and Johnson 2005) and grounded local institutions. For Harvey (1989: 6), 'urban governance means much more than government [since] a broader coalition of forces shapes urban life'. Therefore, for smart development the local government administrative system only plays a facilitative and coordinating role. Effectively, smart development relies on devolved administrative collaboration in a rich network of contracting (firms), social or local partnerships and feedback mechanisms. Political debate determines the mix of public or private finance and development delivery vehicles. Political choice turns on *de jure* or *de facto* sovereignty in various tiers of government and the social contract to mediate competing private or public interests. Finely balanced and context-sensitive, smart administrations consult and reflect before they conserve or appropriately transform places.

Strategy and leadership

Sound planning, unless conceived merely as a reflective forum, involves strategy or shaping the future (Freedman 2014). Smart development strategy must scan its environment to set or alter direction and adjust resources (project portfolios). However, as there is 'no easy path to optimal strategies, and recipes for success are neither universal nor timeless' (Whittington 2001: 6), smart planners employ a mix of approaches. In a classic strategy, leaders drive change. Systemic strategic planning is also deliberative but has multiple targets, accommodates diverse internal processes and is sensitive to external social and cultural considerations. Smart city visions are indicative and alter in line with policy, competition or technology. Independent evaluation and bottom-up wisdom curtail impractical flights of fancy. Wise, servant leaders listen, learn from mistakes and adapt (Greenleaf 1977). In whatever guise, strategy relies on data, integrity, professionalism, robust systems, collaboration, trust, standards, action and feedback. Operational concerns are individual project management.

Information and consultation

To uphold procedural legitimacy (rule of law), project genesis and scrutiny are separate. Smart cities filter out ill-considered measures and wasteful or low-impact projects. Informed multi-criteria judgement distinguishes smart from dumb private or public sector entities. It invokes scrutiny of a sound body of evidence, analysis, discussion, reflection and collaborative meta-cognition by experts. To formulate strategy, make evaluations, screen projects or tenders, manage assets or monitor performance, smart cities systematically collect appropriate real-time, spatial, stakeholder or transactional data. To inform participatory planning, smart planning systems disseminate intelligence to engaged communities. Smart cities act on community feedback to update policy, re-draft legislation or reposition and re-balance service or project portfolios.

Notwithstanding community deliberation, ultimately a high calibre and informed bureaucracy impartially navigates land-use contentions, balancing resilient instrumentalism, protocols and environmental limits. Debate informed by balanced multi-criteria analysis blocks 'white elephants' but fast-tracks projects with robust mandates which pass screening muster.

Once a strategic vision is articulated, smart administrations consult with experts and the affected public to proof plans, review policy, vet measures or scrutinise projects. Procedurally, this involves listening to an inclusive range of stakeholders to temper radical impulses with practical or customary wisdom. Yet, endless analysis or debate can frustrate necessary action. The limits to deliberation are 'irreducible plurality and frequent incommensurability' (du Gay, 2000: 31) as well as ecological limits (Owens and Cowell 2011). Policy, institutions and mechanisms are sensitive to but not bound by tradition and responsive to national priorities or fluctuating international conditions.

However, beyond organisational fit or sound screening mechanics, two critical features distinguish smart from dumb planning systems: 'meta-cognition' for holistic judgements, and policy learning.

Administrative and planning systems

Notwithstanding autocratic concerns or alternate evolutionary mechanisms (localism and incremental, small-scale development), smart urban-scale development needs well-functioning administrative and planning systems. In capable and tightly governed institutions, organisational architecture fits development purpose. If a political consensus emerges for regeneration or construction, the planning process should be competent, fair, transparent but also sensitive to local concerns or cultural settings. Smart planning systems are principled, data-intensive, capably staffed and trusted. Agencies communicate openly with a broad range of project-relevant stakeholders. If a compelling mandate for legitimate development exists, screening considerations involve design, symbolic, operational and financial considerations. Sound management practices entail separation of powers, checks and balances, malfeasance audits, data monitoring and continuous and inclusive stakeholder dialogue. Procedurally, smart urban development relies on fairness, transparency and the rule of law at all tiers of government. Due process tempers idealism and consultation respects legitimate local concerns.

A well-functioning urban system balances local innovation with regulation to circumvent vested interests (Veblen 1919). System constituents include informed strategic leadership, intelligence and feedback systems and competent and adaptable implementation mechanisms. Intelligence screens out noise and collects relevant data to forecast, assess and monitor. To vet policy, regulatory, tendering or planning interventions calls for judgement and evaluation. Analysis helps eschew costly mistakes and feedback accelerates policy learning. It involves continuous deliberation, competent and independent reviews, assessments, feasibility scrutiny and performance monitoring. The main remit of the planning systems excludes actual project delivery. Usually, local councils collaborate with the private sector but outsource the actual construction of affordable homes to the private sector. Practical government functions include land registration, property market regulation, planning law and the oversight of financial markets. Well-functioning systems are transparent, fair and function equitably to meet dwelling needs and respond to demographic, job-market or logistic pressures. Pernicious planning system afflictions include unbridled local populism, corruption, red tape or unwarranted austerity. These infections suppress inspired place-making, delay crucial infrastructure or exacerbate the housing crisis.

Capabilities

Checks and balances, oversight, professionalism, ethics, integrity and administrative competence underpin smart city administration. Its officials, like Plato's guardians, are well educated but with a strong sense of public duty. Fastidious community loyalty quells rent-seeking or other corruptions. Smart development administration converts utopian aspirations into financial signals, practical regulatory measures or useful projects that alter behaviour or improve health, productivity or lifestyle. Careful human resource management strengthens planning and development system and capabilities. Its fruits are servant leaders, principled administrators, competent professionals and organisational learning. Investments to strengthen institutional capacity include training programmes, data governance or decision support systems and technology upgrades. Integrity is central to motivate a skilled workforce, informed by professional standards.

Mechanisms

Politics

Once an inclusive political system settles on the judicious degree of system intervention, consultation and subsidiarity, the smart city deploys or eschews a range of mechanisms to help cost-effectively realise its conservation or development aspirations. Armed with strategic foresight and robust institutions, governments can either shape policy and markets or intervene more directly.

Market shaping

Whilst curtailing ugly developments and blocking unscrupulous operators, smart administrative systems capture unearned land value uplift for infrastructure without dis-incentivising productive entrepreneurs. Policy, regulations, taxation, subsidies, tendering or other instruments align commercial and communal goals. Market shaping incentivises build quality. Developers take a long-term view and investment considerations extend beyond floor space maximisation to design, resilience or energy efficiency. Market-shaping legislative, regulatory, tendering or other instruments range across planning law, building codes, tax bases and rates, lease terms or freehold property rights. Regulated yet efficient capital markets provide long-term development finance for smart projects. Players have professional accreditations and the salience of valuation standards tightens trust in space market operations. Project execution calls for competitive construction firms, effective project management, low-impact delivery mechanisms, monitoring and evaluation.

Direct intervention

As well as gazetting land, monitoring its use, managing education programmes or legislation, governments can directly invest in environmental protection, regeneration schemes and green or low-carbon infrastructure. In practice, direct investment often operates via a controlling equity stake. Environmental projects monitor or treat water, secure or enhance existing natural and semi-natural areas, green corridors, catchments, watercourses, abandoned railways or canals (Davivongs *et al.* 2012). Urban betterment builds appropriate housing, extends public transport or cycleways or landscapes precincts. New university buildings or regeneration of piazzas or other public realm refurbishment can stimulate art or facilitate social interaction (Sorkin 1992). Sacco *et al.* (2014) though caution that culture-led local development initiatives should eschew instrumentalism, over-engineering and parochialism. In other words, let authentic artistic precincts emerge spontaneously and evolve. The book's third part investigates a diverse range of smart development mechanisms, including the greening of Beijing, valuation systems in Dubai, retrofits and green building refurbishment in Germany, UK tenure reform and integrated preservation of archaeological heritage in Istanbul.

Book structure

This volume caters to general readers, students or professionals in real estate, land management or planning. It explores compelling urban issues, questions the 'smart' notion and investigates dimensions of its realisation in varied settings. The book has three parts:

- Visions
- Institutions
- Mechanisms.

Chapters vary but generally include:

- Abstract
- Key terms
- Introduction (chapter importance)
- Literature or conceptual framework
- Data and analysis including figures or tables
- Case studies
- Conclusions
- Policy recommendations
- References or bibliography.

Bibliography

Acemoglu, D. and S. Johnson (2005). 'Unbundling institutions', *Journal of Political Economy* 113: 949–995.

Adams, D. and S. Tiesdell (2013). *Shaping places: urban planning, design and development.* London: Routledge.

Arad, R. and S. Jacob (2014). 'Is beauty an essential consideration in architecture?', *Royal Academy Magazine* Winter 2014, available at www.royalacademy.org.uk/article/debate-is-beauty-an-essential-consideration-in-architecture (accessed 15 July 2006).

Arendt, H. (2003). *Responsibility and judgement*. New York: Schocken Books.

Benevolo, L. (1980 [1975]). *The history of the city [Storia della citta]*. Cambridge, MA: MIT Press.

Benninger, C. (2001). 'Principles of intelligent urbanism', *Ekistics* 69(412): 39–65.

Bodin, J. (1992 [1576]). *Six livres de la République*, Edited by J. Franklin. Cambridge: Cambridge University Press.

Bosker , M., E. Buringh and J. Luiten van Zanden (2008). 'From Baghdad to London: the dynamics of urban growth in Europe and the Arab world 800–1800', *Centre for Economic Policy Research* Discussion Paper 6833.

Campanella, T. (2006) [1899; 1623]. *City of the Sun* in *Ideal Commonwealths: Plutarch's Lycurgus, More's Utopia, Bacon's New Atlantis, Campanella's City of the Sun and a Fragment of Hall's Mundus Alter et Idem.* Edited by H. Morely, 8th Edition. London: George Routledge & Sons.

Dahl, C. (1959). 'The American School of Catastrophe', *American Quarterly* 3: 380–390.

Davis, M. (2006). 'Fear and money in Dubai', *New Left Review* 41: 47–68.

Davivongs, V., M. Yokohari and Y. Hara (2012). 'Neglected canals: deterioration of indigenous irrigation system by urbanization in the west peri-urban area of Bangkok Metropolitan Region', *Water* 4: 12–27.

Doxiadis, C. (1968). *Ekistics: an introduction to the science of human settlements*. New York: Oxford University Press.

du Gay, P. (2000). *In praise of bureaucracy: Weber, organization, ethics*. Thousand Oaks, CA: Sage.

Flyvbjerg, B., N. Bruzelius and W. Rothengatter (2003). *Megaprojects and risk: an anatomy of ambition*. Cambridge: Cambridge University Press.

Forman, R. (2014). *Urban ecology: science of cities*. Cambridge: Cambridge University Press.

Freedman, L. (2014). *Strategy: a history*. New York: Oxford University Press.

GBD MAPS Working Group. (2016). *Executive summary. Burden of disease attributable to coal-burning and other major sources of air pollution in China*. Special Report 20. Boston, MA: Health Effects Institute.

Geddes, P. (1915). *Cities in evolution: an introduction to the town planning movement and to the study of civics*. London: Williams and Norgate.

Giles-Corti, B., A. Vernez-Moudon, R. Reis, G. Turrell, A. L. Dannenberg, H. Badland, et al. (2016). 'City planning and population health: a global challenge', *The Lancet*, September, DOI: 10.1016/s0140-6736(16)30066-6.

Ginsburg, N. (1999). 'Putting the social into urban regeneration policy', *Local Economy* 14(1): 55–71.

Granoff, I., J. R. Hogarth and A. Miller (2016). 'Nested barriers to low-carbon infrastructure investment', *Nature Climate Change* 6: 1065–1071.

Greenleaf, R. (1977). *Servant leadership: a journey into the nature of legitimate power and greatness*. New York: Paulist Press.

Harvey, D. (1989). 'From managerialism to entrepreneurialism: the transformation in urban governance in late capitalism', *Geografiska Annaler. Series B, Human Geography*, 71(1): The Roots of Geographical Change: 1973 to the Present (1989), pp. 3–17.

Hollands, R. (2008). 'Will the real smart city please stand up? Intelligent, progressive or entrepreneurial?' *City* 12(3): 303–320.

Jacobs, J. (1993). *The death and life of great American cities*. New York: Random House.

Kauko, T. (2017). *Pricing and sustainability of urban real estate*. Abingdon: Routledge.

Kratke, S. (2010). 'Creative cities and the rise of the dealer class: a critique of Richard Florida's approach to urban theory', *International Journal of Urban and Regional Research* 34: 835–853.

Leary, M. and J. McCarthy (2013). *The Routledge companion to urban regeneration*. Abingdon: Routledge.

Lewis, C. I. (1955). *The ground and nature of the right, Woodbridge Lectures, V, Columbia University, November 1954*. New York: Columbia University Press.

Lindblom, C. E. (1959). 'The science of "muddling through"', *Public Administration Review* 19(2): 79–88.

McNeill, J. R. (2000). *Something new under the sun: an environmental history of the twentieth century world*. New York: W. W. Norton.

Nussbaum, M. (2000). *Women and human development: the capabilities approach*. Cambridge: Cambridge University Press.

Owens, S. and R. Cowell (2011). *Land and limits: interpreting sustainability in the planning process*, 2nd Edition. Abingdon: Routledge.

Piketty, T. (2014). *Capital in the twenty-first century*. Cambridge, MA: Harvard University Press.

Popper, Karl (1966 [1945]). *The open society and its enemies*, Vol. 1, 5th Edition. Princeton, NJ: Princeton University Press.

Pregill, P. and N. Volkman (1999). *Landscapes in history: design and planning in the eastern and western traditions*, 2nd Edition. New York: John Wiley.

Roy, A. (2009) 'The 21st-century metropolis: new geographies of theory', *Regional Studies* 43(6): 819–830.

Ryan, R. and E. Deci (2001). 'Happiness and human potentials: a review of research on hedonic and eudaimonic well-being', *Annual Review of Psychology* 52: 141–166.

Sacco P., G. Ferilli and G. T. Blessi (2014). 'Understanding culture-led local development: a critique of alternative theoretical explanations', *Urban Studies* 51(13): 2806–2821.

Scott, A. J. (2008). 'The resurgent metropolis: economy, society and urbanization in an interconnected world', *International Journal of Urban and Regional Research* 32: 548–564.

Sen, A. (2009). *The idea of justice*. Cambridge, MA: Harvard University Press.

Shelton, T., M. Zook and A. Wiig (2014). 'The "actually existing smart city"', *Cambridge Journal of Regions, Economy and Society* 8(1): 13–25.

Smith, A. (1759). *The theory of moral sentiments*, available at www.econlib.org/library/Smith/smMSCover.html (accessed 3 February 2017).

Smith, N. (1979). 'Gentrification and capital: theory, practice and ideology in Society Hill', *Antipode* 11(3): 24–35.

Smith, P. and D. Anderson (2016). *Future cities in conversation*, 27 June, available at www.strangehorizons.com/2016/20160627/1SmithAnderson-a.shtml (accessed 7 July 2016).

Sorkin, M. (1992). *Variations on a theme park: the new American city and the end of public space*. New York: Hill & Wang.

Stiglitz, Joseph. (2012). *The price of inequality: how today's divided society endangers our future*. New York: Norton.

Storper, M., and A. J. Scott (2009). 'Rethinking human capital, creativity and urban growth', *Journal of Economic Geography* 9(2): 147–167.

UN (2014). *World urbanization prospects*, available at https://esa.un.org/unpd/wup/Publications/Files/WUP2014-Highlights.pdf (accessed 6 July 2016).

Veblen, T. (1919). *The vested interests and the common man*. New York: B. W. Huebsch.

Veblen, T. (1994) [1899]). *Theory of the leisure class: an economic study in the evolution of institutions*. New York: Penguin Books.

Wadley, D. (2010). 'Exploring a quality of life, self-determined', *Architectural Science Review* 53(1): 12–20.

Whittington, R. (2001). *What is strategy and why does it matter?* 2nd Edition. London: Cengage Learning.

WHO (2010). Environment and health risks: a review of the influence and effects of social inequalities, WHO Regional Office for Europe, Copenhagen, Denmark, available at www.euro.who.int/__data/assets/pdf_file/0003/78069/E93670.pdf (accessed 17 January 2017).

WHO (2014). Global Health Observatory (GHO) data, available at www.who.int/gho/urban_health/situation_trends/urban_population_growth_text/en/ (accessed 6 July 2016).

Wilkinson, R. and K. Pickett (2010). *The spirit level: why equality is better for everyone*, Rev. Edition. London: Penguin Books.

PART I

Visions

1

SMART URBAN PLANNING

*Simon Huston, Arvydas Jadevicius
and Savaş Zafer Şahin*

Abstract

In an unstable world, whether by devolved evolution or *dirigisme*, cities must continuously adapt or perish. Demographic, trade, technological and climatic pressures force land use intensification and low carbon investment, constrained by habitat fragmentation, food insecurity and financial instability. Rather than a utopian destination, smart urban development is an inclusive processual strategic mind-set. Collaborative institutional networks link government tiers with commercial and civil organisations, tempered by local dialogue. Within the networked system, ambition flounders without the appropriate strategy, data, administrative capabilities, trust, professional standards and sustainable finance to block or initiate and execute appropriate projects. Yet urban and institutional peculiarities frustrate generalisable smart city blueprints. Undisputed necessary conditions for smart urban development are a constitutional and legislative framework to guarantee freedom, privacy and due process in a thriving economy with universal access to dignified housing, sanitation, utilities, education and health services. Beyond human rights, dignity and empowerment for human flourishing, paradox vests in three irreconcilable tensions. First is *dirigisme* or evolution in institutional networks. Second is freedom or community cohesion. Third is strategic fragmentation. Productivity, low carbon or humanist imperatives pull in different directions (innovation vs. resilience or hedonic vs. eudemonic well-being). Smart planning contends with fractious politics and property market imperfections. Politically, in current revanchist climates, people question *dirigisme* yet reject laissez-faire. Measured interventions invoke foresight, political calibration and thoughtful design. Well-managed and soundly financed, astute conservation, place enhancement or low carbon infrastructure strengthen resilience. Transformative constraints include land designation and environmental quality thresholds, local identity, culture and funding.

Key terms

Smart, planning systems, adaptation, redundancy, resilience, contention, *dirigisme*, evolution, processual, participatory, horizontal and vertical institutions, sustainable, resilience, Land Administrative System, spatial justice, low carbon infrastructure, partnerships and policy learning

Learning outcomes

- Alternate development political philosophies
- *Dirigisme* vs. evolution
- Vertical vs. horizontal (contractual) and bottom-up (local) institutions
- Diversity of function, form (morphology) and system dynamics
- Historical and geographical context
- Administrative spatial fit
- Multifaceted evaluation
- Collaboration and policy learning
- Alternate vehicles and mechanisms

Introduction

Over the past few hundred years, human activity has transformed the planet but instigated mass extinctions and climate change, and undermined human health (McNeill 2000; Régnier *et al.* 2015; GBD 2016). Air pollution and the polarisation of spatial health outcomes are perhaps the most pernicious impacts of ill-considered development and poorly planned cities (WHO 2010). Smart urban development is not a utopian end state. Rather, it is a continuous, iterative deliberative and inclusive pathway to help negotiate but never fully resolve all land use contention (Owens and Cowell 2011; Childers *et al.* 2015). Notwithstanding its processual and evolutionary character, concrete tasks include to clean up the air, cut fossil fuel dependency and reduce health inequity. However, institutional and built environment diversity precludes formulaic prescriptions. Nevertheless, the smart city invokes appropriate urban design and building density or construction of sustainable, low carbon infrastructure to promote healthier modes of travel (Giles-Corti *et al.* 2016). However, barriers to de-carbonising infrastructure nest within general institutional ones that are particularly pernicious in developing countries (Granoff *et al.* 2016).

Settlements

Human settlements differ greatly in size, setting, form and function (ritual, trade, administrative, military, colonial, industrial or technological). Political, social and economic dynamics vary both spatially and temporally. Organic cities emerge for a variety of sacred, strategic or commercial reasons and evolve. Deliberately shaped or planned cities include Vienna, Letchworth and Welwyn Garden City (Howard 1902 [1898]). The hiatus between planned and organic is often artificial. Akbar planned Fatehpur Sikri in India as a centre for arts, letters and learning (Schimmel 2004 [2000]) but maritime trading hubs like Thera, Carthage, Rome (via *Portus*), Constantinople, Venice or London began as safe havens and evolved into administrative centres and trading hubs (Hoepfner 1997; Norwich 2007; Hughes 2017). Tenochtitlan thrived as a 'lovely forest-Venice' before the Spaniards ransacked it in the early sixteenth century (Dahl 1959: 385).

In the span of human history, permanent settlements are a relatively recent phenomenon, first seen in Anatolia around 10,000 years ago (Göbekli Tepe ~9100 BC and Çatalhöyük ~7500 BC). Even until 1000 BC, mainland Greece remained 'a world of villages' (Morris 2005: 2) although Knossos on Minoan Crete and Akrotiri on Thera flourished until around 1600 BC. But, within a thousand years, cities had spread across the Mediterranean and first-tier ones like Athens, Alexandria, Antioch or Seleucia soon supported populations around a quarter of a million. Since antiquity, although trade, war and disease propelled or destabilised dynamics, overall urban populations grew (Bosker *et al.* 2008). In the 1950s, 746 million people lived in cities compared to 3.9 billion today (UN 2014). In 1960, only 34 per cent of the world's population lived in cities compared to 54 per cent today (WHO 2014). By 2050, the UN forecasts that an extra 2.5 billion people will swell the populations of mainly medium-sized Asian or African cities (UN 2014). Growth will likely extend conurbations, intensify land use, fragment or deplete habitats and accentuate greenhouse gas emissions (Francis and Chadwick 2013). The ramifications of malignant growth include water and food insecurity or civic and financial instability. Its nemesis, 'smart' development purports to eschew the symptoms of urban disease, like squalor, ugliness and sprawl or its associated social evils (poor health, inequality and cultural alienation). However, experts wrestle with its definition. Smart technology, regeneration, rehabilitation, reconciliation or transformation efforts confront philosophical and political contention, complexity and institutional and funding constraints. Conundrums are its eudaimonic or hedonic underpinnings (Ryan and Deci 2001; Wadley 2010), absolute or local sovereignty (Bodin 1992 [1576]), urban visions or institutional arrangements. Argument rages over intervention mechanisms: the merits or pitfalls of state involvement, its scope, scale and geographical productive or consumptive focus (Storper and Scott 2009). Scott (2001) and Kratke (2010) revile localised construction to engineer 'cognitive–cultural' clusters for competitive advantage. Rather, efforts should tackle long-standing causes of social disintegration and polarisation. Strategically, weak regional government, policy flux and planning hamper it. Operationally, the institutions and mechanisms to spark, screen, catalyse, incentivise or control smart projects can malfunction or fail to serve the long-term public interest. Where officials are unaccountable, vested interests dominate (Veblen 1994 [1899]) or funding is sourced from volatile capital markets, ill-considered development can infest landscapes. Despite strategic, operational and tactical contention, smart development likely involves servant leadership (Greenleaf 1977), academic debate, checks and balances, public-spirited administration and widespread consultation.

Visions

Urban utopias have a long history. Imperial Chinese ones, often inspired by Confucius, sought symbolic balance between nature, the state and the cosmos to facilitate human flourishing (Pregill and Volkman 1999). Plato's ideal city relied on the education of its elitist guardians. Thomas More's *Utopia* (1516) and Campanella's *City of the Sun* (1623) described alternate well-intended utopias with nightmarish undertones. In the modern era, Masdar City or Songdo are planned utopias, designed for affluent elites but not marginal artists who question the social order. Montgomery (2008) draws inconclusive links between urban design and prosperity. Paradoxically, over-prescriptive 'smart' urban blueprints can asphyxiate vitality (Sacco et al. 2014). Developers struggle to mimic the convivial morphology and complex social order resulting from evolution (Geddes 1915), seen in Rome's historical core for example.

Nowadays, cities compete to outsmart each other but without settled visions or agreement on institutional frameworks or realisation mechanisms. Urban technological, social or climatic visions proliferate. The term 'smart city' remains vague and ambiguous, yet it somehow provides useful guidance. The 'smart city' and 'sustainable city' compete with alternate utopias for the limelight to contain demographic, competitive and environmental pressures. Other exemplars include the 'Virtual City', 'Knowledge City', 'Broadband City', 'Mobile City', 'Digital City', 'Intelligent City', 'Ubiquitous City', 'Eco City', 'Green City' or Howard's (1902 [1898]) stalwart 'Garden City'. The proliferation of teleological ideals illuminates the existential crisis facing traditional urban planning.

For the Town and Country Planning Association (2017), a Garden City is a holistically planned new settlement that enhances the natural environment whose construction involves:

- Land value capture with community land ownership and stewardship
- Strong vision, leadership yet community engagement
- Mixed-tenure affordable homes
- Local jobs within easy commuting distance of homes
- Beautifully and imaginatively designed homes with vegetable gardens
- Green infrastructure network and net biodiversity gains
- Zero-carbon and energy-positive technology
- Facilities
- Walkable, vibrant, sociable neighbourhoods
- Integrated walking, cycling and public transport networks.

Place and administrative complexity dog smart development. Disagreement hinges on bio-assessment, planning governance and market intervention scale or tools (Owens and Cowell 2011).

Except in authoritarian regimes, squabbling between stakeholders can delay, if not frustrate, the realisation of resilient or creative urban visions. Initially, politicians of different persuasions, planners in various tiers of government, fragmented local communities, small or large developers and local, remote or online financiers are unlikely to share a common vision. Practical participatory tools to help build vision consensus include local meetings, expert focus groups or surveys.

Challenges

The World Resources Institute (WRI) estimates that by 2050, global urban populations will have grown by 60 per cent (Beard *et al.* 2016). To house 2.5 billion urban migrants over the next three decades requires both spatial reconfiguration and policy reform (Kim 2016). Today, many cities even struggle to meet residents' basic needs. Beard *et al.* (2016: 3) estimate that 140 million city dwellers lack clean water. To redress such deficiencies and future-proof cities calls for shared visions, strong coalitions and joined-up policies, to 'unleash a cycle of positive change' for equitable access to core services (land use, water and sanitation, energy, transportation).

In coming decades, most global growth will be urban (Floater *et al.* 2015) yet in the maelstrom of the 'infernal machine', as Bordieu (1998: 100) characterises modern capitalism, planning regimes often seem curiously ill-prepared to tackle looming internal and external challenges. A putative Smart Urban Regeneration (SUR) framework has procedural and multiple teleological dimensions, captured via smart institutions, quality projects and innovative funding. Place-rooted and soundly administered, smart projects balance commercial with public realm considerations (Mehan 2016). Informed transformation balances localism agendas and strategic aspirations for employment, aesthetics, logistics or distributive justice. The private firms or government agencies that formulate, screen or deliver smart developments should be tightly overseen, properly constituted for delivery and procedurally tempered by multiple checks and balances, protocols, standards and regulations within the rule of law. Evaluation involves consideration of likely and cost-effective project delivery and multiple transformative impacts (beautification, pedestrian connectivity, waste management, network connectivity or ecological conservation). Procedurally, to temper ill-considered excess and check corruption, the results of preliminary desktop analysis, consultation, site visits and grassroots dialogues are subject to multiple independent reviews.

Apart from dramatic external pressures, internal constraints can lead to dystopic urban trajectories and bequeath malignant outcomes, involving congestion or a toxic legacy of unstructured sprawl and air pollution such as are endemic in Beijing or Delhi. Dystopic megacities are characterised by planning complacency, poor management, corruption or underinvestment in civic and public amenities. Resentment breeds in slums that abut affluent, gated enclaves. Unstructured urbanisation spillovers manifest in poor health, air pollution, traffic congestion, psychologically stunted children and crime. Such spatial externalities consume 15 per cent of Beijing's gross domestic product (GDP) and cost the United States economy US$ 400 billion annually (Litman 2014).

From a developed world perspective, Kauko (2017) identifies key smart development issues at various scales.

- Building energy efficiency, mix and lifecycle pollution
- Real estate quality, affordability and diversity
- Urban optimal density
- Public transport, accessibility and traffic pollution
- Social cohesion
- Governance and transparency
- Regional and local innovation and production.

Smart planning

Whatever its shape and however realised, smart development impels substantial investment in low carbon connective, utilities and social infrastructure (Granoff *et al.* 2016). Some disruption to status quo is inevitable to tackle access inequality via substantive tax and educational reforms. Finally, smart development sets territorial limits to growth. Its long-term ecological perspective curtails the worst depredations of populist or short-term market proclivities but the cost and political implications are manifest.

- Traditional urban planning has a linear time perspective, in which generalised data about the past is used to forecast the future and decide on the best alternatives. Whereas, in smart cities, planners can access real-time data, mostly in a personalized form, to make real-time changes.
- Urban planning's isolated stance in city governance is also changing. Formerly, urban planning, urban management and implementation of urban plans were all perceived as interconnected but separate activities, happening relatively away from citizens (Shahrokni *et al.* 2015). Yet, smart city considerations necessitate urban planning, urban management and plan implementation occurring simultaneously in real-time, fed by as much data and feedback from citizens as possible. Because citizens' engagement is also changing through new network power structures and new economies, they are becoming active content and service providers rather than passive recipients of municipal services.
- Local governments and the urban planning profession sought to adapt traditional planning by inserting a bundle of extra considerations and tightening operational efficiency. Extra considerations included transportation planning, city-region planning, economic planning, social planning and environmental quality. Efficiency measures included use of informatics and widespread adoption of business managerial techniques. A handful of successful regeneration

projects, new towns or city development regions illustrate the limited legacy of such top-down smart city planning efforts (Stratigea *et al.* 2015).

- Reliance on vertical institutions to make reliable expert forecasts is impractical and risky. In smart city networks, transformation decisions are partly devolved to stakeholders. Plans evolve in response to signals in complex proactive social, commercial and local networks. These political, commercial, social and local networks enable continuous engagement and deliberation. Management of such complex collaborative environments calls for technological adaptation. Urban planning becomes less strategic and more facilitative. At the extreme, planners merely provide infrastructure, information technology (IT) architecture and network integration to coordinate user services and data interactions.

- In practice, many impediments constrain planning system upgrades. Local governments lack the technological capability or cultural inclination for major change. Other smart development barriers are apathy, political fragmentation and ineptitude or chronic budget deficits. Policy flux, depleted infrastructure, public–private partnership shenanigans or misguided initiatives undermine trust. Failed civic leadership and iniquitous education undermine institutional capabilities further (Erie *et al.* 2011). Incremental administrative improvements and IT upgrades are feasible. Practical measures include widening citizen participation in planning deliberation. In time, planners and local councils can roll out new digital platforms such as web-based surveys, data mining, e-participation, citizen-specific decision-making, context-aware information and virtual exploration of environment by communities. Armed with this data, citizens become co-producers of planning (Calegari and Celino 2015).

Institutions

Institutionally, libertarians contest government interventions to shape the future, chancing the vagaries of imperfect markets against risk of tyranny. Arguments turn on libertarian or collective social purpose, intervention scale, participation and bio-centric or instrumental anthropogenic paradigms (utilitarianism).

- Contentious political philosophy aside, the institutional framework for smart development extends spatially beyond city precincts to fit with national constitutional and legal systems (Ekstrom and Young 2009). It encompasses Land Administrative System (LAS), city administrative, urban planning, valuation and project management systems. Smart urban development is neo-evolutionary with processual but inclusive (*corporatist*) as opposed to *exclusionary* anthropology (Blanton *et al.* 1996). It blends judicious and measured processual strategic interventions in collaborative institutional networks. As well as the vertical political system with its government administrative levels, the network encompasses horizontal commercial, non-government institutions and bottom-up inputs from empowered local bodies or citizens. Smart institutions should foster quality growth and curtail its extractive modes. Administratively, its constituents are foresight, policy coordination and well-funded but judicious interventions. It impels capable planning institutions, focused on more compact, connected, resilient and inclusive futures as a prerequisite, but no guarantee of *eudemonic* well-being (Wadley 2010). Rather than indiscriminate output or even *hedonic* well-being, the *eudemonic* focus is competence, autonomy and relatedness of citizens (Acemoglu and Robinson 2013; European Climate Foundation 2010; Geltner and de Neufville 2014; Turner 2014). Requirements include a futures orientation towards resilience and creativity, sensible spatial architecture and disposition towards collaboration. In contrast to extractive ones, smart institutions seek to remedy, not exploit, market failures, and attenuate, not reinforce, structural inequalities (Acemoglu and Robinson 2013). Unlike in *comprador* capitalism, smart urban development is accountable, people-focused and conserves natural systems (Thomas *et al.* 2000). It taps new online technologies and geographical data to capture, model or visualise projects that inform planning and negotiations.

- Collaboration begins with appropriate scales (boundaries) and tight institutional fit (design). Proper governance reduces financial manipulation or fiscal distortion and incentivises projects with conservation, education or health spin-offs. Inclusive institutions, authentic debate, subsidiarity and the rule of law temper extractive proclivities. Smart collaborative institutions negotiate or muddle through (Lindblom 1959) but avoid the quagmire of strategic drift. Integrity, foresight and competence enable them to screen, plan and execute quality projects for urban resilience or enterprise. Resilient settlements can better absorb disturbance or reorganise to retain function, structure and identity (Holling 1973; Forbes *et al.* 2009). Redundancy and a balance of social, economic and environmental capital strengthen it (Wilson 2014). Just as genetic predisposition, trauma exposure or informed treatment engender psychological resilience (Rutter 1985) so, too, urban resilience invokes planning (smart institutions), selective regeneration (quality projects) and system upgrade funding (von Braun and Thorat 2014). Smart institutions employ competent and cooperative staff to generate useful output with positive social and ecological spillovers (Rogers 2012; Turner 2014). Productivity gains come without energy or carbon intensification. Rather, efficiency gains come from distributed energy, transport and information networks.

Foresight and information systems

Strategy is different from strategic planning but has many variants (Whittington 2001). In *dirigisme*, political institutions engineer smart responses to multiple urban challenges. A top-down strategy first articulates planning objectives such as to engineer resilience, cut air pollution, foster creativity or improve well-being. In evolutionary or processual strategic approaches the next step is to collect useful intelligence to understand places (Floater *et al.* 2015) and to celebrate their distinctive historicity, heritage or landscapes. Informed spatial transformations (outcomes) rely on science or architectural and design excellence but need grounded urban intelligence. Archival research, baseline analysis, expert views and structured stakeholder engagement help understand place character (ambience and atmosphere). Comprehensive site diagnostics inform smart institutions on relevant, scientific, commercial and local concerns about contamination or disruptive intensification (habitat loss, blight, noise, emissions, congestion or service stress). In smart cities, decision-support or geographical technologies help stakeholders visualise alternate project permutations to evaluate architectural, connectivity, spatial justice and ecological impacts.

Capabilities

Historically, Venice provides useful guidance for a smart planning effectiveness and resilience (Norwich 2007). Machiavelli considered its mixed system, complicated elections and interlocking councils excellent, but also admired Rome's tribunes who could mobilise the people (Najemy 2010). A smart administrative system articulates institutional scale and scope to balance strategic foresight and 'top-down' leadership (Hemphill *et al.* 2004) with reflection and local dialogue. Strategic leadership, governance and institutional architecture help assure effective, efficient, inclusive and transparent project management. Inspired by the common good, smart planning interventions seek to attenuate spatial injustice without undermining customary or *bona fide* formal property rights or cultural practices. However, conserving or strategically shaping a city requires multiple strategies (Freedman 2014). One approach is to enhance technological capabilities to improve geo-data collection, its management and the dissemination of knowledge. Citizen involvement in Administrative Information Systems (AIS) improves design, provides alternate viewpoints and refines or updates data on urban morphology or infrastructure performance.

SUR's institutional culture is 'managerialist' in the sense that it eschews spectacle and seeks long-term solutions to substantive economic and social problems (Harvey 1989). Governance ensures legitimate and cost-effective delivery of complex projects (Termeer *et al.* 2010). It comprises the formal policies, procedures, and informal culture and norms to focus corporate activity and attenuate agency problems (corruption, nepotism or 'free-riding').

Many iconic construction projects (e.g. Greece's Olympic or Brazil's World Cup stadiums) fail to pass muster against the subsidiarity, spatial justice or transparency criteria but even in tight institutional settings, misconstrued purpose, project complexity or market turbulence can scupper performance (Flyvbjerg *et al.* 2003a, 2003b; Altshule and Luberoff 2003; Van Marrewijk *et al.* 2008). Pragmatism and diplomacy can help institutions navigate complexity, local power politics or vested interests. Institutional fit, good governance and authentic consultation mitigates the risk of outlandish projects, fanciful projections and cost blowouts. Tight governance, financial and spatial LAS transparency and proper tendering oversight cuts waste and roots out corruption or nepotism. It increases competition and broadens private participation in critical infrastructure. Its antithesis is 'patrimony', oligopolistic free-riding and 'plutocratic dystopia' (Piketty 2014).

Institutional design and partnership management facilitate project delivery. Proper spatial, temporal and functional fit help configure institutional and network architecture to match operational requirements. Ekstrom and Young (2009) note that misfit occurs when institutional arrangements ignore ecosystem character, function and dynamics. Spatially, cross-scale misfit occurs where anthropogenic administrative or organisational boundaries diverge from bio-geophysical ones. Catchment management and water security problems are typical. Temporally, urban decision-makers can have a short-term, electoral focus. Functionally, nested organisational concerns can overwhelm foresight or collaboration.

Land, Administrative and other Information Systems

Whilst traditional urban planners rely on paper documents, rules of thumb and formulaic forecasting or scenario modelling, IT-based 'smart cities' employ digital AIS and LAS geospatial data. Real-time (synchronous) data and interactive communication enables planners to engage in active virtual deliberation with other system stakeholders. Smart planning is a continuous interactive process driven by network technologies. Citizens help populate and verify planning database or use information for e-participation (e.g. virtual exploration of proposed developments).

Many obstacles block transformational change, notably political disagreement, lack of funding and institutional weakness. Operational challenges include contested information (Bruijn and Leitjen 2008), fraud, cost escalation or maladroit oversight (Flyvbjerg et al. 2003a, 2003b). Obstacles aside, megaproject outcomes can underwhelm or polarise communities, or rapidly depreciate. Megaprojects like Songdo (Korea), Maasdar (UAE), Skolkovo (Russia) or Dongtan (China) have high opportunity costs and are unlikely to deliver widespread, 'lower level Maslovian sustainability' (Wadley 2010: 19). To deliver these, Güell and Redondo (2012) call for a more tempered approach, involving territorial foresight, debate, local engagement, institutional collaboration, project scrutiny and smart finance. For Batty (2013), social innovation could resolve the 'smart' technological/grandiose or social/grounded paradox.

Urban landscapes are complex and transformation projects pit powerful winners against alienated losers. The dereliction of Bangkok's peri-urban canals illustrates the nefarious long-term ecological consequences of rampant real estate pressure (Davivongs et al. 2012). At play with megaprojects, like the controversial Heathrow Airport expansion, are jobs, landowner uplift bets, hidden commissions, political prestige and multinational corporation profits. For Smith (1979) and Harvey (1989), megaprojects mobilise a coalition of capitalist handmaidens to exploit rent gaps via structural violence and expulsion. For Veblen (1919) and Foucault

(1977), megaprojects are a contrivance by powerful vested interests to over-come urban transformation resistance. Notwithstanding conspiracy, macroeconomic forecasting difficulties; megaproject prognosis must contend with architectural, urban design, institutional and geographic peculiarities. Given these complexities, at the very least then, judgement on the merit or failure of a megaproject must involve multiple ecological, social and commercial considerations.

Smart urban development recognises the interconnection between urban, social, ecological and economic spheres but also requires a robust governance framework, innovation and institutional learning. Unregulated property-led development, on the other hand, can result in sprawl, unstable markets, poor health and depleted ecologies or cultural landscapes. However, preservation, conservation or transforma-tion efforts confront philosophical contention, complexity and institutional and funding constraints. Argument turns on intervention scope or the appropriate balance between public and private realms and the mechanisms to spark, catalyse, incentivise, control and deliver appropriate projects. Future urban visions pit planning utopias against legal precedent, competitive capital markets and practical aspirations of community self-determination. Institutionally, populist impulses, corruption, managerialist bureaucracy or austerity can infect or suppress inspired planning and long-term housing or infrastructure investment. Smart urban development involves multidisciplinary collaboration and widespread consultation. Big data and high-tech decision-support systems (DSS) can be part of the solution but not without administrative competency, civic professionalism and policy learning. Strategic plans should balance logistic infrastructure imperatives with ecological and local considerations. To internalise externalities and balance function with form, developments should consider aesthetics and reflect place character and local views. Urban transformation aspirations include cohesion, green corridors, energy efficient buildings, precinct connectivity, affordable and reasonably stable house prices, resilient utility networks and seamless public transport ones (Banister 2012).

Mechanisms

Project quality

Institutionally, traditional planning confronts alternate policy foci (firm competi-tiveness, local health, school operation). For Healy (2007: 16) 'Clashes between conceptual frameworks and legitimising rationales are commonplace'. The rapidly evolving global economy accentuates stakeholder tensions. Urban regeneration quality is multifaceted but considerations include architecture and design merit, density and housing affordability, public realm enhancement, connective infra-structure (Floater *et al.* 2015). Judgement follows rigorous assessment, planning protocols, expert and local engagement, informed digital information and financial modelling (Nase *et al.* 2013). Data limitations aside, the interpretation of place-making confronts nuanced meanings, complexity, agent network interactions and cultural diversity. The demise of Deepdene palazzo and its demolition in 1967 to make way for drab offices in Dorking, Surrey, provides a salutatory example of crass commercial land transformation, bereft of local place sensitivity and without national policy coherence (Robinson 2012; Jakobsen and Høvig 2014).

Finance

Failure to tackle spatial or market externalities is neither 'smart' nor 'sustainable'. Smart development cuts carbon use and internalises pollution and congestion spillovers but investment requires sustainable finance. Smart development funding draws on both public and private capital. Political and economic instability or budgetary constraints can delay or scupper it. Fiscal austerity and pervasive corporate tax avoidance dampen transformational ambition or feasible socially inclusive aspirations. Without lucrative options, private finance demands risk competitive income streams and payback. To facilitate smart development, it is important to dismantle nested barriers to low carbon investment (Granoff et al. 2016). Policy muddle and flux is one such. Transparency and promulgation or disclosure of credible strategic responses to tackle various urban threats or pressures lowers funding cost (Diamond and Verrecchia 1991; Hall and Hesse 2012).

In imperfect property markets with weak fiscal tax regimes, predatory commercial players hijack government-financed spatial betterment schemes. Development inflates contiguous house prices but inept commercial payback models fail to police or capture public realm uplifts or logistics benefits. Without value capture, regeneration is regressive unless projects deliberately target unprofitable deprived locales. In fluctuating markets, an effective partnership between the public and private sector strengthens the commercial success of targeted projects. Private finance is constrained by the productive or predatory opportunity cost of capital. The vehicle for public sector support can vary but without it, for blighted districts in bearish markets, funding evaporates. Strong public relations de-risking signals include proponent credibility, planning fit, well-designed projects and structured community dialogue. The Greater London Authority (2011) *London Plan* de-risked the Olympic Park, King's Cross and Nine Elms for foreign and domestic investors.

The scale of smart development projects varies from small-scale infill to ambitious forward-thinking investments for Transport Oriented Developments (TODs) or canal restoration with land amalgamation or complex planning, geotechnical and construction issues (Searle et al. 2014). Robust capital and space market intelligence, sound administrative and planning systems cut risks and improve project financial viability. In blighted or poor locales, small-scale improvements and regular maintenance of pathways and public realm help transform market perceptions and unlock commercial potential. However, badly needed projects may never become commercially viable. If compelling long-term social or environmental benefits outweigh commercial considerations, general taxation should fund them (Brookes 2013; Vanolo 2014).

An alternative to earmarking general taxation receipts to fund necessary but unviable projects is to capture land value uplift. The mechanism can be either direct (lease charges or infrastructure connection fees) or indirect, via local higher taxes. Tax Increment Finance (TIF) first demarcates the value-capture project hinterland and then assigns collection rights to the project proponent, usually a special purpose vehicle (SPV). The SPV clarifies project ownership, allocates responsibilities, costs risks and orchestrates construction. Investor due diligence should weed out bad urban infrastructure projects, situated in unpromising sites with fanciful business models, questionable accounting or lukewarm government support. Lack of operational or financial transparency increases funding costs (Easley and O'Hara 2004). In practice, TIF means the local authority effectively cedes an element of its fiscal sovereignty to the proponent.

In the public arena, positive public realm or social improvement 'spillovers' can compensate for a financial deficit. Where substantive public realm investment is necessary, a public–private partnership (PPP) can help (Pattberg and Widerberg 2014). Alternatively, Social Impact Bonds (SIB) can raise finance so bondholders rather than taxpayers fund initial development outlays. Once auditors confirm agreed and social or environmental milestones, the SIB commissioning body (government) redeems the bonds.

SUR partnership effectiveness requires an agreed territorial vision and operational effectiveness. It calls for leadership, collaboration, institutionalisation and local legitimacy rooted in dialogue and community spatial spin-offs – jobs, health, conviviality and spatial justice. Its long-term goals are urban 'resilience' and community 'creativity' but its ethos is public-spirited, administrative and policy-driven. However, multiple and lofty SUR aspirations load development costs on to projects in disadvantaged locales which erodes feasibility. Public funding aside, commercial counterweights are land-gifting, tax breaks, subsidies, project de-risking or TIF. De-risking solutions involve corporate governance, structured community dialogue and a robust payback model. In propitious locales, TIF or social infrastructure bonds can provide alternate funding solutions.

Given the stark distributional backdrop, 'smart' development must address, if not allocative minutiae, then at least the broad procedural mechanics for an inclusive society without compromising enterprise. Practically, SUR sidesteps pedantic semantic quarrels over 'sustainability' or statistical indicators for it, and instead backs catalyst projects to enhance quality of life, cut emissions and boost mobility (Banister 2012). In this regard, pedestrian or dedicated cycle networks would pass muster (Southworth 2005). Smart investment, whether 'hard' (built environment and transport logistics) or 'soft' (institutional upgrades or local up-skilling schemes), should attenuate spatial injustices (Colantonio and Dixon 2010; Couch 1990). For Roberts (2000: 40) it makes 'a lasting improvement in the economic, physical, social and environmental condition of an area'.

Urban regeneration then extends beyond narrow economic development or physical urban renewal. Its proximate pragmatic physical, economic or environmental upgrades improve the daily lives of ordinary people. Within financial constraints and realistic limits, sustainable regeneration improves places, stimulates prosperity and fosters inclusive local capabilities. For Turok (1992: 361), 'unrestrained market-led development' fails to consider locals or the underlying local economy and 'may have detrimental consequences for the economic fabric of cities and for the quality of life of their residents'.

Discussion and conclusion

Smart development is adaptable and evolves in response to multiple pressures. It operates in a networked, collaborative system of vertical, horizontal and bottom-up local institutions. Constraints are philosophical conundrums, political vicissitudes, policy flux, stakeholder diversity and, on the ground, industrial blight, social deprivation and site or engineering challenges. Undisputed necessary conditions

---

CASE STUDY **ÜLEMISTE SMART CITY, TALLINN**

In Europe, according to the Intelligent Community Forum, Eindhoven and Tallinn stand out as leading intelligent cities. Skype was developed in Tallin and, in 2005, Ülemiste Smart City was established near to the airport. Tsar Nicholas II decreed the site's industrial designation, which continued under Soviet military hegemony. Smart City is close to major arterial motorways and a railway station. The Smart City combines historical architecture with modern Leadership in Energy and Environmental Design (LEED) Gold Certified buildings. During the development of the project the old industrial and publicly unknown area was transformed in several steps. A new master plan was developed for Smart City and this has been the basis for transforming the industrial area into a modern urban centre. The old administrative and industrial buildings have been renovated or rebuilt step by step to become modern offices, kindergarten, school, university facilities, parks, recreational areas and restaurants. As of 2015, Ülemiste City has 100,000 sq. metres of modern office space, with 7,000 people working at the Smart City daily. The master plan allows the total office and rental apartment space to increase up to 600,000 sq. metres. As of 2015, Ülemiste Smart City already has a modern street network, sufficient parking conditions, four public bus lines and, by 2017, it will be connected to a brand new tram line, which will connect Tallinn airport, Ülemiste City and the city centre. By 2022–2025, Ülemiste Smart City will also have a Rail Baltica high-speed train terminal, which will be its only train stop in Tallinn and will connect Ülemiste City to other Baltic capitals and will improve the connection between Central and Northern Europe and Germany.

CASE STUDY **UK**

In Britain today, only 2 per cent of commuters cycle to work compared to 19 per cent in Denmark and 27 per cent in the Netherlands. Compared to £24 per person in these European countries, the UK only spends £1.38 per person on cycling infrastructure. Yet, the solution is not simply building more cycling infrastructure but is multifaceted and could take decades to implement. Wardlaw (2014) suggests the following measures:

- Safe cycle parking
- Integration with public transport
- Education of cyclists and motorists
- Legal protection for safe road users, such as strict liability laws
- Bike-sharing programmes
- Restrictions on car ownership and use
- Compact land development to shorten trip lengths.

Arguably, more so than in France, English planners have tempered the worst depredations of industrial blight and sprawl but strangled housing supply. Despite polarisation and policy discontinuity, some remarkable regeneration projects have transformed cities like Liverpool, Manchester and Glasgow (Tallon 2010). Iconic projects notwithstanding, British housing markets remain starkly segregated. Despite demographically induced intensification pressures, pockets of deprivation persist (Meen 2009). To eliminate them, the Urban Task Force (1999) made over 100 recommendations, including design excellence, brownfield development and higher densities. Healey (2007: 227) challenges the implicit *dirigiste* assumption behind such interventions to protect static 'ideal' communities. People have multiple communities in complex, evolving relational geographies. Over the decades, significant public resources such as Community Development Projects or Neighbourhood Renewal Strategies have attenuated territorial injustice.

Arguably, massive development projects like the Olympic Park, Crossrail and King's Cross regeneration accentuate polarisation. Whilst the *London Plan* (Greater London Authority 2011) identified 'polarisation' (*ibid.*: s1.27) and paid lip service to 'promote equality and tackle deprivation' (*ibid.*: s4.61), for Edwards (2009), King's Cross regeneration was 'essentially a business activity aimed at growth and competitiveness'. Despite supposed 'extensive "consultation," local communities felt disenfranchised'. Locals were 'endlessly listened to' but had 'no detectable power to determine the outcome' (*ibid.*: 23). Regionally, UK planning administration presents a confused jumble of district, county or local tiers to frustrate coherent national housing supply. Local resentment centres on the authoritarian imposition of geographically maladapted housing targets. Privileged locales articulate objections most strongly. For Piketty (2014) the root cause of UK polarisation is the wasteful economy of 'patrimony'. Instead of treating the root causes of capital's concentration and malignancy, the regime tinkers intermittently with its symptoms. Rather than progressive tax reforms to curtail evasion or speculative excess, policy fluctuates electorally within a media circus. Robust educational reform falters in the face of entrenched inequality, ministerial posturing and departmental managerialism or chronic ineptitude. Nationally, London's economic dominance festers, undermining affordability and destabilising long-term productivity growth. Regionally, distinctive adjacent towns like Gloucester and Cheltenham reflect atavistic class divisions. Notwithstanding a charade of contrived 'festivals', status differentials (rooted in wealth and housing inequity) fracture local communities and undermine authentic 'dwelling' (Heidegger 1954; Seamon 2000). In the populist imagination, rogue landlords exploit an impecunious and unskilled underclass of renters on zero hour contracts who, in desperation, turn to unscrupulous payday lenders. In stark contrast, the bourgeoisie relish status, overpriced dwellings, trophy spouses, outlandish vehicles or designer baubles. In this extractive narrative, corruption, cronyism and financial malpractice enrich not enterprise or effort. A balanced assessment of the UK built environment backdrop sits between the extreme narratives but wealth inequality remains troubling. According to the Office for National Statistics (ONS 2014), the richest 10 per cent of the UK population controls 44 per cent of the nation's total wealth. In contrast, the poorest half of the population subsists on 9 per cent of the resources (Lucchino and Morelli 2012). Current UK government urban policy is investment-orientated and growth-focused with somewhat less concern for authentic community engagement and distributive justice (Rawls 1971). Policy flux and factional wrangling has left a muddle and a bewildering confusion of policy levers:

- Local Growth Fund (LGF), available for Local Enterprise Partnerships (LEPs)
- The Growing Places Fund (GPF)
- Regional Growth Fund (RGF) Infrastructure Guarantees
- Public Works Loan Board (PWLB)
- Enterprise Zones (EZs)
- Community Infrastructure Levy (CIL).

To conclude, the – admittedly eclectic – review of UK policy context revealed two opposing euphoric or gloomy narratives but impels a considered planning mechanism to address invidious aspects of spatial and social malignancy without undermining the rule of law or sparking nefarious unintended consequences. SUR's socially inclusive aspirations require proper due diligence around partnership structure and public or private funding models. In deprived areas, effective public realm enhancement is expensive. Outlays are either directly recouped from local beneficiaries or they are indirectly recovered from proximate or remote general taxation.

for smart urban development are a constitutional and legislative framework, a thriving economy and universal access to dignified housing, sanitation, utilities, education and health services. Beyond dignity and empowerment for human flourishing, paradox vests in three irreconcilable tensions. First is *dirigisme* or processual evolution in institutional networks. Second is freedom or community cohesion. Third is a productivity or humanist imperative that manifests in alternate innovation or resilience and hedonic or *eudemonic* directions. Contention persists between local identity, environmental quality thresholds, low carbon imperatives, public health and disparate individual impulses. Budget constraints and opportunity costs mean that urban megaprojects likely reinforce privilege but distract from real substantive development challenges like raising productivity and delivering decent, affordable homes and quality public services for the masses. Instrumental development interventions include enhancing streetscapes, improving the public realm and fostering technological innovation.

Administratively, smart development is adaptive and involves a network of resilient public and private sector institutions and local dialogue. SUR seeks to redress spatial injustice, fragmentation and community alienation. However, the approach is not revolutionary but evolutionary. Resilient or smart urban systems are flexible and guided by considerations of the long-term public interest rather than short-term individual/commercial optimisation. Financial responsibility to avoid waste and ecological limits to sprawl and ribbon development notwithstanding, smart systems muddle through with variety, redundancy or excess capacity. Rather than grandiose megaprojects, smart development builds on existing strengths via judicious and selective interventions. Without a final blueprint yet mindful of ecological limits, smart interventions foster sustainable innovation. For example, rather than a glitzy construction project or even low carbon infrastructure, smart interventions to regenerate a dying town could involve cutting retail taxes or sustaining local artists, musicians or entrepreneurs.

SUR confronts many issues; including national policy flux and funding constraints. Municipal administrations, with multiple objectives and often saddled with spatial muddle, drift strategically. Hindrances include collaborative silos, lack of professionalism or meta-cognition and inadequate geospatial systems. Instead of foresight or consultation, short-term gaming by vested interests can dominate decision-making. In smart systems, legitimacy, governance, tendering and milestone reporting transparency help build network trust. Administrative competence, public and private sector collaboration helps to craft policy, refine schemes or initiates and delivers resilient urban regeneration projects. As in the Venetian Republic, the system diffuses power vertically and horizontally, subject to multiple checks and balances, feedback and policy or commercial learning. For public servants, expertise or capability is necessary but insufficient without commitment to the community. Informed deliberation (meta-cognition) in government committees, private companies or local bodies relies on quality LAS data, observation, dialogue and audit of outcomes, outputs and impacts. Ethics, professionalism, the rule of law, trust and standards salience bind commercial contracts, stable and effective partnerships, independent assessments and sound conflict resolution mechanisms. Smart development blends selective and wise *dirigisme* with evolutionary adaptations, shaped in administrative, commercial and local networks. It shuns autocratic or laissez-faire extremes. Smart anathemas are corruption, maladministration, ill-considered decrees, 'white elephant' projects, land grabbing, predatory markets, industrial blight or unchecked sprawl.

Planners need to balance betterment aspirations with practical administrative realities and competing intervention foci. Sober legal, planning and financial considerations tame transformational zeal for urban realm enhancement or spatial equity. Due process, scrutiny, oversight and reflection help avoid ill-considered development. *Ex-ante* evaluation includes modelling and qualitative reflection (meta-cognition) by independent and trustworthy professionals. Independent monitoring evaluates intervention outputs (policy, contracts), construction milestones, local transformative outcomes and community impacts. Measurable targets range across environmental (indicator species numbers, noise, green space, energy use, waste), urban (heritage, density, connectivity, pedestrianisation, cycling, public transport, permeability), social (culture, tolerance, housing affordability, education, health) and economic (business start-ups, financial returns, jobs, tax receipts) outcomes. Eventually, and despite issues of spatial resolution or temporal cut-off, indicators of disease, poverty or crime should decline. In short, smart development rejuvenates communities, strengthens resilience and improves health.

Despite contested political philosophies, multiple imaginings and futures visions, smart development is imperative to tackle chronic urban malfunction and mitigate climate risk. It provides a non-prescriptive collaborative framework for *eudemonic* empowerment and adaptive evolution in collaborative commercial, public and local networks. Foresight is tempered by dialogues and localism to eschew profligate 'white elephants' or the worst debilitating social depredations of laissez-faire.

Stakeholder implications

Governance and planning
- Mix between *dirigisme* and evolution in a network of multiple vertical, horizontal and local institutions
- No final blueprint but adaptable and resilient systems, configured for long-term survival rather than short-term, commercial optimums
- Coherent national–local spatial and climate change policies to integrate plans, local regulations and actions
- Collaborative decision-making at national, regional, metro and local scales
- Institutional capability reinforced by meta-cognition and judgement of well-trained professional administrators, motivated by a strong sense of duty
- Informed subsidiarity with appropriate degree of community/local deliberation and participatory planning

- Strong governance with separation between project champions, case officers and auditors
- Efficient, predictable, fair and cost-effective.

Decision-support technologies

- Considered use of appropriate technologies, linked to sustainable local practices.
- Decision-support systems enrich information field (big data).
- Transparent Land Administration System data dissemination and interaction via smart technologies and mobile platforms.
- Asset management systems to:
  - o Log significant heritage, landscape assets or environmental habitats
  - o Identify geological/contamination/flood hazards.

Ecological integrity

- Ecological carrying capacity limits with strong wildlife and ecological habitat preservation, conservation or adaptation.
- Energy efficient buildings.
- Water resource management.
- Waste management.
- Air pollution monitoring and mitigation.

Identity (sense of place)

- Architecture and urban design celebrate local history, climate, ecology and building traditions.
- Sensitive to place character, landscape and habitat ecologies.
- Simplified culturally sensitive forms and legible public realm (piazzas, playgrounds, parks).

Urban design

- Street-level coherence and aesthetic appeal.
- Diverse neighbourhoods (*quartiers*).
- Appropriate density.
- Central vibrancy (residential).
- Defined and accessible public spaces.
- Mixed-use, compact building design.
- Range of housing opportunities (tenure choice, housing types, locations).
- Adequate supporting infrastructure for accessing a range of quality facilities (community centres, gymnasia, swimming pools, schools, clinics).

Conviviality

- Walkable neighbourhoods
  - o Permeable
  - o Pedestrian friendly
  - o Shaded pavements
  - o Controlled street crossings (speed bumps).

- Human-scale, diverse neighbourhoods and hierarchy of places
  o Individual
  o Friendship
  o Householders
  o Neighbourhood
  o Communities
  o Pedestrian pocket.

Logistics and regional integration
- Urban infill and intensification of existing centres.
- Sprawl curtailed.
- Transit-oriented developments (TOD).
- Integrated, multimodal public transit connectivity.
- Variety of transportation modes.

Bibliography

Acemoglu, D. and Robinson, J. (2013) *Why Nations Fail: The Origins of Power, Prosperity and Poverty*. London: Profile Books.

Adair, A., Berry, J., McGreal, S., Dennis, B. and Hirst, S. (2000) 'The financing of urban regeneration'. *Land Use Policy* 17(2): 147–156.

Adair, A., Berry, J., Hutchinson, N. and McGreal, W. (2007) 'Attracting institutional investment into regeneration: necessary conditions for effective funding'. *Journal of Property Research* 24(3): 221–240.

Altshule A. and Luberoff, D. (2003) *Mega-Projects: The Changing Politics of Urban Public Investment*. Washington, DC: The Brookings Institution.

Atkinson, R. (1999) 'Project management: cost, time and quality, two best guesses and a phenomenon, it's time to accept other success criteria'. *International Journal of Project Management* 17(6): 337–342.

Banister, D. (2012) 'Assessing the reality – transport and land use planning to achieve sustainability'. *Journal of Transport and Land Use* 5(3): 1–14.

Batty, M. (2013) *The New Science of Cities*. Cambridge, MA: MIT Press.

Beard, V. A., Mahendra, A. and Westphal, M. I. (2016) *Towards a More Equal City: Framing the Challenges and Opportunities*. Working Paper. Washington, DC: World Resources Institute, available online at: www.citiesforall.org

Blanton, R., Feinman, G., Kowalewski, S. and Peregrine, P. (1996) 'Agency, ideology and power in archaeological theory: dual-processual theory for the evolution of Mesoamerican civilization'. *Current Anthropology* 37(1): 1–14.

Bodin, J. (1992 [1576]). *Six livres de la République*, Edited by J. Franklin. Cambridge: Cambridge University Press.

Bordieu, P. (1998) *Acts of Resistance against the Tyranny of the Market*. New York: New Press.

Bosker , M., Buringh, E. and Luiten van Zanden, J. (2008) 'From Baghdad to London: the dynamics of urban growth in Europe and the Arab world 800–1800'. *Centre for Economic Policy Research* Discussion Paper 6833.

Brookes, N. (2013) *Emergent Cross-case and Cross-sectoral Themes from the Megaproject Portfolio: An Interim Review*. The Megaproject Cost Action Group.

Bruijn, H. and Leijten, M. (2008) 'Megaprojects and contested information', in *Decision-making on Mega-projects: Cost–benefit Analysis, Planning and Innovation*, Eds: Priemus, H., Flyvbjerg, B. and van Wee, B. Cheltenham: Edward Elgar.

Calegari, G. R. and Celino, I. (2015) 'Smart urban planning support through web data science on open and enterprise data'. *Proceedings of the 24th International World Wide Web Conference*, Florence, Italy, May. Retrieved from http://dl.acm.org/citation.cfm?id= 2742131

Chan, A. P. C. and Chan, A. P. L. (2004) 'Key performance indicators for measuring construction success'. *Benchmarking: An International Journal* 11(2): 203–221.

Childers, D., Cadenasso, M., Grove, M., Marshall, V., McGrath, B. and Pickett, S. (2015) 'An ecology for cities: a transformational nexus of design and ecology to advance climate change resilience and urban sustainability'. *Sustainability* 7(4): 3774–3791.

Colantonio, A. and Dixon, T. (2010) *Urban Regeneration and Social Sustainability: Best Practice from European Cities.* Chichester: Wiley-Blackwell.

Couch, C. (1990) *Urban Renewal.* London: Macmillan.

Cox, R. F., Issa, R. R. A. and Aherns, D. (2003) 'Management's perception of key performance indicators for construction'. *Journal of Construction Engineering and Management* 129(2): 142–151.

Dahl, C. (1959) 'The American school of catastrophe'. *American Quarterly* 3: 380–390.

Davivongs, V., Yokohari, M. and Hara, Y. (2012) 'Neglected canals: deterioration of indigenous irrigation system by urbanization in the west peri-urban area of Bangkok metropolitan region'. *Water* 4: 12–27.

Diamond, D. W. and Verrecchia, R. E. (1991) 'Disclosure, liquidity, and the cost of capital'. *Journal of Finance* 46(4): 1325–1359.

Easley, D. and O'Hara, M. (2004) 'Information and the cost of capital'. *Journal of Finance* 59(4): 1553–1583.

Edwards, M. (2009) 'King's Cross: renaissance for whom?' in *Urban Design, Urban Renaissance and British Cities*, Ed: Punter, J. London: Routledge.

Ekstrom, J. A. and Young, O. R. (2009) 'Evaluating functional fit between a set of institutions and an ecosystem'. *Ecology and Society* 14(2): 16, available at: www.ecologyandsociety. org/vol14/iss2/art16/

Erie, S., Kogan, V. and Mackenzie, S. (2011) *Paradise Plundered: Fiscal Crisis and Governance Failures in San Diego, California.* Stanford, CA: Stanford University Press.

European Climate Foundation. (2010) *Roadmap 2050 a practical guide to a prosperous, low-carbon Europe,* The Hague, available at: www.roadmap2050.eu/attachments/files/Volume1_ fullreport_PressPack.pdf (accessed 15 July 2014).

Finance for Good. (2014) *Social Impact Bonds,* available at: http://financeforgood.ca/about-social-impact-bonds (accessed 24 March 2014).

Floater, G., Rode, P., Friedel, B. and Steering, A. (2015) 'Urban growth: governance, policy and finance'. NCE Cities; Paper 02, LSE Cities, available at: http://files.lsecities.net/ files/2014/12/Steering-Urban-Growth-02.pdf (accessed 2 March 2015).

Flyvbjerg, B., Bruzelius, N. and Rothengatter, W. (2003a) *Megaprojects and Risk: An Anatomy of Ambition.* Cambridge: Cambridge University Press.

Flyvbjerg, B., Holm, M. and Buhl, S. (2003b) 'How common and how large are cost overruns in transport infrastructure projects?' *Transport Reviews* 23(1): 71–88.

Forbes, B. B., Stammler, F., Kumpula, T., Meschtyb, N., Pajunen, A. and Kaarlejärvi, E. (2009) 'High resilience in the Yamal-Nenets social–ecological system, West Siberia Artic, Russia'. *PNAS* 106: 22041–22048.

Foucault, M. (1977) *Discipline and Punish: The Birth of the Prison.* New York: Vintage Books.

Francis, R. and Chadwick, M. (2013) *Urban Ecosystems: Understanding the Human Environment.* Abingdon: Routledge.

Freedman, L. (2014) *Strategy: A History.* Oxford: Oxford University Press.

Freeman, M. and Beale, P. (1992) 'Measuring project success'. *Project Management Journal* 23(1): 8–17.

GBD MAPS Working Group. (2016) *Executive Summary. Burden of Disease Attributable to Coal-Burning and Other Major Sources of Air Pollution in China*. Special Report 20. Boston, MA: Health Effects Institute.

Geddes, P. (1915) *Cities in Evolution: An Introduction to The Town Planning Movement and to the Study of Civics*. London: Williams and Norgate.

Geltner, D. and de Neufville, R. (2014) 'Uncertainty, flexibility, valuation and design: how 21st century information and knowledge can improve 21st century urban development'. *Pacific Rim Property Research Journal* 18(3): 231–276.

Giles-Corti, B., Vernez-Moudon, A., Reis, R., Turrell, G., Dannenberg, A. L., Badland, H., *et al.* (2016), 'City planning and population health: a global challenge'. *The Lancet*, September 2016, DOI: 10.1016/s0140-6736(16)30066-6.

Granoff, I., Hogarth, J. R. and Miller, A. (2016) 'Nested barriers to low-carbon infrastructure investment'. *Nature Climate Change* 6: 1065–1071.

Greater London Authority. (2011) *London Plan: Spatial Development Strategy for Greater London*, available at: www.london.gov.uk/priorities/planning/publications/the-london-plan (accessed 15 July 2014).

Greenleaf, R. (1977) *Servant Leadership: A Journey into the Nature of Legitimate Power and Greatness*. New York: Paulist Press.

Güell, J. and Redondo, L. (2012) 'Linking territorial foresight and urban planning'. *Foresight* 14(4): 316–335.

Hall, P. and Hesse, M. (2012) 'Reconciling cities and flows in geography and regional studies', in *Cities, Regions and Flows*, Eds: Hall, P. and Hesse, M. Abingdon: Routledge.

Harvey, D. (1989) 'From managerialism to entrepreneurialism: the transformation in urban governance in late capitalism'. *Geografiska Annaler. Series B, Human Geography*, 71(1): The Roots of Geographical Change: 1973 to the Present (1989), pp. 3–17.

Healey, P. (2007) *Urban Complexity and Spatial Strategies: Towards a Relational Planning for our Times*. Abingdon: Routledge.

Heidegger, M. (1954) *Bauen Wohnen Denken*. In *Vorträge und Aufsätze*. Pfullingen: Neske. Translated by A. Hofstadter (1971).

Hemphill, L., Berry, J. and McGreal, S. (2004) 'An indicator-based approach to measuring sustainable urban regeneration performance: Part 1, conceptual foundations and methodological framework'. *Urban Studies* 41(4): 725–755.

HM Treasury (2013) National Infrastructure Plan, available at: www.gov.uk/government/publications/national-infrastructure-plan-2013 (accessed 4 April 2014).

Hoepfner, W. (1997) *Das dorische Thera*, Berlin: Gebrüder Mann Verlag.

Holling, C. (1973) 'Resilience and the stability of ecological systems'. *Annual Review of Ecology, Evolution and Systematics* 4: 1–23.

Howard, E. (1902 [1898]) *Garden Cities of Tomorrow*, original title: *To-morrow: A Peaceful Path to Real Reform*. London: S. Sonnenschein & Co., Ltd.

Hughes, B. (2017) *Istanbul: A Tale of Three Cities*, Kindle Edition. London: Weidenfeld & Nicolson.

Jadevicius, A. and Huston, S. (2014) *Cheltenham and Gloucester: A Tale of Two Cities*. Cirencester, RELM working paper, 16 May.

Jakobsen, S-E. and Høvig, O. S. (2014) 'Hegemonic ideas and local adaptations: development of the Norwegian regional restructuring instrument'. *Norsk Geografisk Tidsskrift – Norwegian Journal of Geography* 68(2): 80–90.

Kauko, T. (2017) *Pricing and Sustainability of Urban Real Estate*. Abingdon: Routledge.

Kim, A. (2016) 'Planning inclusive cities: working with the revealed spatial preferences of urban migrants in Asia'. WRR Cities Seminar Series, available at: www.wrirosscities.org/media/video/planning-inclusive-cities-working-revealed-spatial-preferences-urban-migrants-asia (accessed 1 January 2016).

Kratke, S. (2010) 'Creative cities and the rise of the dealer class: a critique of Richard Florida's approach to urban theory'. *International Journal of Urban and Regional Research* 34: 835–853.

Landry, C. (2012) *The Art of City Making*. Abingdon: Routledge.

Lindblom, C. E. (1959) 'The science of "muddling through"'. *Public Administration Review* 19(2): 79–88.

Litman, T. (2014) *Analysis of Public Policies that Unintentionally Encourage and Subsidize Urban Sprawl*, forthcoming, London, LSE Cities. Victoria Transport Policy Institute, commissioned by LSE Cities for the New Climate Economy Cities Program.

López, R., Thomas, V. and Wang, Y. (2008) *The Quality of Growth: Fiscal Policies for Better Results*. Independent Evaluation Group, World Bank Working Paper 2008/6.

Low, S. P. and Chuan, Q. T. (2006) 'Environmental factors and work performance of project managers'. *International Journal of Project Management* 21(1): 24–37.

Lucchino, P. and Morelli, S. (2012) *Inequality, debt and growth*. Resolution Foundation, available at: www.resolutionfoundation.org/media/media/downloads/Final_-_Inequality_debt_and_growth.pdf (accessed 11 June 2014).

McNeill, J. R. (2000) *Something New under the Sun: An Environmental History of the Twentieth Century World*. New York: W. W. Norton.

Mann, C. and Absher, J. (2014) 'Adjusting policy to institutional, cultural and biophysical context conditions: the case of conservations banking in California'. *Land Use Policy* 36: 73–82.

Meen, G. (2009) 'Modelling local spatial poverty traps in England'. *Housing Studies* 24(1): 127–147.

Mehan, A. (2016) 'Investigating the role of historical public squares on promotion of citizens' quality of life'. *Procedia Engineering* 161: 1768–1773.

Millard, J. (2014) *Smart Cities and Social Innovation*. European Social Innovation Research, available at: http://siresearch.eu/blog/smart-cities-and-social-innovation (accessed 15 July 2014).

Montgomery, J. (2008) *The New Wealth of Cities: City Dynamics and the Fifth Wave*. Aldershot: Ashgate.

Morris, I. (2005) *The Growth of Greek Cities in the First Millennium BC*, Stanford University, available at: https://www.princeton.edu/~pswpc/pdfs/morris/120509.pdf (accessed 6 July 2016).

Munns, A. K. and Bjeirmi, B. F. (1996) 'The role of project management in achieving project success'. *International Journal of Project Management* 14(2): 81–87.

Najemy, J. (2010) *The Cambridge Companion to Machiavelli*. Cambridge: Cambridge University Press.

Nase, I., Berry, J. and Adair, A. (2013) 'Real estate value and quality design in commercial office properties'. *Journal of European Real Estate Research* 6(1): 48–62.

Nguyen, L. D., Ogunlana, S. O. and Lan, D. T. (2004) 'A study on project success factors on large construction projects in Vietnam'. *Engineering Construction and Architectural Management* 11(6): 404–413.

Norwich, J. (2007) *A History of Venice*. London: Folio Society.

ONS (2014) Chapter 2: Total wealth, wealth in Great Britain 2010–12, available at www.ons.gov.uk/ons/dcp171776_362809.pdf (accessed 11 June 2014).

Owens, S. and Cowell, R. (2011) *Land and Limits: Interpreting Sustainability in the Planning Process*, 2nd Edition. Abingdon: Routledge.

Pattberg, P. and Widerberg, O. (2014) *Transnational Multi-stakeholder Partnerships for Sustainable Development – Building Blocks for Success*. Amsterdam: IVM Institute for Environmental Studies.

Piketty, T. (2014) *Capital in the Twenty-first Century*. Cambridge, MA: Harvard University Press.

Popper, K. (1945 [1966]) *The Open Society and Its Enemies*, Vol. 1, 5th Edition. Princeton, NJ: Princeton University Press.

Pregill, P. and Volkman, N. (1999) *Landscapes in History: Design and Planning in the Eastern and Western Traditions*, 2nd Edition. New York: John Wiley.

Rawls, J. (1971) *A Theory of Justice*. Cambridge, MA: Belknap.

Régnier, C., Achaz, G., Lambert, A., Cowie, R., Bouchet, P. and Fontaine, B. (2015) 'Mass extinction in poorly known taxa'. *PNAS* 112(25): 7761–7766; doi:10.1073/pnas.1502350112.

Roberts, P. (2000) 'The evolution, definition and purpose of urban regeneration', in *Urban Regeneration: A Handbook*, Eds: Roberts, P. and Sykes, H. London: Sage.

Robinson, M. (2012) *Felling the Ancient Oaks: How England Lost its Great Country Estates*. London: Aurum Press.

Rogers, P. (2012) *Resilience & the City: Change, (Dis)order and Disaster*. Farnham: Ashgate.

Rutter, M. (1985) 'Resilience in the face of adversity: protective factors and resistance to psychiatric disorder'. *British Journal of Psychiatry* 147: 598–611.

Ryan, R. and Deci, E. (2001) 'Happiness and human potentials: a review of research on hedonic and eudaimonic well-being'. *Annual Review of Psychology* 52: 141–166.

Sacco, P., Ferilli, G. and Blessi, G. T. (2014) 'Understanding culture-led local development: a critique of alternative theoretical explanations'. *Urban Studies* 51(13): 2806–2821.

Savindo, V., Grobler, F., Parfitt, K., Guvenis, M. and Coyle, M. (1992) 'Critical success factors for construction projects'. *Journal of Construction, Engineering and Management* 118(1): 94–111.

Schimmel, A. (2004 [2000]) *The Empire of the Great Mughals: History, Art and Culture*. London: Reaktion Books.

Scott, A. J. (2001) 'Globalization and the rise of city-regions'. *European Planning Studies* 9: 813–826.

Scott, A. J. (2008) 'The resurgent metropolis: economy, society and urbanization in an interconnected world'. *International Journal of Urban and Regional Research* 32: 548–564.

Seamon, D. (2000) 'Concretizing Heidegger's notion of dwelling: the contributions of Thomas Thiis-Evensen and Christopher Alexander', in *Building and Dwelling [Bauen und Wohnen]*, Ed.: Führ, E. Munich, Germany: Waxmann Verlag GmbH.

Searle, G., Darchen, S. and Huston, S. (2014) 'Positive and negative factors for Transit Oriented Development: case studies from Brisbane, Melbourne and Sydney'. *Urban Policy and Research* 32(4): 437–457.

Shahrokni, H., Lazarevic, D., and Brandt, N. (2015) 'Smart urban metabolism: towards a real-time understanding of the energy and material flows of a city and its citizens'. *Journal of Urban Technology* 22(1): 65–86.

Smith, N. (1979) 'Gentrification and capital: theory, practice and ideology in Society Hill'. *Antipode* 11(3): 24–35.

Sohail, M. and Baldwin, A. N. (2004) 'Performance indicators for microprojects in developing countries'. *Construction Management and Economics* 22(1): 11–23.

Southworth, M. (2005) 'Designing the walkable city'. *Journal of Urban Planning and Development* 131(4): 246–257.

Storper, M. and Scott, A. J. (2009) 'Rethinking human capital, creativity and urban growth'. *Journal of Economic Geography* 9(2): 147–167.

Stratigea, A., Papadopoulou, C-A. and Panagiotopoulou, M. (2015) 'Tools and technologies for planning the development of smart cities'. *Journal of Urban Technology* 22(2): 43–62.

Tallon, A. (2010) *Urban Regeneration in the UK*. 2nd Edition. Abingdon, Routledge.

Termeer, C. J. A. M., Dewulf, A. and van Lieshout, M. (2010) 'Dis-entangling scale approaches in governance research: comparing monocentric, multilevel, and adaptive governance'. *Ecology and Society* 15(4): 29, availiable at: www.ecologyandsociety.org/vol15/iss4/art29/ (accessed 26 March 2014).

Thomas, V., Dailami, M., Dhareshwar, A., López, R. Kaufmann, D., Kishor, N. and Wang, Y. (2000) *The Quality of Growth*. New York: Oxford University Press.

Town and Country Planning Association. (2017) *Garden City Principles*, available at: https://www.tcpa.org.uk/garden-city-principles (accessed 4 January 2017).

Turner, A. (2014) 'Wealth, debt, inequality and low interest rates: four big trends and some implications', available at: www.cass.city.ac.uk/__data/assets/pdf_file/0014/216311/RedingNotes_Lord-Turner-Annual-Address-at-Cass-Business-School-March-26-2014.pdf (accessed 6 May 2014).

Turok, I. (1992) 'Property-led urban regeneration: panacea or placebo?' *Environment and Planning A* 24: 361–379.

UN. (2014) *World urbanization prospects*, available at: https://esa.un.org/unpd/wup/Publications/Files/WUP2014-Highlights.pdf (accessed 6 July 2016).

Urban Task Force (1999) *Towards an Urban Renaissance*. London: Spon.

Van Marrewijk, A., Clegg, S., Pitsis, T. and Veenswijk, M. (2008) 'Managing public–private megaprojects: paradoxes, complexity, and project design'. *International Journal of Project Management* 26(6): 591–600.

Vanolo, A. (2014) 'Smart mentality: the smart city as disciplinary strategy'. *Urban Studies* 51(5): 883–898.

Veblen, T. (1919) *The Vested Interests and the Common Man*. New York: B. W. Huebsch.

Veblen, T. (1994) [1899]) *Theory of the Leisure Class: An Economic Study in the Evolution of Institutions*. New York: Penguin Books.

Von Braun, J. and Thorat, S. (2014) 'Policy implications of exclusion and resilience', in *Resilience for Food and Nutrition Security*, Eds: Fan, S., Pandya-Lorch, R. and Yosef, S. Washington, DC: International Food Policy Research Institute.

Wadley, D. (2010) 'Exploring a quality of life, self-determined'. *Architectural Science Review* 53(1): 12–20.

Wardlaw, M. (2014) 'History, risk, infrastructure: perspectives on bicycling in the Netherlands and the UK'. *Journal of Transport and Health* (1): 243–250.

Wedding, G. C. and Crawford-Brown, D. (2007) 'Measuring site-level success in brownfield redevelopments: a focus on sustainability and green building'. *Journal of Environmental Management* 85(2): 483–495.

Westerveld, E. (2003) 'The project excellence model: linking success criteria and critical success factors'. *International Journal of Project Management* 21(6): 411–418.

Whittington, R. (2001) *What Is Strategy and Why Does It Matter?* 2nd Edition. London: Cengage Learning.

WHO (2010) Environment and health risks: a review of the influence and effects of social inequalities. WHO Regional Office for Europe, Copenhagen, Denmark, available at: www.euro.who.int/__data/assets/pdf_file/0003/78069/E93670.pdf (accessed 17 January 2017).

WHO (2014) Global Health Observatory (GHO) data, available at: www.who.int/gho/urban_health/situation_trends/urban_population_growth_text/en/ (accessed 6 July 2016).

Wilson, G. A. (2014) 'Community resilience: path dependency, lock-in effects and transitional ruptures'. *Journal of Environmental Planning and Management* 57(1): 1–26.

Winston, N. (2010) 'Regeneration for sustainable communities? Barriers to implementing sustainable housing in urban areas'. *Sustainable Development* 18(6): 319–330.

2

GREEN INFRASTRUCTURE

Will Manley and Sophia Price

Abstract

Future smart cities require 'green infrastructure' (GI). Features of GI are long established including urban parks and landscaping. More recently, GI terminology has become normalised and internationalised. Hence, a significant task for the second chapter is to clarify GI definitions, functions, benefits and emphases. This chapter also provides the necessary background to GI and gives a particular focus to the linkages and connectivity within the GI systems. These connective linkages are attributed a primary function within the context of place and their social, environmental and economic impacts. The ability and requirements to maximise other benefits are then explored, ultimately aiming for connective GI that optimises multifunctional benefits within the parameters of its primary function, be that for example a cycleway or flood management drainage system. GI has a recognised role to play in promoting sustainable approaches to urban development, and in tandem with provision of ecosystem services. We consider the role of GI at different scales, and different opportunities of retrofitting for existing developments or new developments. Rather than providing a best practice guideline, we illuminate key GI issues in relation to ongoing management and future proofing via a case study.

Key terms

Green infrastructure, retrofitting, future-proofing social, environmental and economic impacts

Introduction

As a descriptive term green infrastructure (GI) was conceived in the USA (Benedict and McMahon 2002) with a focus on a strategic approach to land conservation and planning, weighted towards ecological processes. It has some resonance with the world of civil engineering with its comparable 'grey infrastructure' including pavements, roads, railways and so on that link and interlink within and between settlements. Continuing the colour theme, GI can also encompass wetlands, such as watercourses and coastal environments, sometimes referred to more specifically as blue infrastructure. The general term GI has developed and broadened to describe a network of green and blue spaces within urban and rural environments (see Box 2.1). Greenways or green corridors are other common terms applied to linear features of GI (Ahern 1995), whereas the terms green spaces and green belts do not imply that connecting function (Fabos 1995; Jongman and Pungetti 2004; Natural England 2009). GI can also be included, in part, within the 'public realm', with the larger more obvious elements including public parks and sports pitches and, on a smaller scale, roadside verges and street trees; however the public realm also includes 'hard' landscape such as pavements and seating. For GI, whilst connectivity or green *linkages* are often crucial, the parallel with the civil engineering term *infrastructure* is perhaps misleading, but UK policy and local GI implementation practice stress its anthropogenic connectivity role. Box 2.1 summarises key GI components. This is followed by a detailing of the broader benefits of GI, and the concept and benefits of ecological networks are subsequently described. The chapter also considers issues in relation to valuing GI and uses a case study to generate design and management guidelines.

BOX 2.1 **A GREEN INFRASTRUCTURE TYPOLOGY**

- **Parks and gardens** – urban parks, country and regional parks, formal gardens.
- **Amenity green space** – informal recreation spaces, housing green spaces, domestic gardens, village greens, urban commons, other incidental space, green roofs.
- **Natural and semi-natural urban green spaces** – woodland and scrub, grassland (e.g. downland and meadow), heath or moor, wetlands, open and running water, wastelands and disturbed ground, bare rock habitats (e.g. cliffs and quarries).
- **Green corridors/blue corridors** – rivers and canals including their banks, road and rail corridors, cycling routes, pedestrian paths and rights of way.
- **Other** – allotments, community gardens, city farms, cemeteries and churchyards.

GI has multiple anthropic and biocentric spin-offs. An illustration of this could be, for example as a mechanism for addressing climate change mitigation, and adaptation by planting trees along corridors to provide a potential means to cut carbon emissions. An associated provision of cycling and walking routes further helps climate change adaptation, for example, by encouraging more sustainable forms of transport. Additionally, GI tightens physical communication linkages that instil a greater awareness of, and sympathy towards, the aesthetic and quality of well-being within urban landscapes. Increased population pressures and demands for improved quality and potential of living space further increase the demands on GI. However, GI (and green spaces more generally) frequently improves the quality of living space and, consequently, stimulates adjacent property values (RICS 2011; Mell *et al.* 2013). For many urban and suburban populations, the provision of sufficient high quality and more easily accessible GI within new developments can also reduce anthropogenic pressures on protected areas in the adjacent countryside via activities such as dog walking and cycling.

GI spans a spectrum of different scales from the international (for example mammal migration routes across Africa) to the very localised (for example roof gardens). There has also been a gradual movement more particularly towards an understanding and implementation of habitat connectivity initiatives on farmland and within the wider countryside (Watts *et al.* 2005; Donald and Evans 2006; Arponen *et al.* 2013). Such initiatives on farmland in the UK have been predominantly supported by EU-funded agri-environment schemes, which have promoted relatively simple ecological networks using connecting linear habitats such as field margins or hedgerows. These operate more at the individual farm level, but increasingly landscape-scale initiatives incorporating contiguous units are being valued and encouraged (Franks and Emery 2013). The focus of GI within these initiatives and within this chapter is on the provision of linkages and connectivity between the more substantial green spaces, and also includes the recognised importance that can and should be given to key linkages between the urban and the wider rural hinterland. The linkages between the urban fringe and the surrounding countryside can, for example in many parts of the UK, incorporate an increasing proportion of fragmented landholdings, such as industrial units, market gardens, hobby and lifestyle farmers, and equestrian uses, which are particularly challenging to a holistic or strategic approach to GI. An attempt to address urban fringe challenges is recognised, for example, in the development of England's Community Forests (Blackman and Thackray 2007; Mell 2008).

Benefits of GI

The key social, environmental and economic benefits of GI have been highlighted (e.g. Natural England 2009; Roe and Mell 2013; Benedict and McMahon 2012; eftec 2013) and include a varied range of themes that can be appropriately categorised. The key point is that GI is multifunctional – it provides a wide range of benefits (see Box 2.2), and has the potential to do so simultaneously and from

BOX 2.2 **BENEFITS OF GREEN INFRASTRUCTURE**

Social

- Aesthetic appeal – ensuring a more attractive place for people to live, work and visit.
- Informal recreation – opportunities for outdoor relaxation, play and access to nature.
- Improving health and well-being – providing safer, more aesthetic and practical opportunities for exercise.
- Transport routes – provision of sustainable routes for cycling and walking.
- Local food production – in allotments and gardens.

Environmental benefits

- Habitats – addressing fragmentation of habitats and species isolation by providing wildlife corridors and linkages.
- Climate change adaptation – e.g. flood alleviation.
- Temperature buffering – e.g. shading by trees and urban heat island effect.
- Improving air and water quality.

Economic benefits

- Investment – a more attractive area to business investors and potential residents.
- Increase visitor spend – a more attractive area for tourists and visitors.
- Generating employment – attracting new businesses and residents to the area, increasing workplace occupancy rates and increasing the number of jobs in the area.
- Reduction of environmental costs – managing flood risk, storage of water during droughts, reducing urban heat island effect, improving air quality and filtering diffuse pollution.
- Health benefits – reduction of negative impacts on health through improved air quality and surroundings which encourage activity and improve mental health and well-being.
- Promoting food production – enabling increased and local productivity.

Adapted from eftec (2013).

the same area, length or unit of land. Not many GI projects can deliver the full range of benefits, and the objectives required and targeted must be first identified and the GI designed and/or developed to deliver maximum benefits.

The most pragmatic and potentially effective approach to delivering GI benefits is the identification of a primary function or functions which provides the impetus

for the inclusion and ongoing management of the GI within the urban environment. These primary functions are highly likely to be more economically
rationalised, such as drainage and flood management features (Benedict and
McMahon 2002; Ellis 2013) or cycle access routes (Vandermeulen *et al.* 2011).
Multifunctionality is achieved via the secondary functions, for example through
enhancement of the greener elements that can provide socio-cultural benefits and
also promote more natural systems, such as reducing urban heat island effect,
improving air quality, reducing noise levels and filtering diffuse pollution. These
overlapping benefits are a particular advantage of GI and align constructively with
the ecosystem-based approach, enabling the assessment and valuation of the services
provided by GI (see also Chapter 12 for further details). The socio-cultural benefits
of GI, including the contribution to general health, well-being and quality of life
that access to GI can provide, are also becoming increasingly recognised (DEFRA
2011; Benedict and McMahon 2012). GI can also address the discontinuity between
people living in urban areas and nature (Maller *et al.* 2006) and help to meet targets
for the proximity of urban dwellers to green space, for example within the models
of Accessible Natural Greenspace Standards (Pauleit *et al.* 2003; Natural England
2010).

The benefits derived from components of GI may have an identified importance
to the local communities (Bengston *et al.* 2004), which is worthy of protection by
designating specific parts of the GI, for example through the local planning system
(DETR 2012; Planning Practice Guidance 2014).

Ecological networks

In high density developed countries such as the UK with increasingly fragmented
and isolated landscapes, designated protected wildlife sites are critical but not a
panacea. Many less resilient and more sessile species are particularly vulnerable to
changes to their environments. With little potential to recreate large expanses of
contiguous natural habitats, one approach to counter these conditions is to link
high quality sites to provide an increased expanse of suitable habitats, that is to
ensure ecological connections. Such a network of sites, associated buffer zones,
wildlife corridors and smaller wildlife-rich sites that can also act as 'stepping stones'
can be described as an 'ecological network'. The concept and the implementation
of ecological networks have increased in urban and rural biodiversity policies
(Lawton *et al.* 2010; DETR 2012), with their neat summary of the need for more,
bigger, better and joined up, and the potential to integrate this philosophy and
function within GI.

There is a suggested logic and intuitive appeal to biodiversity benefits of
ecological networks, and the ability and greater ease of movement of species.
However, for many species the linear networks and stepping stones provide an
important but lower quality habitat rather than a physical connection between
wildlife-rich nodal habitats (Dawson 1994; Ignatieva *et al.* 2011). More specifically
evidence suggests that the relative quality of individual sites is more important for

plants and invertebrates than networks for dispersal, but that these networks are important for more vagile groups of species such as small mammals (Angold et al. 2006). There is also other evidence demonstrating the variations in use of these by different taxa (e.g. Gilbert-Norton et al. 2010), but this only highlights the importance of identifying the key objectives of components of GI, and where relevant and appropriate any targeting of specific species or groups of species. It can therefore be suggested that for significant sections of biodiversity, more effective resources and efforts would be better directed at management of green spaces such as parks (Miller et al. 2015; DEFRA 2015).

There are opportunities here to include and incorporate innovative design technologies within GI, for example to facilitate movements across barriers such as road networks that have become a more established ecological tool within more rural environments (Corlatti et al. 2009; Clevenger and Wierzchowski 2006). In addition, an important and related challenge to any wildlife friendly management of green spaces and potentially any GI connectivity features requires almost a paradigm shift in mindset within many public and corporate bodies in respect to the priorities of active management of sites. This could range from high levels of management through to specific examples that embrace a non-intervention approach involving minimal management.

Development and the value of GI

The multifunctional benefits of incorporating GI into new developments or retrofitting it into existing settlements are outlined in the previous section, and summarised in Box 2.2. However these benefits are not necessarily realised by those making the financial investment in GI. For example, a local authority which improves the quality of the GI in its ownership in the form of public parks is likely to see an increase in usage of those parks. This should in turn lead to improved mental and physical health outcomes for the park users, with a parallel decrease in the need for medical interventions or days off through ill health. However this will not equate to a financial gain for the local authority, although it could be seen as an important part of that council's leadership and place-making roles. It may develop financial gains for other public sector bodies, such as in this case health authorities and those providing health-related benefits payments. The challenge remains to be able to demonstrate the economic benefits most directly attributable to, for example, a particular government department, agency or business.

For commercial landowners and developers this lack of direct return for investment is potentially even more unbalanced. A developer establishing a large new residential development will only gain financially from some of the multifunctional benefits that might accrue from the creation of high quality GI. Some of these financial benefits are more tangible as outlined earlier, for example properties that are close to well-designed parks are likely to achieve a higher price and may well sell more quickly (Kong et al. 2007; Jim and Chen 2010; Netusil et al. 2014). Specific elements of GI may enable development to come forward on

---

BOX 2.3 GI AND ECOSYSTEM SERVICES

A number of GI features effectively provide operational ecosystem services, that is the services that are provided by nature and benefit people (MA 2005; NEA 2011), and they face a similar costing challenge analogous to that associated with valuing these services. The tools that are available to value some elements of GI are interlinked with the delivery of ecosystem services, and can estimate the economic value of these services. There has been a surge in studies attempting to address the financial issues and practicalities surrounding their valuations (Gómez-Baggethun and Barton 2013; Jones et al. 2016). It is not developed here but the general concept of ecosystem services are addressed further in Chapter 12.

---

sites that would otherwise be considered as un-developable, for example due to flood risk. A significant Sustainable Urban Drainage System (SuDS) incorporated into the site GI may reduce flood risk to such an extent that development becomes feasible and financially viable.

A commitment on behalf of developers to deliver high quality GI may also play a role in shifting the planning balance by providing public benefits to both new and existing communities that are perceived by determining authorities (for example local planning authorities in England) to outweigh the potential harm that the development may cause, thereby ensuring that planning permission is granted, where it might otherwise be unacceptable. Those authorities that have the power to permit development may also need to provide guidance and policy on the quality and quantum of GI that they expect to see in new developments, making clear their expectations and thereby not only ensuring that high quality GI is designed in and implemented from the start of the development process, but also enabling developers to estimate the potential overall costs of development. It may even encourage developers to better understand how the provision of GI can actually reduce some costs, for example flood alleviation features. Local planning authorities in the UK have addressed this in a number of different ways. Some have prepared stand-alone GI strategies, for example Swindon Borough Council (2011), others have adopted supplementary planning guidance on GI to augment the guidance available from government or institutional organisations such as Natural England (2009) or the Landscape Institute (2011). Most have provided clear policy frameworks through local plan policies, often based on those local GI strategies (see Box 2.4).

High quality, well-integrated and carefully designed green infrastructure and landscape provision is crucial to the long-term success of new developments, ensuring that the maximum multifunctional benefits are achieved for those that live in, work at and visit new developments. The spaces in between new buildings, the surrounding areas, and the connections between a new development and the existing townscape or landscape, are equally important to the design of the structures themselves. The detail of the GI and landscape provided on a development site will be related to various factors including the nature of the site itself, and the type, size and impact of the development. Improved GI and high quality landscape is also of great benefit when introduced into existing built areas.

Given the emphasis on the quality of GI in enabling it to achieve the desired multifunctional benefits, it has become increasingly important to be able to evidence that quality, particularly for developers who are promoting their schemes as being of significant value for GI. There are a number of existing benchmarks or standards that developers or decision makers can use, however they are limited in their scope with respect to GI. The University of the West of England and Gloucestershire Wildlife Trust are currently piloting a GI benchmark (Smith *et al.* 2016), which should enable developers and planners to understand how to achieve the best GI options possible. The benchmark works on a points-based system and covers all stages from design through to long-term management.

BOX 2.3 CASE STUDY ACHIEVING EFFECTIVE GI ON DEVELOPMENT SITES – DESIGN AND MANAGEMENT GUIDELINES

Cotswold District Council has incorporated GI into more general design advice (CDC 2016) – the Cotswold Design Code, which is included within the emerging local plan. Once formally adopted this design guidance will have considerable weight in the determination of planning applications. In order to maximise the multifunctional benefits which the Council hopes will be achieved, the guidance emphasises not only the initial design of the GI but also its long-term management and the involvement of the local community. Key principles are outlined in brief:

Key principles

a   National and local standards and best practice
    The amount, type and design of GI should be informed by the appropriate national and local standards, guidance and best practice, including the Accessible Natural Greenspace Standard from Natural England and the national allotment provision recommendations from the National Society of Allotment and Leisure Gardeners.

b   Local character
    The design of newly created elements of GI and landscape should be inspired by and enhance the character of the existing GI, landscape, biodiversity and built environment of the site and the wider area.

c   Existing landscape features
GI design and distribution should be informed by existing landscape, ecological and historical features. For example stone walls, hedgerows, trees and ponds should be retained and successfully integrated into the GI network.

d   Heritage assets
A new development site may include or fall within the setting of historic buildings, structures and archaeological sites. The GI network should be designed, used and managed in such a way as to protect and enhance the heritage assets and their settings, preserving key views and buffer areas.

e   Interface with existing properties
The interface between a new development and any existing adjacent properties should be designed to respect the amenity of existing residents and to ensure that the existing and new developments are well integrated.

f   On-site GI network
This should function as a network of interconnected green (and blue) spaces, which fulfil various functions including: formal sport, recreation, pedestrian and cyclist routes, accessible natural green space, structural landscaping, and wildlife habitat. Most of the elements of the GI should be multifunctional.

g   Distribution of GI across the site
The GI network should be designed to ensure that all residents, employees and visitors have convenient access to green spaces. This should be achieved through dispersal of meaningful and usable areas across the site. Elements of the GI should be of sufficient size to be functional and easily managed. The GI and landscape provision should be located so that it makes best use of and enhances important local views.

h   GI and landscape provision on individual plots
The landscape design of individual plots and the areas immediately surrounding them (e.g. roadside verges) should be of high quality and should reflect the landscape, ecological and built character of the area. Private spaces such as gardens should be of an appropriate size for the dwelling provided, and should be designed to ensure privacy and adequate daylight. Private spaces should be clearly recognisable as such, through the use of suitable boundary treatments.

i   Interrelationship with off-site GI
The on-site GI should be designed to ensure that it links physically with off-site GI to maximise opportunities for ecological connections, footpath and cycle links, continuity of landscape features, etc.

j   Off-site GI enhancements
Where possible enhancements to off-site GI assets should be achieved, for example increasing public access to nearby land, and better management of wildlife sites in the locality.

k   Sustainable drainage solutions

The principle approach to the SuDS infrastructure should be to ensure that as much of it as possible is provided on the surface, mimicking the natural drainage of the site. This will reduce the burden on the existing sewerage system. The SuDS infrastructure should not only serve a drainage role, but also contribute to the visual amenity and the wider environmental performance of the development. Its management should be fully integrated with the management of other aspects of GI.

l   Green features on buildings

Green features (living roofs and walls, bird or bat boxes, etc.) should be incorporated into new and existing buildings.

m   Biodiversity enhancements

Opportunities should be taken within all areas of GI (and the built environment) to enhance biodiversity through species choice, creation of new habitats, land management, etc. There should be linkages with existing biodiversity assets and networks, and increasing access to nature for people.

n   Species choice

Within planting schemes, species choice should be guided by appropriateness to the local area (with an emphasis on native species); suitability for its function (for example winter screening); value for wildlife; and resilience to climate change.

o   Street trees

Wherever possible street trees should be planted to improve amenity and environmental performance. Street trees can also be used to help to define the character of different areas of a development and improve legibility.

p   Road junctions

The landscape design of new or significantly altered road junctions, particularly at visually prominent locations, should be of high quality, reflect the landscape character of the area, help to give a sense of place, and ensure greater legibility.

q   Pedestrian and cycle routes

The walking and cycling network, which will form part of the GI, should encourage 'active travel', in line with the highway user hierarchy principle. On-site routes should link to off-site non-vehicular routes, particularly those that lead to key destinations such as shops, schools and railway stations. These routes should be designed so that they are also available to the existing residents and businesses in the locality, and they should be implemented early in the delivery of the development.

r   Healthy lifestyles

GI should be designed to encourage healthy lifestyles for all, including: encouraging walking and cycling; provision of formal and informal sports facilities; providing volunteering opportunities; and food production.

s   Provision for all sectors of the community
The amount, distribution and type of GI across a site (and any off-site GI enhancements) should be based on an assessment of the needs of the new residents and other users of the site. Consideration should also be given to helping to meet any shortfall in existing provision.

t   Accessibility
The majority of the GI should be accessible, both physically and socially, to all sectors of the community, providing safe, attractive, welcoming and engaging spaces for local people. It should meet the needs of all sectors of the community, including 'hard to reach' groups and those who may require specific provision (e.g. seating for those with limited mobility).

u   Timing of 'construction' of GI
Where appropriate, elements of the GI network should be 'constructed' in advance of built development. Where this is not appropriate, the timing of their 'construction' should be tied to the relevant phase of built development.

v   Long-term management
The management and monitoring of GI should usually be controlled by a management plan. The plan should clearly set out who will be responsible for the management of the GI and landscape provision. Management plans should be implemented in full and regularly reviewed. Where appropriate the local community should be involved in the management of GI.

Adapted from Cotswold District Local Plan 2011–2031 (CDC 2016).

Conclusion

GI is a long-established concept, and an increasingly dynamic field with new initiatives ranging from policy context through to implementation. This chapter assimilated and summarised the generalities of GI and reflected on its substantive benefits. Crucially, the chapter also considered the incorporation of multifunctional GI within built environment planning and development. The chapter noted the limitations of attempts to monetise GI's positive spin-offs, and acknowledged the challenges in cost–benefit analyses of its correlated ecosystem services. The chapter suggests a pragmatic approach to the design and implementation of GI focusing on the more immediate and localised functional benefits that are simpler to justify in financial terms, for example the delivery of ancillary water-related or public access services, and on its connectivity roles. Other more intangible GI benefits can be mediated through smart city plans and are difficult to isolate and quantify but might include reduction in crime, biodiversity protection or more remote impacts such as cleaner air.

Bibliography

Ahern, J. (1995) Greenways as a planning strategy. *Landscape and Urban Planning* 33(1): 131–155.

Angold, P. G., Sadler, J. P., Hill, M. O., Pullin, A., Rushton, S., Austin, K., *et al.* (2006) Biodiversity in urban habitat patches. *Science of the Total Environment* 360(1–3): 196–204.

Arponen, A., Heikkinen, R. K., Paloniemi, R., Pöyry, J., Similä, J. and Kuussaari, M. (2013) Improving conservation planning for semi-natural grasslands: integrating connectivity into agri-environment schemes. *Biological Conservation* 160: 234–241.

Barton, J., Hine, R. and Pretty, J. (2009) The health benefits of walking in greenspaces of high natural and heritage value. *Journal of Integrative Environmental Sciences* 6(4): 261–278.

Benedict, M. and McMahon, E. (2002) Green infrastructure: smart conservation for the 21st century. *Renewable Resources Journal* 20: 12–17.

Benedict, M. and McMahon, E. (2012) *Green Infrastructure: Linking Landscapes and Communities.* Washington, DC: Island Press.

Bengston, D. N., Fletcher, J. O. and Nelson, K. C. (2004) Public policies for managing urban growth and protecting open space: policy instruments and lessons learned in the United States. *Landscape and Urban Planning* 69(2–3): 271–286.

Blackman, D. and Thackray, R. (2007) *The Green Infrastructure of Sustainable Communities. England's Community Forests*, www.roomfordesign.co.uk

Bolund, P. and Hunhammar, S. (1999) Ecosystem services in urban areas. *Ecological Economics* 29(2): 293–301.

CDC. (2016) Cotswold District Local Plan 2011–2031. Submission draft reg. 19, www.cotswold.gov.uk/residents/planning-building/planning-policy/emerging-local-plan/

Clevenger, A. P. and Wierzchowski, J. (2006) Maintaining and restoring connectivity in landscapes fragmented by roads. In *Connectivity Conservation*, eds K. R. Crooks and M. Sanjayan. Cambridge: Cambridge University Press.

Corlatti, L., Hacklander, K. and Frey-Roos, F. (2009) Ability of wildlife overpasses to provide connectivity and prevent genetic isolation. *Conservation Biology*, 23: 548–556.

Dawson, D. G. (1994) *Are Habitat Corridors Conduits for Animals and Plants in a Fragmented Landscape? A Review of the Scientific Evidence.* English Nature Research report, 94. Peterborough.

DEFRA. (2011) *The Natural Choice: Securing the Value of Nature.* London. Department for Environment, Food and Rural Affairs.

DEFRA. (2015) *Information Sheet – Urban Environment.* The Wildlife Trusts, http://www.wildlifetrusts.org/bees-needs/information-sheets

DETR. (2012) *National Planning Policy Framework.* London. Department for Communities and Local Government.

Donald, P. F. and Evans, A. D. (2006) Habitat connectivity and matrix restoration: the wider implications of agri-environment schemes. *Journal of Applied Ecology* 43: 209–218.

eftec. (2007) *Policy Appraisal and the Environment: An Introduction to the Valuation of Ecosystem Services.* Report for Environment Agency.

eftec. (2013) *Green Infrastructure's Contribution to Economic Growth: A Review.* Report for Defra and Natural England.

Ellis, J. B. (2013) Sustainable surface water management and green infrastructure in UK urban catchment planning. *Journal of Environmental Planning and Management* 56(1): 24–41.

Fabos, J. G. (1995) Greenways introduction and overview: the greenway movement, uses and potentials of greenways. *Landscape and Urban Planning* 33(1): 1–13.

Franks, J. R. and Emery, S. B. (2013) Incentivising collaborative conservation: lessons from existing Environmental Stewardship Scheme options. *Land Use Policy* 30(1): 847–862.

Gilbert-Norton, L., Wilson, R., Stevens, J. R. and Beard, K. H. (2010) A meta-analytic review of corridor effectiveness. *Conservation Biology* 24(3): 660–668.

Gobster, P. H. and Westphal, L. M. (2004) The human dimensions of urban greenways: planning for recreation and related experiences. *Landscape and Urban Planning* 68(2–3): 147–165.

Gómez-Baggethun, E. and Barton, D. (2013) Classifying and valuing ecosystem services for urban planning. *Ecological Economics* 86: 235–245.

Ignatieva, M., Stewart, G. H. and Meurk, C. (2011) Planning and design of ecological networks in urban areas. *Landscape and Ecological Engineering* 7: 17.

James, S. M. (2002) Bridging the gap between private landowners and conservationists. *Conservation Biology* 16(1): 269–271.

Jim, C. Y. and Chen, W. Y. (2010) External effects of neighbourhood parks and landscape elements on high-rise residential value. *Land Use Policy* 27(2): 662–670.

Jones, L., Norton, L., Austin, Z., Browne, A. L., Donovan, D., Emmett, B. A., et al. (2016) Stocks and flows of natural and human-derived capital in ecosystem services. *Land Use Policy* 52: 151–162.

Jongman, R. H. G. and Pungetti, G. (eds) (2004) *Ecological Networks and Greenways. Concept, Design, Implementation.* Cambridge: Cambridge University Press.

Kong, F., Yin, H. and Nakagoshi, N. (2007) Using GIS and landscape metrics in the hedonic price modeling of the amenity value of urban green space: a case study in Jinan City, China. *Landscape and Urban Planning* 79(3–4): 240–252.

Landscape Institute. (2011) *Local Green Infrastructure.* London: Landscape Institute.

Lawton, J. H., Brotherton, P. N. M., Brown, V. K., Elphick, C., Fitter, A. H., Forshaw, J., et al. (2010) *Making Space for Nature: A Review of England's Wildlife Sites and Ecological Network.* Report to DEFRA.

MA (Millennium Ecosystem Assessment). (2005) *Ecosystems and Human Well-being: Synthesis.* Washington, DC: Island Press.

Maller, C., Townsend, M., Pryor, A., Brown, P. and St Leger, L. (2006) Healthy nature healthy people: 'contact with nature' as an upstream health promotion intervention for populations. *Health Promotion International* 21(1): 45–54.

Mathews, F. (2016) From biodiversity-based conservation to an ethic of bio-proportionality. *Biological Conservation* 200: 140–148.

Mell, I. C. (2008) Green infrastructure: concepts and planning. *FORUM ejournal* 8(1): 69–80.

Mell, I. C., Henneberry, J., Hehl-Lange, S. and Keskin, B. (2013) Promoting urban greening: valuing the development of green infrastructure investments in the urban core of Manchester, UK. *Urban Forestry & Urban Greening* 12(3): 296–306.

Miller, R. W., Hauer, R. J. and Werner, L. P. (2015) *Urban Forestry: Planning and Managing Urban Greenspaces,* 3rd edition. Long Grove, IL: Waveland Press.

Natural England. (2009) *Green Infrastructure Guidance.* Report no: NE176. Natural England.

Natural England. (2010) *'Nature Nearby' Accessible Natural Greenspace Guidance.* Report no: NE265. Natural England.

NEA. (2011) *The UK National Ecosystem Assessment.* Cambridge: UNEP-WCMC.

Netusil, N. R., Levin, Z., Shandas, V. and Hart, T. (2014) Valuing green infrastructure in Portland, Oregon. *Landscape and Urban Planning* 124: 14–21.

Pauleit, S., Slinn, P., Handley, J. and Lindley, S. (2003) Promoting the natural greenstructure of towns and cities: English Nature's 'Accessible Natural Greenspace Standards' model. *Built Environment* 29(2): 157–170.

Planning Practice Guidance. (2014) *Open space, sports and recreation facilities, public rights of way and local green space.* http://planningguidance.communities.gov.uk/blog/guidance/open-space-sports-and-recreation-facilities-public-rights-of-way-and-local-green-space/open-space-sports-and-recreation-facilities/

RICS. (2011) *Green Infrastructure in Urban Areas.* RICS Practice Standards, 1st edition, information paper. Royal Institution of Chartered Surveyors.

Roe, M. and Mell, I. (2013) Negotiating value and priorities: evaluating the demands of green infrastructure development. *Journal of Environmental Planning and Management* 56(5): 650–673.

Sadler, J., Bates, A., Hale, J., and James, P. (2010) Bringing cities alive: the importance of urban green spaces for people and biodiversity. In *Urban Ecology*, ed. K. J. Gaston. Cambridge: Cambridge University Press.

Smith, N., Calvert, T., Sinnett, D., Burgess, S. and King, L. (2016) *National Benchmark for Green Infrastructure: A Feasibility Study.* eprints.uwe.ac.uk/29514

Soulé, M. E. (1991) Land use planning and wildlife maintenance: guidelines for conserving wildlife in an urban landscape. *Journal of the American Planning Association* 57(3): 313–323.

Swindon Borough Council. (2011) *A Green Infrastructure Strategy for Swindon 2010–2026.*

Vandermeulen, V., Verspecht, A., Vermeire, B., Van Huylenbroeck, G. and Gellynck, X. (2011) The use of economic valuation to create public support for green infrastructure investments in urban areas. *Landscape and Urban Planning* 103(2): 198–206.

Watts, K., Humphrey, J. W., Griffiths, M., Quine, C. and Ray, D. (2005) Evaluating biodiversity in fragmented landscapes: principles. *Information Note – Forestry Commission*, 73.

Whitford, V., Ennos, A. R. and Handley, J. F. (2001) 'City form and natural process' – indicators for the ecological performance of urban areas and their application to Merseyside, UK. *Landscape and Urban Planning* 57(2): 91–103.

3

FOOD, WASTE AND WATER

The urban paradox

Richard Baines

Abstract

Chapter 3 explores the nexus between primary food production, urban food and waste (in particular food and green waste along with waste water management) and outlines strategies that could be considered by urban planners, communities and food supply chains to produce a more holistic food system. In other words, the inclusion of food supply as part of urban planning tools and the development of the concept of 'Foodscapes' that have the capacity to reduce the physical and mental distance between urban and rural communities.

Key terms

Foodscapes, sustainability, food supply, agro-ecological, peri-urban food supply

Learning outcomes

Foodscapes can be seen as a way of bringing several elements of the food system together to create a more holistic and sustainable framework to manage food supply and consumption on the one hand whilst creating a more closed agro-ecological cycle between (primarily urban) consumers and (primarily rural) producers.

- From a rural perspective the Foodscapes concept ensures more of the fertility exported in food is returned through organic waste conversion to nutrients and even bio-energy production.
- For urban communities, foodscapes can facilitate sustainable use of food and green waste along with urban water to produce local food or a valuable export to rural communities who supply food to urban areas.

It is argued that urban planners should embrace the Foodscapes concept and integrate this into urban planning as a vehicle to address food supply, waste management and even urban food production. At the same time, agricultural and food security policies should embrace the Foodscapes concept and the potential for a more sustainable local food supply model.

Background

According to the UK's Chief Scientific Officer, the world is heading towards a 'perfect storm' where an increasing global population will need 50 per cent more food and energy by 2030 along with 30 per cent more water; furthermore, all of this will have to be achieved against the backdrop of climate change and its impacts on food production (Beddington 2009). More recently, events in North Africa and the Near East in 2008 and 2011 have shown how failures to supply sufficient food at an affordable price can lead to civil unrest and even regime change (Lagi et al. 2011). But what has this to do with the development of urban communities?

Global population increase and urbanisation

According to the United Nations, 2008 was the year that globally more people became urban than rural (United Nations, Department of Economic and Social Affairs 2014). Furthermore, between 2007 and 2050, the world population is expected to increase by 2.5 billion, passing from 6.7 billion to 9.2 billion (United Nations 2014). At the same time, the population living in urban areas is projected to gain 3.1 billion people, passing from 3.3 billion in 2007 to 6.4 billion by 2050. Thus, the urban areas of the world are expected to absorb virtually all the population growth expected over the next four decades while at the same time drawing in some of the rural population. As a result, the world rural population is projected to start decreasing in about a decade and 0.6 billion fewer rural inhabitants are expected in 2050 compared to today. Most of the population growth expected in urban areas will be concentrated in the cities and towns of less developed regions. Asia, in particular, is projected to see its urban population increase by 1.8 billion, Africa by 0.9 billion, and Latin America and the Caribbean by 0.2 billion. Population growth is therefore becoming largely an urban phenomenon concentrated in the developing world.

Global food supply

Over recent decades, the food system has become increasingly globalised and dominated by multinational food and food retail companies. As part of this evolution more and more food is crossing national borders, private standards dominate transactions, consumers believe they are getting more choice and value for money, but seasonality has largely been lost (see Baines 2010). This would suggest that more localised or regional food systems are not seen as a major part of food supply; however, a recent meta-study indicates that 60 per cent of irrigated croplands and 35 per cent of rain-fed croplands are within 20km of urban areas (Thebo *et al.* 2014); a distance that is economically viable for local food logistics.

These trends show that the urban landscape is increasingly becoming the dominant human habitat whilst food production mainly occurs in semi-natural or agro-ecological systems outside the city limits. Given that most of the food consumed by these growing urban communities comes from the countryside, how does this trend impact on food, waste and water, and how may this influence urban planning and design?

Foodscapes

Given this spatial separation of food production and consumption, the concept proposed here is to look at ways of linking these two systems from ecological and food security perspectives and, in doing so, introduce the notion of 'Foodscapes' into urban, rural and regional planning (Figure 3.1).

The Foodscapes concept seeks to address the challenges of producing adequate food in rural landscapes and then delivering this to the urban landscape through appropriate logistics. In return for food, cityscapes aggregate food and green wastes (and perhaps even human waste) and deliver this to rural landscapes for fertility building ready for the next cycle of food production. This holistic model can be further refined to include both rural consumption on the one hand and urban food production on the other. Other benefits can also accrue from the basic model; for example, the green waste stream can contribute to solving the increasing urban waste management and decreasing rural soil fertility problems if combined with central or kerb-side segregation of domestic waste. In addition, processing and distribution of wastes can result in increased economic activity, employment and even energy generation from waste. Another advantage of this holistic model within the urban cycle is the ability to utilise rainfall run-off more sustainably for urban food production as long as it is segregated from waste water that is likely to contain pathogenic micro-organisms along with heavy metals and chemicals.

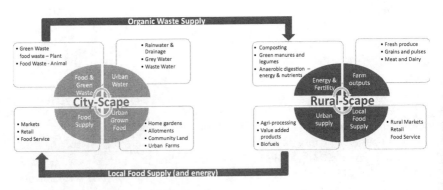

FIGURE 3.1 Foodscape conceptual model

Food, fertility and food waste

Increasing urbanisation results in a number of food-related challenges for urban dwellers that planners need to address; these include: the logistics of food delivery and storage into urban areas; food safety and hygiene from environmental and health perspectives; the design and location of food service and retail outlets; and dealing with food and other green wastes generated by urban living.

At the same time food producers, predominantly in rural areas, are facing their own challenges; these include the net export of land fertility in the form of agricultural raw materials and food to urban areas and the decreasing availability and increasing costs of nutrient inputs, particularly artificial fertilisers, for crop and forage growth. Primary producers face further challenges linked to food supply logistics and trade. In particular many producers suffer from the asymmetry of power resulting from the dominance of the global food supply system over local supply systems where greater volumes of food are travelling longer distances; where private standards, documentation and certification dominate; and where more of the value of food is trapped by actors further along the supply chain as opposed to farmers.

A further consequence of the global food model seeking to make food convenient for consumers is the large amount of food waste. The Food and Agriculture Organisation estimates that almost a third of all food produced globally is wasted; furthermore, the majority of this in developed regions is in the form of food waste (Lipinski *et al.* 2013; Figure 3.2).

Due to the majority of consumers residing in urban areas, food waste has now become a major challenge to urban planners. It is also a significant source of nutrients and organic matter for soil conditioning and fertility building; however, this

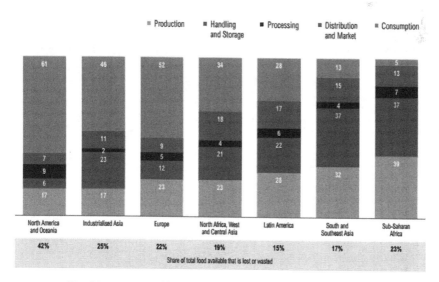

FIGURE 3.2 Food lost or wasted by region and stage in value chain, 2009
Source: After Lipinski (2013).

resource is mainly needed in rural areas. At the same time a significant amount of green waste is generated in urban areas from green spaces and recreational areas that could further contribute to soil conditioning. But what wastes are suitable for this purpose?

Throughout history agrarian societies have relied on organic wastes to rebuild soil fertility for crop growth including the use of green wastes for composting, the incorporation of animal manures and slurry into soils and even the use of 'night soil'. Indeed, night soil has been an important element of fertility building in Asia, including Chinese and Japanese agriculture to name just two. But what is it? Night soil is a Victorian term for human urine and faeces and in earlier periods of urbanisation men and women were employed to collect this waste and dispose of it; however, with the development of sewer systems this unenviable task has largely disappeared. Although there has been significant research into modern sewerage treatment, much of the work carried out on night soil in Asia between the 1500s and 1900s to improve both urban sanitation and agricultural production has been largely ignored in the modern era (Ferguson 2014). Perhaps it is time to reconsider the potential differential use of human waste rather than bulking it with chemical and industrial effluents that add to the challenge of sewerage treatment with only a small return to land in the form of treated sewerage sludge and composts.

Urban water

Since the beginning of the agricultural period, human activities have impacted on the fate of rainfall hitting the land; in particular the proportions that infiltrate, run-off or evaporate. A classical study in the United States noted the impacts of increased impervious layers during urbanisation on water (Arnold and Gibbons 1996; Figure 3.3) where over 50 per cent of precipitation may need to be directed to storm drains. Contrast this with natural ecosystems and even cropland where run-off is typically 20 per cent or less. Although modern urban planning is seeking to increase the area of pervious surfaces to allow more water to infiltrate, there is also the opportunity to think creatively on how to utilise water harvested off urban infrastructure. Such water could mix with the sewerage system to provide volume for transporting waste and pollutants to municipal treatment; however, clean roof water and grey water are relatively safe for urban crop irrigation. Irrespective of the source of water, it is important to consider appropriate treatment before using for irrigation; in this respect the World Health Organisation (n.d.) has developed a robust practical manual for dealing with grey water, waste water and excreta. Another advantage of increasing green space for urban crop production is that a greater proportion of precipitation will infiltrate into the ground.

Urban foodscape innovation

The notion of garden cities is not new; it can be dated back to Sir Ebenezer Howard who proposed it as a method of urban planning that envisioned self-contained

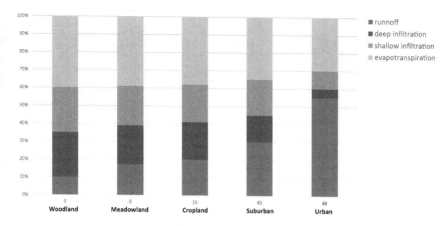

FIGURE 3.3 Impact of urbanisation on precipitation

Source: After United Nations, Department of Economic and Social Affairs (2014).

communities surrounded by green belts of countryside with proportionate areas of residence, industry and agriculture. Indeed in 1899 Howard established the Garden City Association and two cities were built; Letchworth and Welwyn Garden City. More recently, the expansion of Milton Keynes was argued to be a modern Garden City; however, the green space was primarily aimed at recreation and leisure along with meeting the needs of the car!

The Foodscapes concept presented here mirrors the earlier Garden City models but seeks to embed this in the era of globalisation, especially in relation to food supply. This leads us to ask two key questions of urban planning and development in association with their rural counterparts and those responsible for agriculture and food security policies:

1   How can the imported fertility in food delivered to urban areas be returned to food producing units outside of urban areas; and
2   How can food production within urban areas be developed sustainably, based on harvested water and local organic fertility?

These two questions can be addressed by reference to cases in point. In the first case the Revolutionary Government of Cuba faced the challenge of economic and food poverty post the Cuban Missile conflict, the collapse of the Soviet Union and, until recently, US trade embargoes. The agricultural and food systems were redesigned on agro-ecological principles that required greater reliance on fertility building through composting and vermiculture, biological pest control and community supported agriculture. Urban communities by necessity became actively involved in urban gardening and farming with government support. As a result a Cuban version of the Foodscapes concept evolved (see Box 3.1).

BOX 3.1 CASE STUDY **CUBA – REVOLUTION TO URBAN FOOD SECURITY**

Historical context

International relations between Cuba and its neighbours, especially the United States, have been fraught since Fidel Castro took power in 1959, including a US embargo in 1960, broken diplomatic relations in 1961 and the failed US-sponsored exiles invasion at the Bay of Pigs in the same year. Indeed diplomatic relations with the USA were not normalised until recently.

The Cuban Missile Crisis in 1962 created an international stand-off between the USSR and the USA and further isolated Cuba in the region through increased trade sanctions. Initially shortfalls were made up by the Soviet Union in the form of food, fuel and agricultural input subsidies. Indeed Cuba imported from socialist countries: two-thirds of its food stuffs, 57 per cent of its caloric intake, 80 per cent of all proteins and fats, 80 per cent all fertilisers and pesticides, almost all of its fuel and 80 per cent of its machinery and spares.

The Soviet collapse in 1991 led to a severe economic downturn following the withdrawal of former Soviet subsidies worth $4–$6 billion annually along with food and fuel shortages. Post-crisis purchasing capacity was reduced to 40 per cent, fuel importation reduced by two-thirds, fertilisers reduced to 25 per cent, pesticides to 40 per cent and animal feed concentrates to 30 per cent.

From food crisis to foodscape

Agriculture became important with a transition from conventional, high-input mono-crop intensive agriculture to smaller organic and semi-organic farms. In addition, agriculture transportation was reduced or eliminated due to the scarcity of oil. The food crisis incited a massive popular urban food movement. For example Havana with a population of 2.5 million saw significant urban food development including: gardening in and around the home and government action to convert all vacant lots to properties of food production including car parks that had no other use. The Cuban Ministry of Agriculture set up the Urban Agriculture Department and, by 1998, there were over 8,000 officially recognised community gardens in Havana, cultivated by more than 30,000 people and covering approximately 30 per cent of the available land. Ecological agricultural systems were promoted to embrace: biological pest controls and bio-fertilisers (composting and vermiculture), renewed use of animal traction, gardening movements (urban, family and community) and farmers' markets under 'supply and demand' conditions.

Suburban farms have also developed; typically being between 2 and 15 hectares with infrastructure for recycling of waste products, crop growth through intensive cultivation, livestock, efficient use of water and maximum reduction of agro-chemicals. Marketing of products is influenced by the surrounding population and was initially state regulated. In Havana, suburban farms are made up of some 2,000 small private and 285 state farms that cover 7,718 hectares and are highly productive. Now many urban areas are able to supply up to 60 per cent of fresh vegetables needed by communities. In Havana, for example, production of fresh vegetables and herbs has reached 150g and 200g respectively per capita per day, vacant lots are used productively, 100,000 new jobs have been generated and a rich variety of crops have become available to communities. Land use benefits the environment and promotes a nutritional culture while empowering people to build communities, and it has stimulated local markets.

There have been challenges to sustainability of the system, these include food theft and high irrigation demands allied to insufficient research on toxic elements in water that may contaminate leafy green vegetables.

Summary

This case study demonstrates how communities and administrators can work together to develop functional foodscapes, even under international political isolation. The urban model developed addresses many of the elements outlined in the conceptual model; however, there are areas that need further innovation. These include: improving the safety of urban waste water, especially for irrigation of leafy greens, and further linkages between the urban and rural food systems.

The second case focuses on the Transition Town movement which provides a model of grassroots community actions to address the challenges of climate change and globalisation though more local trading and community relationships, including food production and consumption (see Connors and McDonald 2010). In particular, Bristol, in the South West of England, is driving a local food movement strategy (see Box 3.2). Although at the early stages of development 'Bristol Food Network' has achieved a great deal as a Community Interest Company (CIC) in seeking to transform Bristol into a sustainable food city (Bristol Food Network n.d.).

Conclusions

Most urban areas have transitioned from a local food supply system where urban communities and neighbouring rural communities were interdependent on each other to a more globalised food system with a resulting loss of local and regional

BOX 3.2 CASE STUDY **BRISTOL: WORKING TOWARDS A SUSTAINABLE FOOD CITY**

Bristol Food Network CIC supports, informs and connects individuals, community projects, organisations and businesses who share a vision to transform Bristol into a sustainable food city. Key objectives include: promoting and encouraging people to cook from scratch, grow their own and eat more fresh, seasonal, local, organically grown food; championing the use of local, independent food shops and traders to help keep Bristol's high street vibrant and diverse; promoting and encouraging the use of good quality land in and around Bristol for food production; promoting and encouraging the redistribution, recycling and composting of food waste; advancing education about the part that food, nutrition and lifestyle can play in meeting the needs of disadvantaged individuals, families and groups in the community to encourage social inclusion and cohesion; promoting community-led food trade such as cooperatives, buying groups, Community Supported Agriculture (CSA) and pop-up shops; and building network expertise in food and sustainability that allows access to and creates opportunities for local people within Bristol.

Partnerships and projects

In Bristol, three partnership structures help create opportunities to link up businesses, community groups and individuals in a city where food systems are highly complex.

1   The Bristol Food Policy Council takes a strategic view in relation to building and embedding a resilient food system for Bristol including: production, processing, distribution, retail, catering, consumption and waste disposal. The Council meets at least four times a year and has published the Charter and the Good Food Plan. It is developing ways of measuring progress.

2   The Bristol Food Network (CIC) is the practical delivery organisation that aims to support, inform and connect individuals, community projects, organisations and businesses that share a vision to transform Bristol into a sustainable food city. It produces a bi-monthly Bristol local food e-newsletter update and coordinates much of the communication within the wider city food networks including bi-monthly networking sessions.

3   The Bristol Green Capital Partnership (CIC) is an independent leadership organisation whose aim is to make Bristol 'a low carbon city with a high quality of life for all' and who played an active role in Bristol's 2015 European Green Capital Programme. It has assisted or funded a number of strategic tasks including support in 2010 for the 'Who Feeds Bristol' baseline study and in 2015 it co-hosted an event to develop a 'Good Food Action Plan'.

Applying a foodscape model to Bristol

Both formal and informal networks operate around the city, many of which focus on specific aspects of the food system such as: providing cooking skills in local neighbourhoods; redistributing edible wasted food to community organisations; bringing chefs together to look at ways of improving sustainable food sourcing; helping urban growers cooperate; organising seed swaps, networking and open day events on urban food growing sites; helping food businesses find new food waste collection solutions; connecting up healthy school initiatives with school gardens and school meals; campaigning for the safeguarding of best value agricultural land for food; and promoting independent food businesses.

The coordination of these activities by the Bristol Food Network creates the environment to define Bristol's foodscape and has the potential to realise a sustainable food city linked to the green halo of farmland surrounding the city (Figure 3.4).

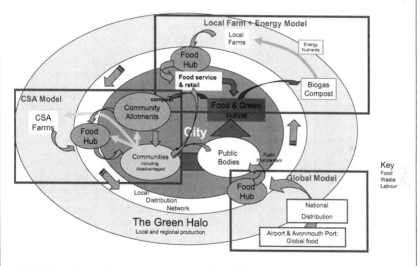

FIGURE 3.4 Bristol foodscape (conceptual model)

food sovereignty. Indeed, cities could be considered as clusters of people who do not produce their means of sustenance and who are technically, socially and spatially separated from those who do produce this sustenance for them (Ascher 2010).

The Foodscapes concept described here provides an opportunity for urban communities to reconnect with their food systems in a more agro-ecologically resilient way; it also provides a vehicle for community cohesion and partnerships. For urban planners it partly solves the dual problems of food wastes and urban rainwater management. Although food is currently not part of the planning tool, the inclusion of foodscapes may be fundamental to the design and development of more sustainable city-scapes that are co-dependent on rural landscapes for a more food secure and resilient future.

Stakeholder implications

Implementing the Foodscapes concept will have significant impacts on the development of urban communities as well as neighbouring rural communities. These can be summarised as follows:

Communities – Foodscapes provides urban communities the opportunity to reconnect with their food system and to contribute to a more sustainable and resilient local food system. This connection can further drive community partnerships in business, health, nutrition and education. Foodscapes also contributes to meaningful employment and social connections within and between urban and rural communities.

Planners – By building food systems into the planning toolbox, Foodscapes provides the framework to address the interrelated problems of food supply, organic waste management and waste water as all are essential elements of food production and supply chains. Urban planners have the option of building food production into the urban landscape, thus making green spaces productive whilst still facilitating recreation and leisure along with landscaping. Organic waste management now becomes an opportunity as opposed to a cost which can be utilised through urban agriculture or exported to peri-urban and rural farms. Furthermore, by diverting green waste and rainwater from the sewerage system, the size and scale of sewerage treatment can be optimised.

Developers – Foodscapes provides a different challenge to developers ranging from integrating water harvesting and storage systems and waste segregation facilities to the development of rooftop, vertical and urban food production facilities. There would also be opportunities to develop integrated waste facilities capable of recouping fertility and energy from green and food waste along with the need to develop storage, wholesale and retail outlets for food produced within the city-scape or imported from surrounding rural areas.

Investors – This diversity of activity around the Foodscape concept provides an ideal opportunity for investors and investment funds that prefer to support sustainable, ethical and social businesses and development. As the model seeks to develop resilient food systems, then any investment is also likely to be more resilient and profitable in the longer term.

Professionals – Foodscapes draws together a wider range of professionals to address urban development. In addition to urban planners, the professional network should also include: agriculturalists to advise on urban, peri-urban and local rural production systems; health and nutritional professionals to advise on food consumption and nutrition; environmental professionals to advise on food safety and environmental challenges, especially associated with urban agriculture and composting; and waste management professionals to address the logistics of waste management, recycling and use.

As Foodscapes is a holistic concept, then it goes without saying that the above stakeholders would have to be prepared to not only offer their specialist and professional expertise but to also listen to and integrate the views of other professionals and citizens into urban development. This would logically lead to a more participatory and multidisciplinary approach to strategic urban development in synergy with associated rural communities.

Glossary

Biogas The production of methane and carbon dioxide from anaerobic fermentation with the potential to provide energy, nutrients and soil conditioners.

Community Interest Group (CIC) A business or organisation model orientated towards community benefits. Any profits are reinvested back into the community.

Community supported agriculture A system of food production where producers and consumers share in the risks and rewards of food production and consumption.

Composting The partial breakdown of organic matter under aerobic conditions in order to provide nutrients and soil conditioners for crop growth. Effective composting should reach temperatures above 60–70°C to reduce pathogenic organisms.

Foodscapes A food system defined by its linkages between production and consumption and by the social, ecological and technical links of the component parts. Key to the concept is the importance of recycling nutrients.

Food supply chains The links between businesses and individuals from primary production to food service and retail.

Food waste Waste from the food supply system which can be further divided into uncooked plant waste, animal waste and cooked food waste.

Globalisation Internationalisation of business where production is located in least cost regions and outputs are delivered to the highest value markets.

Green waste Plant waste largely made up of green material with some woody elements. Not containing any cooked food wastes.

Grey water Water that has been used for washing, etc. but not containing any excreta.

Urban and peri-urban agriculture Formal and informal food production carried out within city limits and in suburbs respectively.

Vermiculture Composting using worms.

Waste segregation Separation of waste materials into component groups such as food and green wastes, plastics, metals and glass.

Waste water Grey water that also contains excreta; water that needs treatment before releasing into the environment.

Discussion question

Discuss how the Foodscapes concept provides a framework for urban planning to address key urban development challenges and creates links with rural communities.

Key issues

In considering the above question, the following key issues should be addressed:

- Foodscapes is a holistic concept that brings together urban and rural communities around a more sustainable local food system where:
  - o Food productive capacity is maintained in rural areas due to recycling and processing of food and green wastes – the nutrient loop
  - o There is the potential to harvest bio-energy from processing
  - o Planners have the option to design or retrofit food production into urban landscapes using local fertility building and urban water.
- Foodscapes provides opportunities for urban waste segregation and productive use.
- By separating rainfall and roof drainage from foul water and sewage, the former can be used productively and the latter is reduced in volume.
- Urban and peri-urban food supply can contribute to food and nutrition security as well as social cohesion within and between communities.
- Foodscapes encourage urban planners to think beyond the city limits while encouraging more consumer and agro-ecological policies to evolve.
- The importance of strategic food supply should be a formal aspect of urban planning and Foodscapes provides the model.

References

Arnold, C. A. Jr and Gibbons, C. J. (1996) Impervious surface coverage: the emergence of a key environmental indicator. *Journal of the American Planning Association* 62(2): 243–258. http://dx.doi.org/10.1080/01944369608975688

Ascher, F. (2010) *Novos principios do urbanismo: novos compromissos urbanos: um lexico*. Lisbon, Portugal: Livros Horizonte.

Baines, R. N. (2010) Quality and safety standards in food supply chains. In Mena, C. and Stevens, S. (eds) *Delivering Performance in Food Supply Chains*. Cambridge: Woodhead Publishing.

Beddington, J. (2009) Food, energy, water and the climate: a perfect storm of global events. London: Government Office for Science. http://webarchive.nationalarchives.gov.uk/20121212135622/http:/www.bis.gov.uk/assets/goscience/docs/p/perfect-storm-paper.pdf (Accessed 20 January 2017).

Bristol Food Network. (n.d.) www.bristolfoodnetwork.org/about/

Connors, P. and McDonald, P. (2010) Transitioning communities: community, participation and the Transition Town movement. *Community Development Journal* 46(4): 558–572.

Ferguson, D. T. (2014) Nightsoil and the 'Great Divergence': human waste, the urban economy, and economic productivity, 1500–1900. *Journal of Global History* 9(3): 379–402.

Lagi, M., Bertrand, K. Z. and Bar-Yam, Y. (2011) The food crises and political instability in North Africa and the Middle East, August 15. Available at SSRN: https://ssrn.com/abstract=1910031 or http://dx.doi.org/10.2139/ssrn.1910031

Lipinski, B., Hanson, C., Lomax, J., Kitinoja, L., Waite, R. and Searchinger, T. (2013) Reducing food loss and waste. Working Paper, Installment 2 of Creating a Sustainable Food Future. Washington, DC: World Resources Institute. www.worldresources report.org (Accessed January 2017).

Thebo, A. L., Drechsel, P. and Lambin, E. F (2014) Global assessment of urban and peri-urban agriculture: irrigated and rainfed croplands. *Environmental Research Letters* 9(11): 114002.

United Nations, Department of Economic and Social Affairs, Population Division. (2014) World Urbanization Prospects: The 2014 Revision, Highlights (ST/ESA/SER.A/352). https://esa.un.org/unpd/wup/publications/files/wup2014-highlights.Pdf (Accessed January 2017).

World Health Organisation. (n.d.) Dealing with waste water, grey water and excreta. http://apps.who.int/iris/bitstream/10665/171753/1/9789241549240_eng.pdf?ua=1

4

PLACE AND COMMUNITY CONSCIOUSNESS

Negin Minaei

Abstract

Lately, concepts of smart cities, intelligent cities, digital cities, resilient cities, sustainable cities and creative cities have been widely discussed among planners and designers. Governments of many developed countries have commenced strategically researching and planning for adoption of these concepts. In principle, these concepts are broadly similar but their precise meaning unclear. Ultimately, smart cities aim to achieve high levels of liveability and sustainability for urban inhabitants. However, they cannot progress without people's awareness and cooperation. The case for planning strategically for sustainable futures seems indisputable. Many global cities and metropolises have already benefited from new infrastructure and technologies associated with intelligent and smart planning. However, most urban areas in regional conurbations suffer from inadequate transport and digital infrastructure while both play fundamental roles in keeping a town or a city sustainable and smart.

The fourth chapter investigates the notion of 'urban consciousness' and its relationship with 'smart cities', and quality of life. The chapter seeks answers to the following questions: What is urban consciousness? To what extent have urban populations achieved consciousness?

- What is the role of a smart city in creating a conscious environment?
- How can we create a smart and conscious atmosphere for urban inhabitants?

Key terms

Smart city, urban consciousness, sustainability, information technology (IT), smart infrastructure

Learning outcomes

- Explore notion of a 'smart city', and its qualities (intelligent systems, infrastructures, creative industries, resilience, sustainability).
- Explore relationship between 'smart cities' and residents consciousness.

The issue

With the global urban challenges emerging due to increased urban population and climate change (UN 2015; BSI 2014; Schaffers *et al*. 2012; Owens and Cowell 2011), thinking about sustainable options is critical. Creating smart cities is a good response as they incorporate well-advised approaches towards the environment. 'Smart cities' can help in saving the planet. It is impossible to debate smart cities and consciousness without requiring people's engagement. One indicator of sustainable conscious planning is the documents that incorporate 'smart planning' considerations. Another is the scale of procedural planning or participatory deliberations between citizens and various tiers of government. In the aftermath of participatory formal dialogue, a shared 'smart city' consciousness spreads. Unresolved questions remain and include:

- Is the smart city about styles of living (e.g. eco-friendly living) or about digital awareness and smart mobility?
- Is urban consciousness about conscious planning and designing of a city to shape it more sustainably or about sharing an awareness and ongoing dialogue between citizens and their governments?
- Which are more important, smart planners or conscious city dwellers?
- To what extent is the knowledge about the current city concepts and trends (intelligent city, liveable city, smart city, resilient city, sustainable city and creative city) transferred to the public?

New research bodies have started exploring the possible answers and have raised public awareness and citizen's engagement in city affairs. Here is an example: TRADERS (Training Art and Design Researchers in Participation for Public Spaces) is a FP7 (under the EU's Seventh Framework Programme for Research) and Marie Curie Multi-ITN project which is developing a method to study public spaces and improve private/public partner engagement by employing the underexplored techniques of intervention, play, multiple performative mapping, data-mining and modelling in dialogue.

Technology leaders such as IBM, Cisco, Siemens and others are planning for smart cities that inspire a breakthrough (Townsend 2013: 104). Global cities and world cities are undergoing continuous evolution. To retain their rank in the global hierarchy, most of them now seek to become a smart city since they already have the required infrastructure and the prerequisites from the earlier concepts of resilient and sustainable cities. Creating smart cities has become an emerging business and there is a good market for it. Global Smart City Industry is the name of a report that analyses the worldwide markets for smart cities (PRNewswire 2016). Booz Allen predicted that urban information and communication technologies (ICT) and telecommunications will attract global investments of over $30 trillion within 30 years. He believes that smart communities are potential hubs with the power of drawing young professionals and companies (the 'Human Capital Theory' (HCT) of Jane Jacobs). Creative and highly educated people are the driving forces in regional economic growth. They are interested in diverse inclusive places which attract more talented people, bringing in growth and job opportunities, and simultaneously improve the real estate conditions such as property refurbishment, entertainment and leisure (BIS 2013: 13; Florida 2005: 33).

Urban consciousness in smart cities

Consciousness is a debatable philosophical topic. Chalmers states consciousness is impossible to define. Dennett (2003) believes since we are a package of billions of small robotic cells with no information on who we are, we are not able to comprehend our own consciousness. Others believe all humans are aware of their own consciousness. Searle believes consciousness is a subjective matter, neither an objective materialistic paradigm nor a scientific concept. In common terms, it means awareness and knowledge of feelings are only in humans and not in any kind of robots.

Miller refers to Durkheim and Simmel and argues a large city can be a single cognitive system (2006: 17), albeit a multilayered complex one which impacts us and our urban consciousness at different levels:

At personal level Our health and well-being are the result of internal and external interactions with the environment. Climate, physical signs and urban furniture or the way-finding task interact with memory and cognitive abilities (Minaei 2016) as our brain constantly interacts with its surroundings and continuously makes real-time adjustments based on the environmental information it receives. *It offloads some of its cognitive duties to the environment* (Miller 2006: 5) hence the environment should be designed properly with enough signals and clues to render the space comprehensible (Minaei 2014).

At social level The assumption is that cognition is a social and cultural process (Hutchins 1995: 354). Varieties of influence in the city and in urban life can produce different kinds of consciousness in urban dwellers (Simmel 1969, cited in Hutchins 1995: 354). These consist of atmosphere (lively, happy or depressed), place attachment and sense of belonging, urban identity, the 'Image of the City' and social responsibility (maintenance, protection, etc.). Cities are places for endless social contacts, consumption and leisure, and integrated mechanisms to adopt.

All definitions, characteristics and examples in this chapter are viewed from the consciousness lens and selected based on their part in the conscious interactions between people (the inhabitants) and place (the city and The City). 'Being more content aware through devices connected to internet, can this mean we are more conscious, or cities are more conscious of what is happening, where and how?' (Dennett 2003).

While in 1984, only 1,000 Internet devices were connected to the Internet, 6 per cent of things are connected to it today; 50 billion devices are expected to be linked by 2020. At the end of 2012, 4,000 hexabytes of data were stored in the Cloud just by Amazon and Facebook. Data is growing exponentially by about 20 petabytes of data per day; that is more data than was created by all humans in the previous 5,000 years (Dennett 2003). As Townsend says, we've all built the Internet together, it is the most participatory construction project in human history (2013: 106). Most businesses in the developed world, including health care, agriculture, industry and manufacturing, are now digital businesses; the transformation is bigger than just technology. A new intelligent network is required to accommodate it.

Goleniewski (2001: 8) predicted moving from an era of human-to-human interactions to an era of machine-to-machine interactions. During the 1980s, the main attempt of IT and telecommunications (TC) was to create a new order for financial institutes, insurance companies, services, governments and even retailers, and encourage people to use TC to make services faster and cheaper. IT and TC were the two main drivers of globalization in cities, accelerating the process (Minaei 2004). Nine years later, information and communication technologies are mentioned as the new economic forces for urban growth changing the way cities compete (Bakici et al. 2013: 136) and are the foundations of a smart city.

Smart city, definitions, domains and characteristics

Having the issue of a lack of a clear definition for smart cities, the International Telecommunication Union (ITU), a specialized agency of the United Nations (UN), analysed 116 definitions for smart cities in a comprehensive report (ITU 2014: 14–53). Ben Letaifa (2015: 1415) quotes Hall's definition for both intelligent and smart cities:

> A city that monitors and integrates conditions of all of its critical infra-
> structures, including roads, bridges, tunnels, rails, subways, airports, seaports,
> communications, water, power, even major buildings, can better organize
> its resources, plan its preventive maintenance activities, and monitor security
> aspects while maximizing services to its citizens.

She says intelligent cities are the historical version of smart cities with a top–bottom approach and sees smart and intelligent cities as different from creative cities due to the balance they create between technology, institutions and people as the core structure of the city. In her view, the top-down and bottom-up participation approach is the main difference between those three cities. A city like Montreal that employs the community-based bottom-up approach can be categorized as a creative community where technology drives what is needed for people and people are at the core of structure (Ben Letaifa 2015: 1414–1415). Slightly differently, Richard Florida suggests the 3T concept (technology, talent and tolerance) as the shaping structure of its creative cities (Florida 2005: 6). In Florida's view it is the diversity, openness and tolerance for the diversity which make creative cities more attractive and economically more successful.

Smart cities are based on information networks, communication grids and technologies which are built around users' quality of life. They aim to optimize resources toward sustainable development (Bakici et al. 2013: 137). Citeos (2015) indicates the core of smart cities is smart connected streets and street lights to collect information and send it to citizens to have real-time data and make the city alive. According to Neirotti et al. (2014), based on their empirical analysis of 70 cities with developed projects or best practices in the 'smart city domains', a smart city optimizes the exploitation of assets, tangible (natural resources, energy distribution networks and transport infrastructure) and intangible (intellectual capital of companies, human capital, etc.). Natural resources and energy, transport and mobility, buildings, living, government, as well as economy and people are the six main smart cities domains. Slightly differently, Giffinger et al. (2007) introduced the six 'common indicators of smart cities' including 'smart' (people, economy, governance, mobility, environment, and living) (cited in Ben Letaifa 2015: 1416). Hamnett, in a review of Deakin and Al Waer's book *From Intelligent to Smart Cities* (as cited in Moore 2014), mentions that smart city goes beyond technology; it concentrates on environmentally sustainable outcomes and aims to achieve a higher level of quality of life. Others believe environmental and social features of the intelligence aspect are as important as the technology aspect (Anttitoiko et al. 2014, Halpern 2005, cited in Yigitcanlar 2015: 29); the reason people have a non-negotiable role in smart cities.

The UK government definition is:

> a smart city enables and encourages the citizens to participate more as active members of the community by selecting a healthier and more sustainable lifestyle, providing feedback on the quality of services or the road conditions, state of the built environment, volunteering for social activities or supporting minority groups.
>
> (BIS 2013: 8)

In this definition, conscious citizens are at the core of the smart city not the IT infrastructure. The role of citizens and their participation in smart cities have been highlighted. Carl Piva from Sweden mentioned five principles for a smart city: 'existing smartness level, integration of ICT (Information and communication technology), use of effective data, leadership and management, and participation of stakeholders and citizens' (cited in Huang 2016).

Smart city characteristics

Sustainable digital infrastructures

Smart cities invest in their infrastructures which may not be visible. They improve their IT and broadband networks with high-speed Internet everywhere. The UK Department for Culture Media and Sport (DCMS) had a £1.2 billion investment programme on ultra-fast broadband in addition to £150 million to have high speeds and wireless Internet access via the Mobile Infrastructure Project for homes and businesses to improve coverage and quality where it did not exist (BIS 2013: 25).

Open data provide good interactions between cities, their citizens and their infrastructures. Real-time data can be used by citizens to make chronic problems visible and create new pressure for long-term repairs (Townsend 2013: 276). Having a transparent society is the common vision most intelligent and smart cities have, such as Open Data in the UK and Open Government in Canada. Open Data has been promoted by governments as a tool to provide business opportunities by establishing an accessible environment of a pooled public, commercial and social media data (BIS 2013: 15). This is the case for global and capital cities like Toronto or London where necessary infrastructure and business opportunities are available. In Toronto, the City Hall has a specific Open Government webpage where people access the city information, records, reports, progress reports, statistics, economic and service values, and business information for small and medium-sized enterprises (SMEs) in addition to Toronto public service by-law, council and committee decisions, polling results and revenues. Anyone can find data to develop software, a website or an application. Citizens are involved in decision-making and developing city policy (City of Toronto 2016). Similarly in the UK, e-infrastructure Leadership Council, Digital Economy Programme, Urban Prototyping (UP) London, Digital City Exchange (DCE) and Liveable Cities are examples of information economy and smart cities research which have been funded by Research Councils UK (BIS 2013: 16). Undoubtedly, in the era of intelligence, open data increase public awareness, provide chances for experimentation and contribute to citizens' urban consciousness. In the developing world, lack of data and communication is still the major setback preventing cities from functioning well; conversely, ICTs have the problem of access to abundant data and instantaneous communications (Townsend 2013: 283), not mentioning privacy and data security.

Sustainable mobility

An intelligent transport system monitors the existing traffic constantly and predicts congested areas. Citizens are informed on a real-time basis. Upgrading the systems of transport modes to accept contactless bank cards (used in London) and smart tickets is another step to make mobility even smarter and its data collection easier (BIS 2013: 22). Transport data such as real-time train and bus information services and applications around road congestion, traffic information, finding the best fares and personal navigation services are amongst the most popular (BIS 2013: 23). Equipping public transport with broadband and offering free Wi-Fi (in China) encourages people to use public transport. Florida: 'To smart people connectivity matters.' Nowadays most young professionals prefer to have hands-free transits to check their emails, texts on phone, work, or even rest while commuting (Florida 2005: 168).

The City of Toronto has invested in a bike share scheme which in 2016 had 200 stations (2,000 bikes) serving about 10,000,000km travel in Toronto, with 1,700 tons of carbon dioxide offset. By 2017, 500 stations with 5,000 bikes will be provided. The cycling network plan works on improving road safety and building new bike lanes. Another cost-effective solution is to encourage flexible working hours or remote working to ease traffic congestion by expanding the peak travel times over the 24 hours and improve employees' well-being.

Smarter decision-making in urban planning and in daily life

A smart city needs flows and networks of people, goods and money. It has a multilayered structure including electricity, water and transport interacting with one another. It is affordable with a health care system, safety and security for all which results in a high quality of life for the happy and healthy population (VINCI Energies 2015). High quality of life in a community can make jobs appear more attractive by 33 per cent. Talented people consider the quality of life as the second most important factor when selecting their living/working place, hence developing creative environments so that talents' new ideas can grow is the best policy for economic growth (Florida 2005: 165).

Smart energy, connected buildings and districts

The European Union in the Zero Energy Buildings (ZEB) principle envisioned all buildings ought to be nearly zero-energy from 2020. This is achievable by using more Renewable Energy Sources (RES). Buildings can reduce two-thirds of their energy consumption applying ZEBs (Kylili and Fokaides 2015: 87, 94). A 40 per cent reduction in greenhouse gas emissions is encouraged by both the European Strategic Energy Technology Plan (SET-Plan) and the Smart Cities and Communities Initiative (SCCI). Key Performance Indicators (KPIs) have been defined to monitor and assess the progress of the this initiative in different categories including advanced ICTs, transport, electricity, smart electric grids, buildings, and heating/cooling systems (Kylili and Fokaides 2015: 89).

The Environmental Protection Agency (EPA) reports that since 1972 carbon dioxide emissions have risen by 90 per cent globally and that is the reason the Union of Concerned Scientists estimates by 2050, 80 per cent of US power will use RES to decrease carbon dioxide emissions (Lee 2016). Project Neutral in

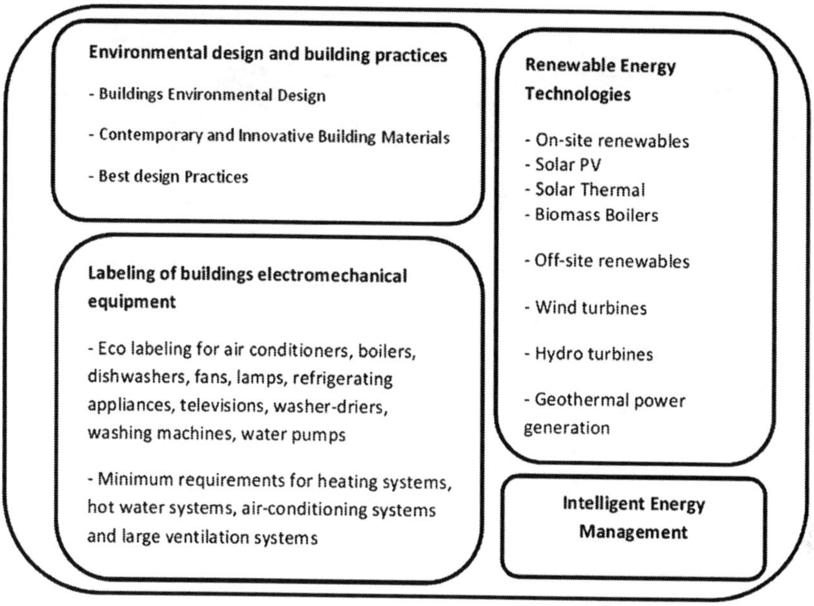

FIGURE 4.1 ZEB aspects as integral part of smart cities
Source: Kylili and Fokaides (2015: 91).

Toronto informs people how to meter their energy consumption and calculate their carbon footprint and consciously engages them to live eco-friendly. The UK aims for a 40 per cent reduction of greenhouse gas emissions by 2020 by sustainable use and production of energy (BIS 2013: 30). The two suggested initiatives are:

1   **Smart city heating** This approach employs low carbon solutions for heating buildings sustainably which can be achieved by Heat Networks to strengthen cities' energy resilience as well as tackling fuel poverty (BIS 2013: 25).
2   **Smart city electricity distribution grid or 'smart grid'** The smart city requires a sustainable energy source to function therefore controlled use of energy is vital (BIS 2013: 26).

Smart grid includes some major technologies: 'energy storage devices, advanced superconducting transmission cables, smart substations and smart transformers, Advanced Metering Infrastructure (AMI) and Home Area Networks (HANs)'. AMI and HANs directly affect building operation (Syal and Ofei-Amoh 2013) but AMI is more important since it is a dynamic interactive system for power and real-time data exchange (NETL 2007, Roncero 2008, cited in Syal and Ofei-Amoh 2013) which makes it intelligent and conscious to optimize electricity consumption. Figure 4.2 shows the linkage of AMI and HAN to the smart grid.

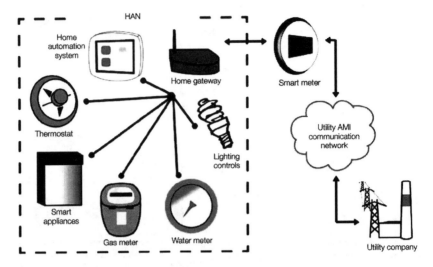

FIGURE 4.2 Linkage of AMI and HAN to a smart grid

Source: Syal and Ofei-Amoh (2013: 285).

Smart actions, urban consciousness, and changing behaviours in daily life

Sassen (2011) believes cities are intelligent; they talk back to us as a survival mechanism but urbanizing technology is not using interactive technologies. She talks about people who reappropriated a dangerous public space called 'River Park' in New York by buying dogs to protect themselves while walking so gradually the image of the park changed to a safe neighbourhood.

Professor McCarney[1] (2015) explained the neighbourhood resilience hubs in Toronto to build trust and improve social bonds in communities so people help each other in hard times. The idea came from American cities to encourage kind communities in the age of climate change and extreme weather; with their health impacts on the vulnerable population, creating a culture of kindness in communities to secure resilient neighbourhoods and communities is a must. Being smart is about planning all prerequisites and future needs of cities in advance and planning the right response by investing in the needed infrastructure. As part of this strategy creating a transparent data set which both government and people can have access to was suggested. Having access to the real up-to-date data can increase consciousness in a community. People see the potentials and problems and feel being part of a reliable community. This will generate feelings of attachment to the city which is the reason to emphasize the bottom-up approaches, community and participatory planning and people at the core of the smart city; it is about peoples' awareness, understanding, participation and investment to keep a city liveable, smart and conscious.

Changing peoples' behaviours using technology, Internet and social media can benefit our societies. There are two good examples from the book *Smart Cities* (Townsend 2013) to explain how new technological innovations and conscious citizens can save our cities and ultimately our planet. The first, 'Dontflush.me', is an example of a system developed by Leif Percifield, a designer from New York, which saves costs by linking an intelligent system to the city's conscious inhabitants. New York has a single network of drains for both rainwater and sewage like many other old cities. Normally treatment plants process the combined outflow before it is released into adjacent canals, however when it rains heavily the treatment plants cannot hold the deluge from flowing into the streets, hence a combination of run-off and raw sewage is emitted into the city's canals: approximately 27 billion gallons annually.

> Percifield's gadget links an 'Arduino' to a proximity sensor and a cell phone which sends an alert via the Internet to a network of bathroom-based light-bulb overflow-warning indicators to inform people not to flush toilets during overflow occurrences therefore reduce the discharge of sewage.
>
> (Townsend 2013: 130)

Another example is called 'Botanicalls' which was developed for a class on sustainability in 2006. It turned social networks and the Internet of Things (IoT) to the challenge of gardening by adding a tiny computer to a moisture sensor inserted among roots of a plant and connecting it to the Internet via a network adapter. The moisture readings were uploaded to a cloud-based web server and the software designed by the students analysed the data and triggered a call for help when dryness was detected. It was linked to a Twitter account and a phone system so the plant's friends could follow its Twitter posts to be informed of its water requests. They could exchange messages among themselves to coordinate care (Townsend 2013: 126).

Transport for London and the Department for Transport (DfT) launched an aspiring programme to change behaviour and reduce the transport network's demand during the London Olympics. Some 30 per cent of Londoners altered their travel patterns so the network could be shared with visitors and game lovers. DfT has invested on the London 2012 success, and started examining the impacts technology and changing travel behaviour can have in lightening the burden of the public transport in congestion periods, in carbon reduction and ultimately in economic growth (BIS 2013: 25).

Policy recommendations

Rometty (2011) explains the three main driving forces of smart cities: growth of the middle class, decay of the infrastructure (most smart cities are old) and competition among cities. She mentions the importance of 'coopetition' meaning cooperating and competing and suggests five solutions to build a smarter city:

1 Data collection and data management from every system possible, e.g. observing the interaction of people with urban spaces to distinguish cultural patterns for the right actions.

2 Using functional integrated systems such as a real-time crime centre which could decrease crimes in New York City by 30 per cent.

3 Optimization by linking transport system to the IT systems; for public transport, based on the information in one's smart cards, street lamps adjust the time for one to pass (in case of an elderly or disabled person who needs more time). Using powered walkways and seat prediction in buses can also increase the value of the public transport.

4 The city should be seen as a comprehensive complex system which needs a comprehensive view.

5 About 30 per cent of the world's produced food is wasted annually. In 2050, it is predicted that 70 per cent more food would be required; the problem is lack of collaboration between farmers, goods distribution and proper economic systems. Rometty thinks collaboration among city, businesses and citizens can solve the food problem.

6 Having resilient neighbourhoods.

Installing satellite sensors and advanced imaging cameras for large-scale management activities has proven successful. Video surveillance and 'Interactive Voice Response' (IVR) systems in cities are suggested for public safety, crime prevention, citizen protection and crowd control and will soon be widespread in developing cities (Dorfman 2015: 5; Breetzke and Flowerday 2016: 1). Sensors, cameras and satellite services can enable cities to collect comprehensive data, control better and plan for disaster management, but it is also important to consider what the IoT can do to an individual's privacy as a basic human right (Kitchin 2016a).

Stakeholder implications

Designers

Navigating between the virtual and the physical word is a real challenge for designers (Townsend 2013: 272). Designers need to think about the layers of infrastructure underneath and above the city, sufficient biking and walking passages, beautiful healthy walkable neighbourhoods to prevent creating 'obesogenic' environments (research shows the number of takeaways in an area has an impact on obesity as those who live or work near these areas are more likely to eat unhealthy food (BBC 2014)). Designers should think about healthy spaces where traffic pollution is non-existent because air pollutants are associated with suppressed lung growth, heart disease, asthma, fetal brain growth damage and the onset of diabetes. Air pollution from traffic and industry leads to 3 million premature deaths annually (Tinker and Levitt 2016). Designers should create friendly and happy environments, defensible neighborhoods where people can watch the neighborhood, protect it and trust their neighbours; it is important specifically in cosmopolitan societies. Figure 4.3 illustrates a smart city with its interrelated links to other factors.

McCarney (2015) mentions a new concern of planners in Los Angeles is to create a culture of kindness. Being kind and offering a helping hand in times of crisis can add to the resilience of communities. Living in natural environments with trees can cause the good in people; hence using this element to empower our cities to trusting communities is suggested. Research shows people who live in areas with more parks are more helpful and trusting regardless of their ethnic background, culture and income level.

There are new approaches in design which consider social interactions. Design thinking is a new field which can transform social services sector to approach social change but with a design mind-set. It is said it can make a whole community more resilient by boosting their collaboration and improving their ability to comprehend each other (United Way Toronto and York Region 2015).

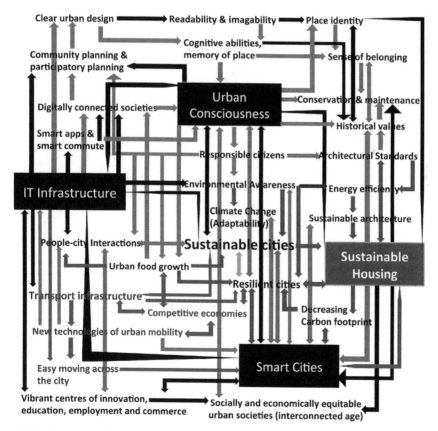

FIGURE 4.3 System thinking and the complex smart city concept

Source: Minaei (2016).

Planners

Keeping the process of planning fluid is more important. Master strategies replace master plans so plans can be updated in shorter intervals (Townsend 2013: 276). Planners need to think about societies, culture, investments, job creations and the interactions of all these because they know how social interactions are important to a city. In Robert Park's opinion, great cities are melting pots of cultures and races. The newer social types are the results of interactions that happen in these cities (Florida 2005: 27). Neuroscientists believe our evaluations are influenced by memory, culture and images of our former experiences. Adam Smith states that human conscience comes from social relationships. Our actions are guided by the natural empathy that is produced by being among other people and is an essential part of our well-being (Montgomery 2015: 39). 'Exposure to cultural information can change the way our brains function by dredging up images and feelings that alter the way we experience things' (Montgomery 2015: 91). There is a statistically positive relationship between the 'Melting Pot Index' and the 'Tech-Pole Index' which Florida (2003: 40) relates to the HCT theory as the main driver of production. A diverse society requires a variety of lifestyle amenities and these amenities lead to city growth and therefore attract talented people.

CASE STUDY **SAN DIEGO, US**

World's Smart Cities: San Diego (2015), a National Geographic Channel documentary, features San Diego (the eighth largest US city) as a smart leading city of the twenty-first century where technology, talent and innovation collaborate and create a novel urban environment. San Diego was selected for its smart public planning, local innovators, strong technology sector, green practices, cultural diversity, liveability, economy and business climate, educational institutions, leadership and strong sense of community and an unparalleled quality of life. City scores on four global categories of innovation, green ranking, quality of life, and digital city were considered for rankings.

Conclusion

Smart cities have already taken concrete steps towards addressing climate change by thinking of strategies and policies to increase citizens' awareness and to improve their city's resilience, sustainability and intelligence. They are reinvesting in augmenting IT, TC and digital infrastructure, planning for smarter solutions on data privacy, data protection and data security (Kitchin 2016b) to create public trust, designing and building smart grid infrastructures to collect real-time data to improve citizens' experiences. Improving transport infrastructures particularly public transport modes with clean buses or trains, designing safe walking or biking passages, encouraging bike share schemes to have pedestrian-friendly cities, electrifying transit and improving electrical infrastructure significantly impact on carbon footprint reduction through electric cars, buses and trams as well as smart cars and driverless cars. Having world-class academic research institutes, labs, R&D centres of innovation or 'living labs initiatives' for learning, testing, researching the implementation of new technologies, and user innovating at a large scale can attract talented people and world-class investors which keep the city growing. That can impact their climate resiliency and strategic plan to approach self-sufficient cities while decreasing their carbon emission by using RES as well as greener agriculture systems, planning and designing for a functional urban agriculture. The first step is increasing public awareness through free education and training of eco-friendly living by city councils and volunteer organizations to encourage preserving public gardens and promote gardening in an actively engaged community to lead to conscious community planning. Preserving historical buildings and protecting the cultural identity, transforming shrinking areas into smart districts by defining and investing in regeneration projects can create meaningful urban spaces for outdoor activities. Having a higher quality of life through outdoor activities and enough cultural amenities attract citizens' engagement and the so-called creative class who prefers communities with high-quality experiences of diverse cultural amenities and not entertainment amenities (Florida 2005: 35, 90). That will lead to an inviting city with conscious inhabitants tolerant about diversity which will draw growth and economic opportunities.

Notes

1   Director of the Global Cities Institute at the University of Toronto, president and CEO of the World Council on City Data, speaker at the 'Talk Transformation: Preparing Toronto for an Extreme Weather Future – Community and Infrastructure Resilience' event in Toronto, December 2015

Bibliography

Bakici, T, Almirall, E & Wareham, J 2013, 'A Smart City Initiative: the Case of Barcelona', *Journal of Knowledge Economy*, vol. 4, pp. 135–148. DOI 10.1007/s13132-012-0084-9

BBC, 2014, 'Who, What, Why: What Is an "Obesogenic" Environment?' *Magazine Monitor*, 28 May, www.bbc.com/news/blogs-magazine-monitor-27601593

Ben Letaifa, S 2015, 'How to Strategize Smart Cities: Revealing the SMART model', *Journal of Business Research*, vol. 68, pp. 1414–1419. http://dx.doi.org/10.1016/j.jbusres.2015.01.024

Breetzke, T & Flowerday S V 2016, 'The Usability of IVRs for Smart City Crowdsourcing in Developing Cities', *Electronic Journal of Information Systems in Developing Countries*, vol. 73, no. 2, pp. 1–14.

British Standards Institution, 2014, *Smart City Framework – Guide to Establishing Strategies for Smart Cities and Communities*, BSI Standards Publications, PAS 181:2014.

Bull, R & Azennoud, M 2016, 'Smart Citizens for Smart Cities: Participating in the Future', *Proceedings of the Institution of Civil Engineers – Energy*, vol. 149, no. 3, pp. 93–101. www.ice virtuallibrary.com/doi/abs/10.1680/jener.15.00030

Citeos 2015, 'What Is a Smart City', online video, 24 August, viewed 22 December 2015, www.citeos.com/video?v=Br5aJa6MkBc

City of Toronto 2016, Open Government, Accessing City Hall, City of Toronto, viewed 13 July 2016, www1.toronto.ca/wps/portal/contentonly?vgnextoid=4550b9ca56ccf410 VgnVCM10000071d60f89RCRD

Dennett, D 2003, 'The Illusion of Consciousness', TED, viewed February 2015, www.ted. com/talks/dan_dennett_on_our_consciousness#t-196539

Department for Business Innovation & Skills (BIS) 2013, *Smart Cities, background paper*, London: BIS.

Dorfman, A 2015, *The Vision of Smart City Networks Becomes Reality*, LinkedIn, viewed June 2016, www.linkedin.com/pulse/presidentobamasvisionsmartcitiesbecomesreality avidorfman

Florida, R 2003, 'Cities and the Creative Class', *City & Community*, vol. 2, no. 1, pp. 1–17.

Florida, R 2005, *Cities and the Creative Class*, London: Routledge.

Goleniewski, L 2001, *Telecommunications Essentials*, Upper Saddle River, NJ: Addison-Wesley.

Huang, X 2016, 'Innovation Trends for Smart Cities, My review of ZOOM Smart Cities Conference', *Smart Cities Online,* viewed 29 July, www.smartcitiesonline.net/?cat= 67&lang=en

Hutchins, E 1995, *Cognition in the Wild*, Cambridge, MA: MIT Press.

ITU 2014, 'Smart Sustainable Cities: An Analysis of Definitions', viewed 15 August 2016, www.itu.int/en/ITU-T/focusgroups/ssc/Documents/Approved_Deliverables/TR-Definitions.docx

Kingwell, M 2008, *Concrete Reveries (Consciousness and the City)*, New York: Penguin Group.

Kitchin, R 2016a, 'Continuous Geosurveillance in the "Smart City"', *Dis Magazine,* viewed 4 August 2016, http://dismagazine.com/dystopia/73066/rob-kitchin-spatial-big-data-and-geosurveillance/

Kitchin, R 2016b, 'Getting Smarter about Smart Cities: Improving Data Privacy and Data Security', Data Protection Unit, Department of the Taoiseach, Dublin, Ireland. www. taoiseach.gov.ie/eng/Publications/Publications_2016/Smart_Cities_Report_January_ 2016.pdf

Kylili, A & Fokaides P A 2015, 'European Smart Cities: The Role of Zero Energy Buildings', *Sustainable Cities and Society*, vol. 15, pp. 86–95.

Lee, F 2016, 'Renewable Energies that will Revolutionize Sustainable Cities of the Future', *Sustainable Cities Collective,* 6 July, viewed 7 July 2016, www.smartcitiesdive.com/ ex/sustainablecitiescollective/renewable-energies-will-revolutionize-sustainable-cities-future/1215022/

McCarney, P 2015, 'International Standards on Global City, Indicators for Sustainable Cities, Resilient Cities And Smart Cities', speech presented at the 'Talk Transformation: Preparing Toronto for an Extreme Weather Future – Community and Infrastructure Resilience', Toronto, 24 September.

Miller, B J 2006, 'Cognition and the City', *Proceedings of the Annual meeting of the American Sociological Association, Montreal Convention Center, Montreal, Quebec, Canada.* www. allacademic. com/meta/p103566_index.html, viewed, http://crisisfronts.org/wp-content/ uploads/2008/09/brian-miller_cognition-in-the-city.pdf

Minaei, N 2004, 'The Effect of Globalization (Information Technology and Telecom-munication) on Physical and Conceptual Aspects of Cities: Case Study of London', PhD thesis, Islamic Azad University, Science and Research Branch.

Minaei, N 2014, 'Do Modes of Transportation and GPS Affect Cognitive Maps of Londoners?' *Transportation Research Part A: Policy and Practice*, vol. 70, pp. 162–180.

Minaei, N 2016, 'Reflections of Transport Mode and GPS on the Experience of Urban Atmosphere, Case Study of London', in *Proceedings of the Sensing the Place – Experiences & Wayfinding, ARCC and the Feeling Good Foundation*, London, 27 April, viewed 2 May 2016, www.arccnetwork.org.uk/health_wellbeing/feeling_good_in_public_spaces/sensing_the_place_experiences_wayfinding_in_a_changing_climate/#.VyeJ%E2%80%A6

Montgomery, C 2015, *Happy City*, London: Penguin Books.

Moore, T 2014, 'From Intelligent to Smart Cities', *Australian Planner*, vol. 51, no. 3, pp. 290–291. DOI:10.1080/07293682.2013.810163

National Geographic Channel. 2015, *World Smart Cities: San Diego*, online video, 24 June, viewed December 2015, www.youtube.com/watch?v=LAjznAJe5uQ

Neirotti, P, De Marco, A, Cagliano, A C, Mangani, G & Scorrano, F 2014, 'Current Trends in Smart City Initiatives: Some Stylised Facts', *Cities*, vol. 38, pp. 25–36.

Owens, S & Cowell, R 2011, *Land and Limits: Interpreting Sustainability in the Planning Process*, 2nd edition, Abingdon: Routledge.

PRNewswire 2016, 'Global Smart City Industry', *The Business Journals*, 22 March, viewed 22 July 2016, www.bizjournals.com/prnewswire/press_releases/2016/03/22/BR53050

Rometty, G 2011, 'Smart Cities Rio', 10 November, viewed January 2015, www.youtube.com/watch?v=DKmj7nlQndg

Sassen, S 2011, 'Who Needs to Become Smart in Tomorrow's Cities, The Future of Smart Cities', online video, lift France11 with Fing, 6–8 July, http://videos.liftconference.com/video/2895375/saskia-sassen-the-future-of-smart-cities

Schaffers, Hans, Komninos, N, Pallot, M, Aguas, M, Almirall, E, *et al.* 2012, 'Smart Cities as Innovation Ecosystems Sustained by the Future Internet' [Technical Report], <hal-00769635>, https://hal.inria.fr/hal-00769635, 2013

Syal, M. G. M & Ofeo-Amoh, K 2013, 'Smart-Grid Technologies in Housing', *Cityscape*, vol. 15, no. 2, Mixed Messages on Mixed Incomes, pp. 283–288.

Tinker, P & Levitt, T 2016, 'How Air Pollution Affects Your Health – Infographic', *The Guardian*, 13 July, viewed 13 July 2016, www.theguardian.com/sustainablebusiness/2016/jul/05/howairpollutionaffectsyourhealthinfographic

Townsend, A M 2013, *Smart Cities, Big Data, Civic Hackers, and the Quest for a New Utopia*, New York: W.W. Norton.

UN, 2015. *World Population Prospects, The 2015 Revision (Key Findings and Advance Tables)*, New York: United Nations.

United Way Toronto & York Region, 2015, 'Designing a Blueprint for Social Change', 15 January, http://imagineacity.ca/2015/01/15/designing-a-blueprint-for-social-change/

VINCI Energies, 2015, 'What Is a Smart City?' online video, 24 August, viewed November 2015, www.youtube.com/watch?v=Br5aJa6MkBc

Yigitcanlar, T 2015, 'Smart Cities: An Effective Urban Development and Management Model?' *Australian Planner*, vol. 52, no. 1, pp. 27–34. DOI: 10.1080/07293682.2015.1019752

PART II
Institutions

5

LOCAL INNOVATION IN EMERGING CREATIVE ECOSYSTEMS

Onur Mengi and Koray Velibeyoğlu

Abstract

Globally, most future economic growth will occur in regional cities, but infrastructure and employment are often inadequate. In short, default development approaches may focus on shaping the urban form (infrastructure/housing) at the expense of the institutional and intangible factors driving jobs growth, such as creativity, innovation and sector productivity. Effective local partnerships can help counter the limitations of default approaches to urbanism. Analyses reveal that the wedding wear sector in Izmir, Turkey, in particular is in many respects unique, with great potential as an emerging cluster due to its inherited knowledge and know-how, yet still lacks in design considerations, and is in desperate need of promotion, advertisement and cost-effective returns. Therefore, the main lesson is that enabling interactions between the local government, NGOs and firms both inside and outside the cluster can promote smart development. At policy level, establishment of an incubator within a four-leg structure is an effective local development partnership. In a practical Turkish context, Chapter 5 investigates the institutional and partnership management arrangements to facilitate such urban innovation hubs for creative ecosystems.

Key terms

Local partnerships, investment, infrastructure, sustainability, feasibility, urban development, diagnostics, analytics, risk

Introduction

Unprecedented demographic pressures drive regional city growth, but environmentally, socially and commercially sustainable cities need appropriate infrastructure.

Smart development involves not only the construction of appropriate dwellings, but also the strengthening of the productive capability of local industries (Rudlin & Falk 2014; Elkington 1997). Smart planning systems promote modern physical and technological infrastructure, and support the productivity of local firms. In short, smart urban development involves considered and well-designed physical transformation, supported by social and innovation measures. Woo (2013) calls for 'a new compass to guide bold policy directions, change incentive structures, reduce or phase out harmful subsidies and engage business leaders in a vision for an innovative, new economy' (Woo 2013: 1). One institutional mechanism that can encourage such smart aspirations is a Local Development Partnership (LDP) as a workable alternative to fragmented and conflicted planning landscapes. Such local partnerships played a role in transforming Wilhelmsburg and Hamburg in Germany from blighted industrial locales into thriving ones, replete with affordable and energy-efficient housing. LDPs target Placed Based Enterprises (PBE), which catalyse sustainable urban development (Shrivastava & Kennelly 2013). PBE not only adopt triple bottom line (TBL) performance evaluation but are also locally rooted, structurally embedded in unique geographical locales. Distinct place identity is conditioned both by urban form and socially constructed historicity. Ecosystems are important for business sustainability; firms operating in non-descript locales or soulless construction wastelands are more at risk. Without a rooted 'sense of place' or an 'identity', workers and end-users become alienated, unable to connect to the *genius loci* or form any emotional bonds with place (Low & Altman 1992). Ultimately, in such bleak landscapes, productivity and consumer confidence suffer.

To investigate whether urban ecosystems with a LDP and PBE focus enhance resilience, this chapter focuses on inner-city fashion clusters. Specifically, the Turkish case study investigates the wedding industry in the Mimar Kemalettin Fashion District of Izmir, Turkey. Formally, the research question is: 'What are the planning implications of adopting an ecosystem management approach to nurture fashion clusters?'

To date, most cluster research has focused on the defence, automotive, aeronautical or biochemical and technology sectors. Scrutiny of the Turkish fashion sector in the Mimar Kemalettin Fashion District provides new insights for regional cities as they struggle to compete in the global marketplace. Often, misguided policy or ill-considered local interventions can create rather than solve problems. Top-down approaches tend to disregard the potential of local innovation, and ignore actual global trading realities.

Through the 1990s, new specialized services, information technology, innovation and design, cultural production and international mega-projects have led a major shift in the spatial and social structure of cities; cities have shifted and expanded and consequently evolved new spatial as well as social organization patterns. 'Whereas the dominant industries of the nineteenth and twentieth century's depended on materials and industry, science and technology, the industries the twenty first century will depend increasingly on the generation of creativity through innovation' (Landry & Bianchini, 1995: 2). Creativity has been both the driver and the component of this new economy. However, in general, rather than residing in only certain industries, creativity is a central and increasingly major input into all sectors where design and content constitute a base for competitive advantage in global economic markets. The definition of sector here becomes confusing if creativity is associated only with the arts, and not other sectors of production such as science or technology. Rather, creativity should be approached at the macro-level, defined in a cross-sector and multidisciplinary way, mixing elements of 'artistic creativity', 'economic innovation' as well as 'technological innovation'. KEA (2006) regards creativity, especially in industry environments, as 'a process of interactions and spill-over effects between different innovative processes' (p. 41). In fashion, a creative industry in the global context, many nations experience economic benefits as a participant in the global fashion industry system. Fashion is often found in the form of clusters as ecosystems containing dynamic interactions between firms, and which are surrounded by other ecosystems in the environment. Therefore, the ecosystem emerges around the diversified members and interactions, as well as evolution dynamics. Interaction and evolution, major characteristics of such ecosystems can be developed by the local institutional partnership.

Within such a context, local innovations in sectoral cluster ecosystems in general, as well as the wedding wear cluster in particular, must more adequately be incorporated along with the issues of system design. This can lead to vitality of ecosystem, its sustainability and the issues diversity, redevelopment and evolution dynamics, enabled networks, potential interactions and collaborations for competitive global market, investment of diverse human capital for a better value chain production, particular urban planning interventions based on organic approaches to clusters, and realizing a creative environment specific to our case. These key policy implications are highlighted, since the ecosystem approach to sectoral clusters within given urban places of any scale is unique and comprehensive, and is key to managing and sustaining the existing local innovation.

Literature and conceptual framework

The ecosystem approach is very common in the literature on creative ecology, business ecosystem and ecosystem management (see Moore 1993; Pirot et al. 2000; Argote et al. 2003; Dvir & Pasher 2004; Shorthose 2004; Iansiti & Levien 2004a, 2004b; Teece 2007; Hearn et al. 2007; Duxbury & Murray 2010; Chen et al. 2010; Chan 2012; Winden et al. 2012; Kannangara & Uguccioni 2013). However, the institutional management of creative industry clusters as ecosystems is still not touched upon. As this present study suggests, there has been no comprehensive and concrete study on the overall dynamics of the creative industry clusters that investigates the contexts in which a key role is played by local diversity, interactions, competitions, growth and survival. Therefore, the ecosystem approach based on the application of appropriate scientific methodologies of recent studies is undertaken and adapted to the creative industry clusters. The intersections of previous ecosystem approaches are reviewed to enable a better understanding of the creative industry clusters (Figure 5.1).

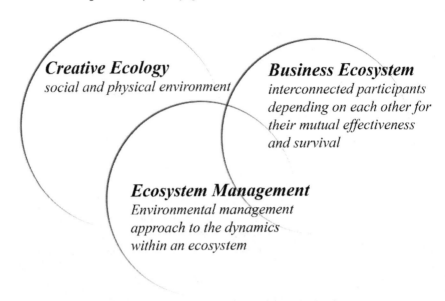

Creative Ecology
social and physical environment

Business Ecosystem
interconnected participants
depending on each other for
their mutual effectiveness
and survival

Ecosystem Management
Environmental management
approach to the dynamics
within an ecosystem

FIGURE 5.1 The foundation of the research approach of the study

Local innovation: buzz and pipelines

The dynamics of clustering vary. Some scholars treat clustering as a purely geographical phenomenon based on spatial or geographical proximity (Gordon & McCann 2000; Feldman 2000). Other schools of thought stress not geographical 'space' but 'place', regarding it as a social rather than a physical phenomenon. Torre and Rallet (2005) highlight proximity, claiming that 'the purpose of examining geographical proximity is to determine whether one is "far from" or "close to"' (p. 49). Similarly, Asheim and Gertler (2006) refer to this type of geography as neo-regionalism. In this respect, neo-regionalists attempt to extend the idea of the region to encompass interactions and processes for knowledge, creativity and innovation. Storper and Venables (2002), Owen-Smith and Powell (2002) and Bathelt *et al.* (2004) build upon these discussions, strengthening the spatial and non-spatial perspectives by incorporating the concepts of local buzz and global pipelines. These two concepts represent two scales for local innovation within the clusters.

Local buzz

Storper and Venables (2002) identify the conception of 'buzz', referring to the spatial dimension of information and communication ecology created by face-to-face contacts, co-presence and co-location of people and firms within the same industry and place or region. This type of interaction consists of specific information and continuous updates of this information, planned or spontaneous learning processes, and also shared cultural traditions and habits within a particular field. These all stimulate the establishment of conventions, and also other institutional arrangements. In a similar way, so does Owen-Smith and Powell's (2002) concept of 'local broadcasting' and Grabher (2002) uses the concept of 'noise' to refer to the same notion.

> Participating in the buzz does not require particular investments. This sort of information and communication is more or less automatically received by those who are located within the region and who participate in the cluster's various social and economic spheres . . . It is almost unavoidable to receive information, rumors and news about other cluster firms and their actions. This occurs in negotiations with local suppliers, in phone calls during office hours, while talking to neighbors in the garden or when having lunch with other employees and so on.
>
> (Bathelt *et al.* 2004: 13)

Being in close proximity, co-located and visible creates potentials for interpersonal translation of important news and information among the cluster actors and firms. Also, Maskell *et al.* (1998) suggest that the trust exists in localities as something inherited, and easily available to any known-member.

Global pipelines

Owen-Smith and Powell (2002) employ the non-spatial stand, and propose the term 'pipeline' to describe contexts that include the channels used in such distant relations and contacts. The findings of their study on the Boston biotechnology community finds that, while knowledge spillovers and creation may be more effective within rather than across regional network borders, physical distance is not the only influence. They argue that the creation and information accumulation results not only from the local and regional interaction but is often obtained through strategic partnerships of inter-regional and international reach. These pipelines often enable planned, systematic and decisive knowledge flows rather than undirected and spontaneous local buzz. Unlike the local buzz between cluster firms, this approach lacks the benefit of shared trust in inter-regional and international environments. Rather, the formation of global pipelines with distant partners requires time and involves costs (Bathelt *et al.* 2004).

Regarding the mutuality of these two concepts, local buzz and global pipelines feed each other and they are complementary. The more firms of a cluster create trans-local pipelines, the more innovation about products, industries, markets and technologies are channelled into internal networks, and the greater the dynamism of local buzz. Explaining the reasons for circulation of economic activities is vital to recognize how, for certain industries, local innovation, as well as capital and labour are attracted to particular places. Particular sectors with high potential of local innovation differ from each other because they are not driven only by the regional geographies of specialization, whether spatial or non-spatial, but also by the externalities that arise from the diversity of these places.

CASE STUDY **FASHION SECTOR IZMIR, TURKEY**

Due to the recent developments, a view is emerging that, within the fashion industry in Izmir, the wedding wear sector has local innovation potential for increased profits and economic growth. The fashion industry reveals some path-dependent development for Izmir, yet is still evaluated as an emerging creative industry, in spite of various weaknesses (Mengi & Velibeyoglu 2013). Over three years, it was observed that the wedding wear sector was evolving into an emerging cluster (see IZKA 2009, 2010, 2013). This aspiration draws attention to the existing sector and its ecosystem, and the wedding wear sector and Mimar Kemalettin Fashion District in particular has been chosen as a case study area. A survey was used to explore the interaction potentials and evolution dynamics for local innovations in the existing cluster.

For analysis of the wedding wear cluster, all 266 firms located in different parts in the Mimar Kemalettin Fashion District were contacted regarding a survey for the final field study. Among 266 firms, 28 were not manufacturers, but retailers, and were excluded from the survey. In all, 132 firms agreed to participate in the survey.

The case study area of this study covers the wedding wear firms located around the Mimar Kemalettin Fashion District and adjacent area. Several mappings and use of illustrative methods show how locationally the wedding wear sector is distributed and clustered in this district. According to these findings, wedding wear firms are mostly concentrated around the core of the area. According to the final investigation, the area currently contains 266 wedding wear firms, including retailers, wholesalers and manufacturers. Their locational distribution as a cluster in Mimar Kemalettin Fashion District is given in the Figure 5.2.

Interaction potentials of the cluster

For the sharing of creative ideas and knowledge among firms through interaction in different scales, Figure 5.3 illustrates that in-house sharing, in other words, internal interaction within firms, is more common than interactions with others within the ecosystem. However, Figure 5.3 shows that wedding wear companies have a higher level of interaction with similar companies inside the cluster compared to those outside the country. Also, considering the standard deviation of 1.2 for the international scale of interactions, firm characteristics vary: 35 per cent report good relations with foreign companies, while 32 per cent have weak relations.

Specifically for the internal interactions within the ecosystem, in order to keep informed about the recent sector developments in the cluster, face-to-face relations is the medium employed the most, followed by trade fairs and other formal relations with other firms. However, compared to other mediums,

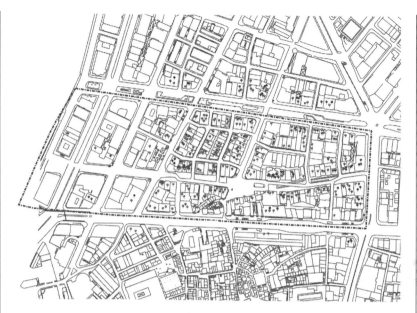

FIGURE 5.2 Locational distribution of wedding wear firms on site

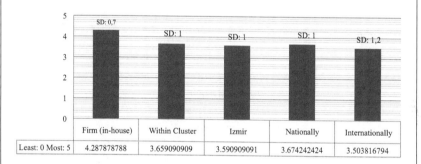

FIGURE 5.3 Interaction in different scales

there is a higher level of disagreement over the role of the Mimar Kemalettin Association and informal conversation about the current conditions of the sector in this ecosystem (Figure 5.4).

Evolution dynamics of the cluster

In terms of the development attempts through mutation, the major outcomes indicate the importance of the role of national new networks, competitiveness, increased sales, knowledge mutation among workers with share and exchange, knowledge accumulation within firms, international new networks and increased trade activities. When comparing outcomes, firms reported new

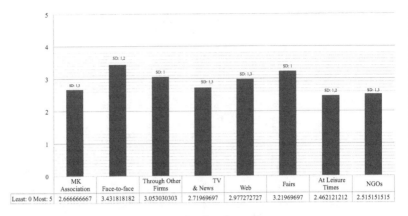

Least: 0 Most: 5	MK Association	Face-to-face	Through Other Firms	TV & News	Web	Fairs	At Leisure Times	NGOs
	2.666666667	3.431818182	3.053030303	2.71969697	2.977272727	3.21969697	2.462121212	2.515151515

FIGURE 5.4 Mediums of interaction for the cluster

connections and networks through other wedding wear firms in the country as most important, followed by new designs and collections differentiating them from the other firms, as well as increases in their sales (Figure 5.5).

In terms of innovation through crossover, via hiring employees with experience in the sector, there is a different degree of benefits according to whether they are from inside or outside the cluster. The knowledge crossover through inside the cluster is greater than from the outside. Respondent firms generally agree that hiring an employee experienced in their particular cluster contributes considerably in terms of design and collection. This also provides increased sectoral knowledge transfer to the firm through a new employee and also results in more efficient manufacturing processes. Also the arrival of a new employee hired from the cluster is generally considered to result in an increase in sales. According to the survey, firms are undecided about the knowledge contribution of a new employee from the sector, but from a different locality, with standard deviation of 1.2 (Figure 5.6).

All firms agree that research and development is a major driver of ecosystem evolution, whether for new technologies, materials (fabrics and textiles) or

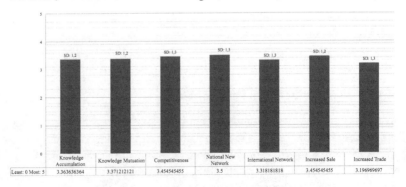

Least: 0 Most: 5	Knowledge Accumulation	Knowledge Mutuation	Competitiveness	National New Network	International Network	Increased Sale	Increased Trade
	3.363636364	3.371212121	3.454545455	3.5	3.318181818	3.454545455	3.196969697

FIGURE 5.5 Outcomes of knowledge mutation and in-house development

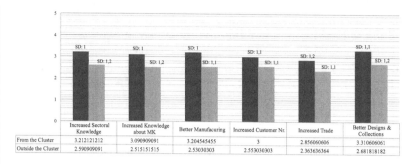

	Increased Sectoral Knowledge	Increased Knowledge about MK	Better Manufacuring	Increased Customer Nr.	Increased Trade	Better Designs & Collections
From the Cluster	3.212121212	3.090909091	3.204545455	3	2.856060606	3.310606061
Outside the Cluster	2.590909091	2.515151515	2.53030303	2.553030303	2.363636364	2.681818182

FIGURE 5.6 Outcomes of knowledge crossover through hiring

new wedding wear apparel; however, the scope for e-commerce is limited (Figure 5.7).

The case study research of the wedding wear sector in Izmir illustrates that ecosystem dynamics for creativity and smart development are closely related to interaction between firms as well as the process of evolution within the spatial unit. Therefore, smarter urban development is closely associated with creative ecosystem dynamics and underlying policies.

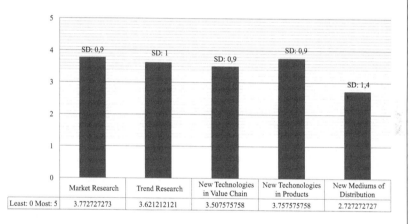

	Market Research	Trend Research	New Technologies in Value Chain	New Techonologies in Products	New Mediums of Distribution
Least: 0 Most: 5	3.772727273	3.621212121	3.507575758	3.757575758	2.727272727

FIGURE 5.7 Areas of research and development for evolution

Policy recommendations and stakeholders implications

The Turkish fashion sector case study articulated an emerging ecosystem approach aimed at accentuating the commercial benefits of clustering; however, whether the conclusions are relevant for other geographical or sectorial contexts is uncertain. The ecosystem approach highlighted the importance of internal wedding wear sector organizational operational dynamics with specific implications for spatial planning policy in Izmir.

Development and evolution dynamics

For evolution of the sector towards more creative and fashion-based industry, it is important to introduce new business models which both create and capture creative and cultural value. Such creative industry policy should encourage a wide-ranging research and development for the sector. Research and development with the possible cooperation of university fashion design departments is a possible innovative approach. Such cooperation also brings vitality and promotion to the district. Besides, a particular focus on the relationships between publicly funded associations, universities and the private sector can result in greater planning in the development of local innovation. However, resilience becomes an issue within such development efforts, especially which occurs around the ecosystem. In this respect, it is important to take into account the amount of change that the ecosystem can adopt while maintaining its structure and function, and the extent to which the ecosystem is capable of operating itself without the need for excessive external forces, and the degree to which the ecosystem develops the capacity to learn and adapt in response to disturbances.

An ecosystem evolves over time through new knowledge emerging from mutation of internal populations, or its incorporation of new knowledge and creativity pollinated from outside sources. For policy implications, it is crucial to take into consideration how and when evolution relies on outside knowledge crossover. Creativity and knowledge are vital, and their diffusion among firms through spin-off creation, labour mobility and cooperation networks suggest strategic opportunities for creative industry clusters.

Enabled organic networks

Networks create a form of community and identification. They are also a form of validation and legitimation of protection and vitality, within which ecosystems can ensure their survival.

Creative industry clusters tend to provide different types of network. The specific formal organization is one level of the network, which is considered useful, but rather inflexible and strict. At another level, ecosystems comprise social links to less formal shared experiences on different scales. Also, within large ecosystems, it is not possible to argue that members form smaller and closer networks of their own as a direct outcome of their social interactions. These smaller scale networks are relatively organic, satisfying more directly to the needs of ecosystem, and integrating members more effectively. Especially, for the local innovation in creative industries informal networks are very difficult to access, in contrast to formal ones. Therefore, a top-to-bottom approach may not be effective in constructing such organic network-based ecosystems of creative clusters.

Interactions and collaborations

Connectedness and interdependence realized through interactions and collaborations are the key operating principles in such ecosystems. Creative industry clusters benefit by understanding their existing place conditions, specifically, their links both internal and external to ecosystem, and the interdependence between firms inside and outside the cluster, and also between firms and the related sectors in their environment. Export capability of this ecosystem provides a place in the global creative ecology. Policy implications should also emphasize the mutual interdependence and interconnectedness in an attempt to promote wedding wear manufacturing at both regional and national level, as well as strengthen relationships with other industrial ecologies nationally and globally. For instance, development opportunities are provided by the internal interaction at firm and district level and external interactions with the related sectors within the creative industries, and collaborations with the services and cultural sectors. Of particular relevance are services that promote the integration of the creative class including professional, technical and creative knowledge skill sets from the design, information technologies and engineering fields. On the other hand, promotion and commercialization strategies supporting these interactions require management capacity in various disciplines, as well as a capacity to combine these disciplines in collaborative ways. It is already acknowledged that creative and design professionals are highly embedded in all industry sectors, and the wedding wear sector is no exception. Collaborations among public–private partnerships and associations as well as external interactions are also crucial drivers for such implications. Lastly, the social relationships based on trust can build local cultures of production as part of collaborative business relations, even in a competitive environment, and support longer term individual and collective cognitive image of ecosystems.

Investment in human capital

The new economy that has triggered the urban restructuring has also enabled a global competitive arena, which has become a place where human capital is the core component, in the form of creativity, know-how and/or tacit knowledge. As mentioned in the vast literature, human capital is one major driver of the creative industry formation known as the creative class. The diversity of the inner organizations of industries and the occupational distributions, also diversity of the cluster itself that consists of these firms and also contains the physical variety, human diversity and product variety are regarded vital to the environments. More importantly, carefully managed socio-economic urban spaces like the sectoral cluster can, from the human capital perspective, contribute to integrating creative workers and craftsmanship into everyday culture, enabling wider local innovation opportunities.

Along with the policy implications, it has been underlined that education and training activities, and facilitation of learning and communication among key actors can create long-term benefits for the vitality of such ecosystems. The creative class is the essential player, and although the present research does not investigate the issue of creativity, the potential future areas for exploration include investment in the creative workers which is outside the scope of the current study. Therefore, policy implications should consider creative workers as central to success, and acknowledge that the capacity to produce and observe new ideas is an outcome of their particular creative education and training, which is one of the underlying mechanisms of growth of an ecosystem of creative clusters. Policy should therefore address how to foster creative human capital within the expanding creative workforce of creative industry clusters.

More organic planning

The stage in the cycle of clusters appears vital and needs to be taken into consideration by evaluating the existing structure of creative clusters. The strengths and potentials of clustering of creative industries are associated with their stage in the development process: dependent, aspirational, emergent and mature. The ecosystem of the cluster in this study can be said to be emerging due to the existing infrastructural investment from the public sector and rapidly developing local and regional markets in which demand has become significant and the market is expanding towards international levels. Understanding the emerging characteristics of the ecosystem requires a long-term visioning, rather than being an addition to urban discourse, and the creative and cultural factors should be considered as grounded in the urban context through a less mechanistic approach. There are no fast-track solutions in the development of ecosystems. However, it should be acknowledged that in current urban planning practice, the integration of organizational and spatial dimensions to the ecosystem approach in terms of local innovation management is still more conceptual than practically orientated. Instead of the pure organizational ecosystem management, appropriate protocols and mechanisms of planning should be determined and, as far as possible, incorporated at the policy level.

The issue of sustainability

As the concept of sustainability for systems has matured, growing emphasis has been given to the interconnection to social dimension with economy in the framework of local innovation. Generally, in sustainable thinking, diversity has become crucial, and as a major component of this ecosystem both as an input and output, it is the ultimate aim of fashion production. This is applicable to the wedding wear sector, as a sub-set, in which diversity is lacking yet increasingly recognized as a means of achieving success for the ecosystem as a whole. However, in future interventions, there is a need to balance the extent of the possible innovation policies for the district along with growing interest in the creative industry based development, especially considering the uniqueness of the sector.

At the organizational level, the development of skills of workers affiliated with individual functional units can grow very fast, while employing their primary knowledge in processes. Also, an appropriate distribution of knowledge populations is important to the success of the ecosystem. Without adequate levels of knowledge or necessary and required knowledge such ecosystems will not survive. However, at the same time, the failure of individual members should not affect the overall effectiveness and operations of the ecosystem, due to the resilience capability and viability of the ecosystem.

Resilience describes how a given system responds to the demands and pressures of external conditions; whether physical or non-physical. In order to acknowledge the resilience of the ecosystem, possible investigations need to be done from different aspects, as illustrated in Figure 5.8. First, for resistance, it is important to assess the amount of change that a system can undergo while retaining the same controls on structure and function. Second, it is necessary to check to what extent the system is capable of organizing itself without becoming disorganized or dominated by external environment. Finally, there is a need to examine the adaptive capability of ecosystem to prepare for unexpected events, respond to disruptions, and recover from them by sustaining its operations at the desired level.

The faster a system returns to steadiness, the greater its viability. There is an implicit assumption of resilience in the system; without resistance, there would be no presumed return to the pre-disturbance state, but rather, with the help of resilience, an adjustment to some new equilibrium level, which could be better

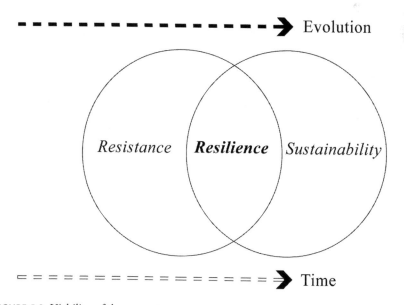

FIGURE 5.8 Viability of the ecosystem

or worse than the previous state. To attain long-term sustainability, the ecosystem must be resilient and be prepared to respond to changes that produce disturbances, but enable evolution over time.

Local innovation efforts for Mimar Kemalettin Fashion District

As a part of the overall creative environment, allocating resources for wedding wear clusters and underlining the unique craftsmanship of the manufacturing process may contribute to the social sustainability of this ecosystem while producing greater economic returns for the city. The planning efforts for the district implemented in 2000 resulted in improved performance of the wedding sector cluster without entirely changing the sociality of the district. Developing a sharing culture is important for better creative and knowledge transfer and exchange. A certain degree of competition among firms is beneficial for development and promotes survival skills. However, supporting collaborative competition and limiting the conflictive competition increases the vitality of the ecosystem.

Efforts for planning creative environments, in particular, for Mimar Kemalettin Fashion District, may indicate a more intentional basis for actions in comparison to a top-down approach. This involves participatory planning and priority-setting, integrated planning, and horizontal coordination across municipalities and other local NGOs. It is important to maintain a certain degree of proximity between the ecosystem and related sectors in order to enable effective external interactions within this ecosystem. This increases the competence to derive and adapt innovations from similar sectors through crossover, and also internal mutation, through the flow and exchange of creativity, knowledge and innovations.

Urban environments with a unique historical value, that attract fashion designers to work and live, and that provide high accessibility, low transaction costs, as well as a high quality built environment with a certain social mix, and an alternative, creative atmosphere, have enabled the creative industries to take advantage of all these properties and accumulate in certain places. The quality of the atmosphere is also underlined in the theory of creative cities. This involves three factors: first aesthetic attributes, which provoke visual attraction, unexpectedness and spontaneity; and, second, place which brings co-location, clustering as well as attachment, comfort and authenticity. The final factor is social life with enabled networks and with individual and physical diversity. The presence of the historical and cultural context around the Mimar Kemalettin Fashion District should also be stimulated to strengthen this spatial environment. Furthermore, its identity needs to be sustained in all relevant policy-making decisions.

Mimar Kemalettin Fashion District already embodies the history, meaning and local knowledge of wedding wear production. But while spatial imaginaries are powerful, amenities are lacking. Thus, the interventions for restoration or rehabilitation of spaces for repurposed uses would overcome such deficits. While some projects are proposed for public uses, others are more profit-orientated. This district is a unique environment, full of architectural significance and locational opportunities, yet possible development scenarios on creative space-making, such as marketing, theming, creation of facilities aimed at costumers, all carry the risk of destroying the authenticity of this area. Thus, the decisions towards promoting the local assets and dynamics should be considered the key to unlocking access to global markets for this ecosystem to compete.

Planning interventions based on an organic approach to such a cluster should necessarily assist the value creation chain of firms and provide urban services, including hard and soft infrastructure. Developing infrastructure appears as an important strategy in providing mobility and access from major connecting roads, airports and ports both to production, products, human capital and other resources of the given ecosystem. The interventions should also consider the physical environment and amenities, enable access to human capital, access to broad networks and markets across Izmir and Turkey. They should allow a diversified industrial structure since all creative industry clusters, regardless of the sectoral indication, are interconnected with other creative sectors, and build diversity of skills for different stages of the fashion production value chain. Additionally, such interventions need to enable openness to flow and exchange of creative ideas and know-how through the existing local buzz of the Mimar Kemalettin Fashion District. Also needed are zoning in relation to adjacent land uses in the city and other urban policies that promote recreation and entertainment. Less significant advantages of such clustering are its proximity to the bus terminal and airport. Also, its former proximity to the Izmir International Fair has disappeared due to recent developments and the relocation of the fair outside the city centre. Therefore, these should be considered in future planning and any decision on the relocation of the cluster.

It is also essential to highlight human creativity as an integral part of this existing ecosystem. Thus, the ultimate goal of creative environment interventions must be to address strategies for planning, design and the required physicalities to allow creativity to be attracted, retained, developed, shared and exchanged in and around the district. In order to achieve this, policy-making needs to involve careful observations of potential members of the ecosystem and their particular involvement in the processes of wedding wear manufacturing as a sub-set. The branding efforts also need to be elaborated with the physical planning of this ecosystem. Respectively, the success of sustainability through social content of this ecosystem is likely to depend on the long-term functioning of creative value accumulation through workers.

The careful examination of the existing ecosystem brings about the instability of markets, the need for distinction as a competitive advantage, the issue of product differentiation for the niche market as a wedding wear sector. Otherwise, any creative industry formation will dissolve in the long run without well-planned production, distribution and consumption, and, more importantly, without creativity and good design as competitive advantages. However, creative and cultural essence in such a sector is, without doubt, embedded in existing socio-cultural and institutional structures.

Development of a local innovation incubator and its operational management

As part of such an intended creative environment of the wedding wear sector, a possible creative incubator in Mimar Kemalettin Fashion District would be developed. The operations of this incubator could be designed as a system within a four-leg structure, as illustrated in Figure 5.9.

First, the structure of the incubator would consist of multiple actors. Initially, this incubator might be operated by the Izmir Metropolitan Municipality or by a non-profit cooperative, or a combination of public and profit-based organizations. A possible non-profit actor is the Aegean Clothing Manufacturers Association (EGSD), in collaboration with Izmir Chamber of Commerce (IZTO) and Aegean Exporters Association (EIB), both of which are already involved with the sector. Considering the current situation of wedding wear production in Izmir, other than its cluster in Mimar Kemalettin Fashion District, there are further actors likely to be part of the operations of the incubators. Additional actors who could be incorporated are the independent designer boutique owners located in Karsiyaka and Alsancak, who currently produce wedding wears on demand, and relatively larger manufacturing firms based in MTK (Textile Manufacturers Site in Izmir). Such multi-actor based structure would also diminish the further need for monitoring, or any performance indicator requirements for maintenance of the creative environment.

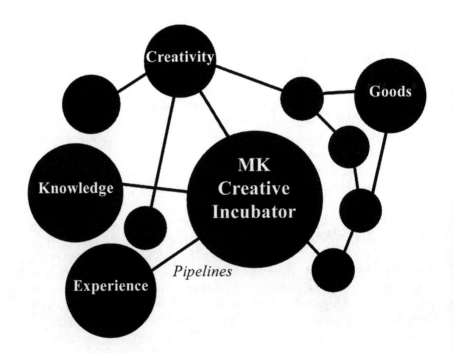

FIGURE 5.9 Content of the possible pipelines

Second, collaborations with education centres, institutions and, more importantly, with the relevant design and engineering departments of universities could provide benefits through shared specializations via such a platform. Particularly, such a structure becomes a major mechanism for the recognition of talent and professionalism. The incubator allows opportunities for individuals to be perceived and recognized at the right spot, to interact with professionals, fashion designers and the social environment. With the additional help of independent designers, such talent management could be carried out. This could allow the constant and continuous creativity and knowledge exchanges, and much cooperation with designers that the sector currently lacks. Such a centre can also serve as a node where the education, research, archiving and displaying of wedding wear designs can be carried out, and encourage the new members into the ecosystem.

Third, the proposed incubator might serve as a hub to the wedding wear manufacturing that has currently unstable regional conditionals for evolving into a more organized structure with a creative value chain. That covers not only its production, but also distribution and marketing considerations. The incubator works as a showcase where established and emerging wedding wear brands display their products, promote themselves, and participate in the competition in a more collaborative environment. On the other hand, for marketing, it could overcome the readjustment of the existing ecosystem caused by the relocation of Fair Izmir, and function as a new centre for fairs, exhibitions, corporate events and conferences, and provide all the necessary equipment. That eventually would lead the cluster to form a more comprehensive identity.

Finally, the necessary pipelines could also be built both regionally and nationally through such an incubator. This incubator may function as a pump functioning through the established pipelines. The current networks fail to create sufficient global pipelines, providing only for the goods and product transfer of the sector. Rather, knowledge and creativity should be incorporated with experience flows in order to build more effective pipelines and reach non-spatially to other geographies. Also, the flow needs to operate in both directions. This might also offer a platform of non-spatial links for creators and enable connection, production and networking among non-creative and creative workers. More trans-local pipelines bring more information and news about products, industries, markets and technologies through internal networks, and this boosts the already existing local buzz in Mimar Kemalettin Fashion District.

Conclusion

This study aimed to reveal the underlying behaviour beyond the local innovation dynamics within the case of emerging creative district in Izmir, Turkey. The Mimar Kemalettin case study illustrated the importance of ecosystem dynamics and particularly shows how local innovation interaction and evolution can scale up new economy sectors. The fundamental role for policy makers therefore is to shape and create contexts in which creative industry clusters can grow and evolve. In this respect, policy makers can establish the drivers, such as education and investment, to create a sustainable pattern of operation for different stages, attracting relatively large companies to the district. On other occasions, drivers may be needed to overcome the obstacles to better performance, such as research incentives for the distribution, or grant programmes through NGOs, while taking into consideration the long-term effects of the sector on the entire ecosystem. The key principal here is to regard the ecosystem as a system-based perspective facilitating the long-term local innovation opportunities for the sector. Adequate establishment of the global pipelines that connect the clusters non-spatially to further geographies should also be taken into account, while designing such systems.

In information-rich industries like fashion, creating a local buzz, informed by global trends and producer pipelines is equally important for smart and for creative development. Local buzz transmits tacit knowledge which underpins incremental local innovation. Diverse loose spaces in historical districts, like Mimar Kemalettin Fashion District, are an ideal milieu to nurture circles of entrepreneurial improvement. Global pipelines, as neo-regionalists argue, are a part of relational capital and harbinger of novel knowledge and radical innovation. Therefore, synergy management framework can help to maximize the positive implications of learning processes that contribute to domestic/local policy change. In the case of wedding wear cluster, specifically tailored policy implications should be directed at its own ecosystems, both at the organizational and spatial levels. Regarding its organizational structure, that takes account of the unique context, innovation policies are also crucial to the existing business culture.

This chapter concludes that smart urban development aimed at enhancing urban economic dynamism and resilience requires heritage conservation and judicious alterations to urban form, in conjunction with smart policies, and with collaboration implemented by institutions.

References

Argote, L., McEvily, B., & Reagans, R. (2003). Managing knowledge in organizations: an integrative framework and review of emerging themes. *Management Science*, 49(4), 571–582.
Asheim, B., & Gertler, M. (2006). The geography of innovation: regional innovation systems. In J. Fagerberg, D. Mowert, & R. Nelson (Eds) *The Oxford Handbook of Innovation* (pp. 291–317). Oxford: Oxford University Press.
Bathelt, H., Malmberg, A., & Maskell, P. (2004). Clusters and knowledge: local buzz, global pipelines and the process of knowledge creation. *Human Geography*, 28(1), 31–56.
Chan, Y. M. (2012). *Study of Creative Ecology and Cultural Policy for Sustainable Urban Development in Local District of Hong Kong* (Doctoral dissertation). University of Hong Kong, Pokfulam, Hong Kong.
Chen, D. N., Liang, T. P., & Lin, B. (2010). An ecological model for organizational knowledge management. *Journal of Computer Information Systems*, 50(3), 11.
Duxbury, N., & Murray, C. (2010). *Creative Spaces*. Los Angeles: Sage.
Dvir, R., & Pasher, E. (2004). Innovation engines for knowledge cities: an innovation ecology perspective. *Journal of Knowledge Management*, 8(5), 16–27.

Elkington, J. (1997). *Cannibals with Forks: The Triple Bottom Line of 21st Century Business.* Oxford: Capstone.

Feldman, M. P. (2000). Location and innovation: the new economic geography of innovation, spillovers, and agglomeration. In G. Clark, M. Feldman, & M. Gertler (Eds) *Oxford Handbook of Economic Geography* (pp. 373–394). Oxford: Oxford University Press.

Gordon, I., & McCann, P. (2000). Industrial clusters: complexes, agglomeration and/or social networks? *Urban Studies,* 37, 513–532.

Grabher, G. (2002) Cool projects, boring institutions: temporary collaboration in social context. *Regional Studies,* 36: 205–214.

Hearn, G., Roodhouse, S., & Blakey, J. (2007). From value chain to value creating ecology: implications for creative industries development policy. *International Journal of Cultural Policy,* 13(4), 419–436.

Iansiti, M., & Levien, R. (2004a). Strategy as ecology. *Harvard Business Review,* 68–78.

Iansiti, M., & Levien, R. (2004b). *The Keystone Advantage: What the New Dynamics of Business Ecosystems Mean for Strategy, Innovation, and Sustainability.* Boston, MA: Harvard Business School Press.

IZKA (Izmır Development Agency) (2009). *İzmir ve İlçeleri İstatistiki Analiz Raporu.* İzmir Kümelenme Stratejisinin Geliştirilmesi Projesi.

IZKA (Izmır Development Agency) (2010). *İzmir Kümelenme Analizi.* İzmir Kümelenme Stratejisinin Geliştirilmesi Projesi. İzmir ve İlçeleri İstatistiki Analiz Raporu. ISBN: 978-605-5826-04-8.

IZKA (Izmır Development Agency) (2013). *İzmir mevcut durum analizi 2013.* ISBN: 978-605-5826-12-3.

Kannangara, S. N., & Uguccioni, P. (2013). Risk management in crowdsourcing-based business ecosystems. *Technology Innovation Management Review,* 3(2), 32–38.

KEA (2006). *The Economy of Culture in Europe.* Brussels: KEA European Affairs.

Landry, C., & Bianchini, F. (1995). *The Creative City.* London: Comedia.

Low, S. M. & Altman, I. (1992). *Place Attachment: A Conceptual Inquiry.* New York: Plenum Press.

Maskell, P., Eskelinen, H., Hannibalsson, I., Malmberg, A., & Vatne, E. (1998). *Competitiveness, Localised Learning and Regional Development: Specialization and Prosperity in Small Open Economies.* London: Routledge.

Mengi, O., & Velibeyoglu, K. (2013). Wedding wear cluster in Izmir: how does the creative knowledge ecosystem self-operate? In T. Yigitcanlar & M. Bulu (Eds) *Proceedings of the 6th Knowledge Cities World Summit, KCWS 2013* (pp. 21–35). Istanbul, Turkey: Lookus Scientific.

Moore, J. F. (1993). Predators and prey: the new ecology of competition. *Harvard Business Review,* 71(3), 75–83.

Owen-Smith, J., & Powell, W. W. (2002). *Knowledge Networks in the Boston Biotechnology Community.* Paper presented at the Conference on Science as an Institution and the Institutions of Science, Siena.

Pirot, J. Y., Meynell, P. J., & Elder, D. C. (2000). *Ecosystem Management: Lessons from around the World: A Guide for Development and Conservation Practitioners.* International Union for Conservation of Nature (IUCN).

Rudlin, D. and Falk, N. (2014). *Uxcester Garden City.* Wolfson Economics Prize URBED, Manchester. Available at http://urbed.coop/sites/default/files/20140815%20URBED%20Wolfson%20Stage%202_low%20res3.pdf

Shorthose, J. (2004). Nottingham's *de facto* cultural quarter: the Lace Market, independents and a convivial ecology. In D. Bell & M. Jayne (Eds) *City of Quarters: Urban Villages in the Contemporary City* (pp. 149–162). Farnham: Ashgate.

Shrivastava, P. & Kennelly, J. (2013). Sustainability and place-based enterprise. *Organization & Environment*, 26(1), 83–101.

Storper, M., & Venables, A. J. (2002). *Buzz: The Economic Force of the City*. Paper presented at the DRUID Summer Conference on 'Industrial Dynamics of the New and Old Economy – Who is Embracing Whom?' in Copenhagen, Denmark.

Teece, D. J. (2007). Explicating dynamic capabilities: the nature and microfoundations of (sustainable) enterprise performance. *Strategic Management Journal*, 28(13), 1319–1350.

Torre, A., & Rallet, A. (2005). Proximity and localization. *Regional Studies*, 39, 47–59.

Winden, W., Braun, E., Otgaar, A., & Witte, J. (2012). *The Innovation Performance of Regions: Concepts and Cases*. European Institute for Comparative Urban Research (Euricur). Erasmus University Rotterdam. Retrieved 5 June 2014, from http://urbaniq.nl/news/2013/03/new-report-regional-innovation-ecosystems-sweden-china-and-netherlands

Woo, F. (2013). Regenerative urban development as a prerequisite for the future of cities. *The Guardian* online, sustainable business section, accessed on 30 October 2013 at www.theguardian.com/sustainable-business/regenerative-urban-development-future-cities

6

MEGAPROJECT SCREENING AND MANAGEMENT

King's Cross, Olympic Park and Nine Elms

*Reyhaneh Rahimzad, Simon Huston
and Ali Parsa*

Abstract

Urban megaprojects are complex, unique and, almost invariably, contentious. Some megaprojects radically transform but others backfire and not only waste resources but damage the environment. At play with megaprojects, like the controversial Heathrow Airport expansion, are jobs, tax receipts, landowner uplift bets, hidden commissions or legitimate profit and prestige or disgrace. Sceptics condemn megaprojects as a cover or a contrivance by powerful vested interests to exploit rent gaps via structural violence and expulsion. Notwithstanding controversy, macroeconomic turbulence, policy flux, project complexity and practical difficulties vary in diverse architectural, urban design, institutional and geographic settings. Given political contentions, at the very least then, judgement on the merit or failure of a megaproject must involve multiple ecological, social and commercial consequential but also procedural considerations. From the literature and expert dialogue, the research developed a draft multi-criteria project evaluation framework with five key megaproject success drivers:

- Robust planning
- Smart institutions
- Quality project
- Project management
- Sustainable funding.

London, with its global status, provides a rich milieu to analyse megaprojects using the assessment framework. The research investigated three case studies and details the first but summarises the others:

- King's Cross
- Olympic Park
- Nine Elms.

Key terms

Urban megaproject, urban regeneration, stakeholders, consultation, vision, public realm

Introduction

Urban megaprojects are complex, unique and, almost invariably, contentious (Flyvbjerg 2003; Owens and Cowell 2011). Yet the challenge to accommodate and provide sustainable utilities for billions of new urban migrants is serious (Dellesky and Da Silva 2016). Some megaprojects radically transform but others backfire. Promised amelioration can turn into dust (literally in the case of the Aral Sea irrigation schemes). In the urban field, without proper consultation, in South-East London the 1960s facelift to Deptford High Street backfired (Ginsburg 1999). The dereliction of Bangkok's peri-urban canals illustrates the nefarious long-term ecological consequences of rampant and ill-considered real estate development (Davivongs *et al.* 2012). At play with megaprojects, like the controversial Heathrow Airport expansion, are jobs, tax receipts, landowner uplift bets, hidden commissions or legitimate profit and prestige or disgrace. For Smith (1979) and Harvey (1989), megaprojects mobilise a coalition of capitalist handmaidens to exploit rent gaps via structural violence and expulsion. Veblen (1919) and Foucault (1975) would likely consider megaprojects contrivances by powerful vested interests to overcome urban transformation resistance. Notwithstanding conspiracy, bungling involves macro-economic forecasting difficulties, planning and institutional politics with multiple stakeholders, project complexities and funding constraints. Given these political contentions, at the very least then, judgement on the merit or failure of a mega-project must involve multiple ecological, social and commercial considerations. The literature and preliminary dialogue with, arguably compromised, 'experts' generated a draft multi-criteria project evaluation framework. Five key drivers underpin megaproject project success:

- Robust planning
- Smart institutions
- Quality project
- Informed project management
- Sustainable funding.

London, with its global status, provides a rich milieu to investigate the usefulness of the assessment framework. The research investigated three projects:

- King's Cross ('KX')
- Queen Elizabeth Olympic Park ('QEOP')
- Nine Elms ('9E').

Methodology and data

The research used an explanatory deductive methodology. The five key dimensions of the draft framework structured the evaluation of three London megaprojects. Archival material in the public domain and primary qualitative data from both a questionnaire and over 30 stakeholder interviews provided assessment evidence. For each megaproject, the research scrutinised project web pages to first generate a sample frame and then systematically selected key respondents. Interviewees included senior planners, architects, lawyers, council members and developers. NVivo software facilitated the analysis of qualitative information. For this chapter, we selectively present excerpts from these interviews to illuminate discussion. Detailed analysis from the King's Cross project is presented but only summaries of the results from the other two megaproject case studies are given.

CASE STUDY **KING'S CROSS**

Overview

King's Cross ('KX') is a £1 billion plus, iconic mixed-use central London regeneration project, anchored on two historical railway stations and the British Library. In the Victorian era, KX was an important industrial neighbourhood and transport hub. Its central position linked the commercial East of the city with West End retail and entertainment precincts. However, its transport advantages could not offset the post-war demise of the railways. At the end of the 1960s, KX was characterised by industrial blight. Without formal functions, its dilapidated disused buildings, railway sidings, warehouses and contaminated land provided a rich milieu for a shadow, night-time economy of insalubrious strip-clubs, brothels, bohemians and artists. Lower-income social housing dominated its housing stock. Blight and social deprivation kept the area stigmatised with market rents below commercial uplift potential (Highest and Best Use). Recession thwarted 1980s maladroit regeneration attempts but in booming post-global financial crisis London property markets, investor sentiment turned and Argent,[1] via its KX unit trust, realised KX's potential. Project finance involved equity contributions by the original landowners (London Continental Railway) and several pension schemes.

The consortia built shiny new buildings on site but also restored some historical ones. It transformed King's Cross into a high density commercial precinct and London transport hub (see Figure 6.1), linked to six London

FIGURE 6.1 King's Cross
Source: The authors (2016).

Underground lines, two national mainline train stations and an international high-speed rail link to Paris. Google and High Speed One pre-let space and now Google sits as an anchor tenant, facing its bitter rival Apple across the Thames at Nine Elms ('9E').

Robust planning

Aside from its commercial objectives of making an adequate return for its investors, KX aimed to deliver:

- Regeneration and area enhancement
- Public realm connectivity
- Railway and station repair and refurbishment of historic buildings.

To evaluate KX, Argent deployed a full spectrum of conventional assessment techniques, including Environmental Impact Assessment, Transport Modelling, Community Impact Assessments and Economic Evaluations.

> We used all those standard tools. The key one was Excel spreadsheet, which allowed us to monitor activities of project moving forward, combined with the GIS. They are plot-by-plot basis you can say exactly what legal arrangement they have what infrastructure allowance are plot by plot. [D][2]

Smart institutions

In our proposed multi-criteria explanatory framework, smart institutions are the second driver for megaproject success. Smart institutions foster quality growth and curtail its short-term, predatory or extractive modes. Requirements include a futures orientation towards resilience and creativity, sensible spatial architecture, and disposition towards collaboration. To mitigate Agency Problem risks (Eisenhardt 1989), institutional arrangements, responsibilities and oversight must be clear and governance arrangements tight. Internally, the KX partnership has three boards. Externally, audits and review provide additional assurance. Beyond the usual internal controls and external auditing, Islington and Camden councils host quarterly KX group meetings to monitor construction impacts. Notwithstanding the diatribes of Edwards (2009) against the effective alienation of vulnerable locals from development consultation and their eventual physical displacement, on balance, the archival and interview evidence suggests that, from a commercial perspective, KX's corporate institutional arrangements were sound.

Quality project

As mentioned, over the years the KX precinct had deteriorated. The KX plan reversed its decline and, although the scheme has its detractors, notably Edwards (2009), it certainly has some very positive aspects. For a start, despite site intensification, the KX master plan leaves 40 per cent of the 67 acre precinct as quality public realm. *Ceteris paribus* and English weather notwithstanding, this space has health and social spin-offs. The deliberate avoidance of automobile parking slots increases the pedestrianised public realm. Multiple bus stops and many cycling bays nudge people towards adopting healthier transportation choices and facilitate connectivity. However, beyond transport practicalities, megaprojects need an authentic identity. According to Argent's public relations brochures, various staged events infuse the KX project with cultural meaning. Examples include seminars to diffuse the local history. Concerns remain about the authenticity of over-engineering top-down cultural precincts (Sacco *et al.* 2014). More tangible is the preservation of Victorian era, industrial land, steam-train and other heritage around the area. The idea was to have a mixed-use development and to retain cultural integrity.

> The most important thing we ever produced was this idea of vision of what KX is going to become as part of A to Z of city of London when it finished. Basically, we drew this from start point. [D]

> One other thing to consider is that they didn't want shiny shiny new buildings. [M]

> For the building efficiency we have achieved brand new BREEAM awarded (86%) new construction for our 1 per square office. We do have some carbon emission targets to achieve for example all the building has to be 50% better than part L. so having consideration for building envelope the passive design the use of facade natural ventilation and etc. [L]

Based on historical analysis, visual aesthetics and interview comments, KX seems to pass muster as a quality development project.

Project management

Risk management is a key aspect of successful project management. The main KX risks were:

- Delivery risk in terms of actually completing infrastructure and buildings on time, on budget and up to standard.
- Market risks, mitigated by the mix of product which is manipulated to suit market conditions.
- Cash flow to ensure liquidity, mitigated by robust financial analysis and continuous monitoring.

> In 2008–9 when we started the KC we started lots of public sector uses, university and affordable housing and student housing so that way we attracted lot of market rezzy we got at the moment. [M]

> It's much more about identifying the important factors and how to de risk the project and create the best economic circumstances so that the project will succeed. We do that by cracking to social infrastructure. [D]

Sustainable funding

The final criterion in the megaproject evaluation framework is a sustainable funding model. Financial viability is a necessary condition for project success. As the project developed, KX adjusted its funding structure.

> The big challenge about KX and a lot of regeneration projects is because there are a lot of infrastructure and you can't borrow against it as they doesn't producing any income so you have to fund the infrastructure through cash resources or equity. [M]

> Keeping the cash flow was our major risk. The business model evolves and it is updated. If you stick to a particular model the chances will be limited as the more commercial will come forward. It has to be a very dynamic process and constantly updated. [D]

All initial funding comes from equity capital provided by Argent and the major partners, British Telecom Pension Scheme that bought the site from London Continental Railway and DHL. After an initial affordable housing grant, KX received no subsidies. The proponents built the infrastructure first and flexed phasing of other stages to suit circumstances and market conditions. Bank finance funded individual projects until leasing enabled development loans to be converted to lower risk investment ones. Place-making was an important, if instrumental and subsidiary, objective to help mitigate project political and market risks.

> KX now on this tipping point that we have now enough money in our bank sheet so we can now go out to banks and say this is a nice service plot and we need money for direct construction. [M]

> Investment priority for landowners was mainly capital appreciation and income, however Argent has wider interest such as urban real estate and place making also the developers' expertise were important because they wanted to prove they can manage such project and achieve success. [P]

Argent uses its own equity and debt rather than sourcing funds from third parties. It sells off plots piecemeal but retains control of infrastructure. Later, the focus is on leases and operating income (cash flow). Unlike many other urban megaprojects, KX stands on its own commercial feet without governmental grants. New investors come with different criteria and hurdle rates, depending on the development stage and risk. In terms of business planning, it is very creative and dynamic.

> All of the investors have to make an appropriate return, which depends on risk and also the risk during the project as it changes. At fountain/high end of the project you expect high expectation for the return than from the back end of the project where the risk profile would reduce. [M]

Synthesis

The research used the same explanatory mixed methodology to analyse three London megaprojects (King's Cross, Queen Elizabeth Olympic Park and Nine Elms). Over several months in 2105, the researchers queried over 30 senior managers, senior public policy makers, architects, planners, lawyers and social entrepreneurs. However, for its synthesis and conclusion this monograph only abridges the results of the QEOP and 9E case studies. The results are organised by the draft framework's five key megaproject success dimensions: planning, institutions, project, management and funding.

Robust planning

Structure of the project board

For robust planning, the project proponent and collaborator responsibilities must be clear. The structure of the project boards differs among the projects. For KX, the Argent Group board monitored performance. Arup managed the operational side of the project and oversaw engineers and other practitioners. The London Legacy Development Corporation (LLDC) ran QEOP single-handed and owns most of the venues. For 9E a Strategy Board included leaders of councils and landowners.

Project visions and objectives

All three projects had similar regeneration objectives: site transformation (betterment) and value uplift. Initial infrastructure construction, housing densification and public realm enhancement altered market perceptions, catalysed precinct value uplift and stimulated economic activity at a wider scale. In addition, QEOP hosted the Olympic Games. Having a credible and inspiring project vision significantly de-risks projects. Due consideration to unique site heritage is critical. KX sought to blend culture, identity and well-being in one place so that the site offered more than a business office district. Central to realising such a vision was well-designed and active public spaces. For QEOP, the Olympic Games vision involved the radical transformation of an impoverished neighbourhood. The main vision behind 9E was to enhance London's prominence as a world city.

Smart institutions

Stakeholder's coordination at state, regional and local level

KX continuously negotiated with landowners, the local council and the City of London. It was all about making a joint venture and trust between the partners. For QEOP, formal borough arrangements facilitated project coordination. Memorandums kept stakeholders informed. For 9E, regional coordination dominated but there was an informal partnership with the council leader.

Project milestones and feedback

Project oversight depends on the structure of the partnership and the project board. All three projects deployed the standard planning tools and techniques (e.g. EIA, transport modelling and community impact assessments). In the KX development, the Argent Group board queried data on new buildings, tenants, environmental issues, finances and so on. Arup controlled engineers and other project practitioners.

Every single project was BREEAM rated for environment assessment. In the KX project, signing leases is a key performance indicator. Community feedback provided supplementary risk feedback. Other tangible milestones were buildings completion or occupation. For QEOP, the LLDC regularly updated different KPIs to the board and its mayoral chair as well as relevant local authorities. For 9E, initially and as part of monitoring vision impact assessments were widely conducted. To inform performance monitoring, the Nine Elms group sent annual business and action plans to the Strategy Board. For 9E, commercial milestones involved buyer contract signatures or bank confirmation to that effect. Key 9E social milestones involved jobs, work placements or school visits.

Community consultation

Smart institutions remain close to the needs of the people they ostensibly serve and, therefore, regularly consult with local residents. KX respondents considered that communication with Islington and Camden council was open and transparent, while those who worked on the QEOP complained that the public accountable body was slow and unresponsive. Unusually, for 9E the public sector had no land interest. According to some respondents, the downside of this arrangement was a lack of public sector responsiveness to infrastructure and structural issues. On the surface, community consultation appeared well organised in all three projects. Whether these arrangements were merely a 'tick box' exercise or represented a real transfer of power to locals is questionable. For KX, the Argent Group formed a panel of local people and residents. In 2004, Argent published feedback from its community consultation. Locals favoured prioritising young people as project beneficiaries. Local people wanted a clean, safe and accessible site for them and an opportunity for their kids. The second phase of the process was getting locals involved with the activities (e.g. creation of the new access centre). For QEOP, the LLDC broadly communicates with local community about the project through face-to-face meetings, as well as the formal meetings around the park. The park panel is comprised of community members and community organisations. QEOP also constituted a youth panel with the idea of young people getting involved in shaping of the park's future. Local people were involved very actively in development of the park by running events and so on. The LLDC worked in partnership with people who ran venues or who brought universities, businesses and communities to the park. In both the boroughs of Lambeth and Wandsworth, 9E organised community consultations. Each developer also hosted consultations. Delivery teams organised annual open days to disseminate information about the development. Consultations took place at accessible local venues during weekdays on evenings or weekends with views and comments submitted as a part of the planning application. Once the planning application was made, people were informed of any changes.

Quality project

Baseline investigations and job creation

A baseline investigation helps to develop projects in line with *spirit of place* but also provides data to gauge subsequent performance. All three megaproject proponents conducted baseline investigations. In KX this involved environmental impact assessment and analysis of air quality and historical context. All three megaprojects conducted baseline assessments of the local employment because job creation was their main substantive contribution to local community. Internet sources suggested that, in the long term, KX would generate 35,000 workplaces. For QEOP, the touted employment spin-offs were 15,000 permanent jobs of which around 8,000 were local people. For 9E, employment projections inflated baseline models by an extra 20,000 jobs of which 20 per cent were reserved for local people in job schemes or work experience/apprenticeships for students.

Cultural meaning of place

Of course as well as their economic boost, megaprojects should also enhance the urban realm and enrich culture. KX ran many free events to disseminate local history. QEOP attracted large cultural institutions to East London, such as London's Victoria and Albert museum and theatres. 9E offered the local community a cinema screen, a pleasure garden, horticultural, farmer, and flower markets and cultural activities. Whether these cultural trappings will inspire real insurgent artistic talent remains unanswered.

Project management

Stakeholder management

To be successful, mega urban proponents must tightly manage projects but the case studies illustrate that the number of partners involved can vary. Viability is safeguarded if a proponent coordinates stakeholder activity via robust institutional arrangements and regular communication. KX involved a property unit trust in which Argent Group and other investors injected capital but the original landowners retained a controlling interest. For QEOP, the LLDC operated alone. In contrast, 9E had a large number of partners, including the landowners, the boroughs of Lambeth and Wandsworth, the Greater London Authority and Transport for London.

Risk management

For KX delivery, risk remained a concern but excellent construction teams spread this risk while a diversified product helped mitigate KX market risk. For QEOP, a dedicated project manager cut delivery risk. In 9E, the biggest risk was completion delay that would undermine confidence, lose tenants and dent values. Mitigation involved flexible development scenarios, ensuring a tight infrastructure and design fit and, when domestic demand faltered, selling plots to overseas investors.

Sustainable funding

The main initial funding sources for KX came from the BT pension fund and Argent itself (via reserves, asset sales or debt). Later, bank finance funded individual building construction on a project-to-project basis until residential sales or commercial leasing payback. Interviewees involved in the KX project indicate that there was not a clear business model and payback structure for the project, because the business model has to be dynamic, constantly updated and opportunistic. If you stick to one model, your chances will be limited. For QEOP, the main funding sources were public, either Olympic Games budgets or City Hall. The Mayor of London's interest was an enduring post-Olympic business and socio-economic legacy. The National Lottery and consortia with private developers raised operational finance to manage the park. Curiously, no Tax Increment Financing mechanism captured land value uplift. In 9E, Section 106 funding and Community Infrastructure Levy schemes did recoup some of the uplift for the community. For 9E, landowners funded initial works, with off-plan sales part-financing later construction stages but each developer had their own specific long-term business model. In all three studies, investors took a long-term view (20–30 years). Whilst all projects promulgated urban beautification credentials, in reality financial gain remained a significant incentive.

Conclusion

Interviewees rated the importance of the five putative alternate megaproject success drivers on a five-point Likert scale. For *robust planning*, respondents considered a clear project vision the most important factor. Interviewees rated megaproject vision either 'very important' (28) or at least 'important' (four persons). Respondents considered public realm enhancement [*quality project*] the second most significant megaproject success factor with 21 respondents rating it as 'very important'. Surprisingly, despite the fact that all three projects conducted baseline investigations, it was not a highly rated aspect of *project quality*. Three respondents even dismissed the activity but, on average, most considered it at least 'moderately important' and 19 interviewees rated it with mark 3. Also surprisingly, in terms of *project management*, clear articulation of Key Performance Indicators (milestones) scored an average mark close to 3 (15). Possibly interviewees, aware of the uncertainties that usually exist in the early stages of the projects, were reluctant to let ill-considered performance matrices bog projects down. Rather, KX interviewees stressed the need for dynamic and constantly updated business models. Respondents considered all other project management facets, as articulated in the draft framework, 'important'. Respondents rated other specific elements of *robust planning* (strategic goals, design excellence, fostering creative industries, heritage preservation or articulating local identity) similarly. For *institutions,* tight governance and partnership management were important. For *project management* important considerations were stakeholder coordination, efficiency, risk management and authentic community consultation. For *funding,* respondents looked for a clear and viable payback model. In short, this chapter used an explanatory framework with five dimensions (robust planning, smart institutions, quality project, project management and sustainable funding) to assess three key London urban megaprojects at King's Cross, Queen Elizabeth Olympic Park and Nine Elms. The research conducted over 30 expert interviews with senior managers, senior public policy makers, architects, planners, lawyers and social entrepreneurs involved in these projects. The interviews suggest that the most important factor for megaproject success is a clear vision with credible public realm enhancement, rooted in urban design and place understanding.

Note

1   Argent (Property Development) Services LLP is a limited liability partnership of English urban development professionals. In 2015, it formed a partnership ('Argent Related') with a US real estate investment firm. Recently, an Australian superfund bought 25 per cent of Argent.
2   To cite interviewees anonymously, the research attributes quotations with a simple initial rather than a name (e.g. [D] (bottom of p. 110) and [D], [M] (bottom of p. 111)).

References

Davivongs, V., Yokohari, M. and Hara, Y. (2012) 'Neglected canals: deterioration of indigenous irrigation system by urbanization in the west peri-urban area of Bangkok Metropolitan Region', *Water* 4, 12–27.
Dellesky, C. and Da Silva, L. (2016) *Sustainable and Livable Cities Initiative: Pioneering Impact,* Working Paper, October. Washington, DC: World Resources Institute. Available online at: www.wrirosscities.org/research/publication/sustainable-and-livable-cities-initiative-pioneering-impact
Edwards, Michael (2009) 'KX: renaissance for whom?', in Punter, John (ed.) *Urban Design, Urban Renaissance and British Cities.* London: Routledge.
Eisenhardt, K. (1989) 'Agency theory: an assessment and review', *Academy of Management Review* 14(1), 57–74.
Flyvbjerg, B., Bruzelius, N., and Rothengatter, W. (2003) *Megaprojects and Risk: An Anatomy of Ambition.* Cambridge: Cambridge University Press.
Foucault, M. (1975) *Surveiller et punir.* Paris: Gallimard.
Ginsburg, N. (1999) 'Putting the social into urban regeneration policy', *Local Economy* 14(1), 55–71.
Harvey, D. (1989) 'From managerialism to entrepreneurialism: the transformation in urban governance in late capitalism', *Geografiska Annaler. Series B, Human Geography,* 71(1), The Roots of Geographical Change: 1973 to the Present (1989), pp. 3–17.
King's Cross Partnership (2016) Principles for a human city: www.kingscross.co.uk/media/Principles_for_a_Human_City.pdf

Owens, S., and Cowell, R. (2011) *Land and Limits: Interpreting Sustainability in the Planning Process*, 2nd Edition. Abingdon: Routledge.

Sacco P., Ferilli, G. and Blessi, G. T. (2014) 'Understanding culture-led local development: a critique of alternative theoretical explanations', *Urban Studies* 51(13), 2806–2821.

Smith, N. (1979) 'Gentrification and capital: theory, practice and ideology in Society Hill', *Antipode* 11(3), 24–35.

Veblen, T. (1919) *The Vested Interests and the Common Man*. New York: B. W. Huebsch.

7

REVITALISATION AND DESIGN IN THE TROPICS

Tale of two cities – Singapore and Kuala Lumpur

Marek Kozlowski and Yusnani Mohd Yusof

Abstract

The major issues affecting contemporary cities are climate change and rapid urbanisation. To avoid disaster and chaos, cities around the world are in the process of rethinking their strategies and promoting the management of land in a responsible manner. Issues such as the promotion of brownfield infill development and revitalisation of the old urban fabric, provision of pedestrian-friendly environments, mixed uses, higher densities, public transit systems, local identity, public safety and security, and return to traditional streets have been identified as essential prerequisites for future smart urban development.

In the face of growing competitive globalisation, one possible response is to revitalise decayed parts of cities. The revitalisation includes existing historic precincts, post-industrial areas, abandoned waterfronts and large parts of decaying inner cities. Liverpool and London Docklands regenerations provide vivid examples. This trend is common in all the main and even medium size cities around the world including the Southeast Asian region and has also become one of the precepts of Intelligent Urbanism and Smart Growth.

This study reviews the approach to urban revitalisation in two major cities of the Southeast Asian region: Singapore and Kuala Lumpur.

Singapore was a major trading hub and naval base for the British Empire whose lacklustre defence in 1941 weakened British power. After independence from Britain in 1965, Singapore embarked on a rapid development spree. Singapore's top-down central planning initiatives introduced a comprehensive city-wide revitalisation strategy which targeted historic areas and the urban streets. Vast areas of the traditional urban fabric were demolished to pave the way for new modern buildings. On the positive side, Singapore's ambitious transformation plan included intensive tree planting, to foster a 'tropical garden city' ambiance. Opposite the island state, on the Malaysian mainland, planning was less centralised. The Kuala Lumpur Metropolitan Region (KLMR) includes ten different municipalities. Governance fragmentation resulted in a spate of uncoordinated development, and the belated construction of a massive highway system made KL known as the 'Los Angeles of Southeast Asia'. Planners thoughtlessly replaced traditional urban fabric with modern structures, often contrary to the principles of sustainable and tropical climate responsive architectural and urban design. However, in recent years local authorities have introduced design and planning measures, focusing on restoring traditional buildings, revitalising old river corridors and creating tropical public spaces.

Key terms

Urban revitalisation, urban design, smart growth, climate responsive design, regional planning

The issue: the notion of urban revitalisation

The contemporary urban challenges such as rapid urbanisation, changing demographics, the adverse impacts of climate change, traffic congestion and social polarisation pose threats to the physical, social and economic functions of our cities (Newman, Beatley & Boyer 2009; Leary & McCarthy 2013). In response to these challenges, some cities are developing strategies to improve urban resilience. The notion of 'resilience' can be traced to leading cities in Western Europe including Barcelona, Amsterdam and Copenhagen (Gehl 2010; Newman *et al.* 2009; Beatley 2000). It involves infrastructure upgrades, socio-economic measures, conservation, environmental protection and improved energy and waste performance. Regarding urban transformation, one aspect of resilience is the revitalisation of the obsolete and decayed urban areas identified as of low social, environmental or economic value (Tallon 2010).

According to the Australian Oxford Dictionary (1989), revitalise means 'to imbue with new life and vitality'. Revitalisation is a response to obsolescence or diminished utility which reflects the reduction in the useful life of a capital asset. Attempts to revitalise decayed parts of the city must address and remedy obsolescence of buildings as well as the entire economic life of the building stock (Tiesdell, Oc & Heath 1996; Carmona, Heath, Oc & Tiesdell 2010). Mainstream authors argue that the obsolescence results from the mismatch between urban fabric current services and contemporary needs. Progressive authors like Fainstein (2008) and Harvey (2007) argue that capitalists exploit these 'rent gaps'. Whatever the underlying dynamics, revitalisation can reconcile the mismatch engendered by the decay of physical fabric and socio-economic change. Analysing the revitalisation of historic precincts, Tiesdell *et al.* 1996 assert that the physical fabric may be adapted to contemporary requirements through various modes of renewal which include refurbishment, conservation, or by demolition and redevelopment. Regarding economic activity, revitalisation can also arise from changes in occupation, with new uses replacing the former ones. Although a physical revitalisation creates an improved urban environment and physical public realm, a comprehensive economic revitalisation is also required as the activities and uses within buildings are the major financial contributor to the maintenance of the improved physical public realm. The authors also stress the importance of social revitalisation as the vitality of the area is of crucial importance in maintaining a healthy balanced and vibrant urban environment.

According to Tallon (2010), dimensions of urban regeneration include economic, social, cultural, physical, governance and environmental factors. Consequently, a successful urban environment should involve a combination of successful physical, economic and social strategies. As previously implied, revitalisation should be considered in its physical, economic and social dimensions. Sustainable and contemporary urban revitalisation is more often understood as recycling and reuse of existing buildings and abandoned spaces and the conservation of historic precincts.

In some countries such as the United Kingdom, the term urban regeneration is a general term for revitalising blighted urban areas (Peiser 2007). For Peiser (2007), retail and housing revitalisations are the primary components of property-led regeneration which has been commonly encountered in the USA and the UK. In the USA the private sector includes many small local developers who take leading roles in revitalisation projects. In the UK the central government plays a substantial role in local redevelopment financing and policy (Peiser 2007).

An in-depth analysis of 11 major cities in Southeast Asia and East Asia conducted by Hamnett and Forbes (2011) revealed that global aspiration and the rapidly expanding economies drove fast development.

In Southeast Asia, countries such as Singapore, Malaysia and Indonesia grew their economies from very primitive infrastructures (Marshall 2003). Since the 1960s, cities in the region have developed very rapidly. Within one generation, these nations transformed from typically rural communities to predominantly urban communities. However, the pace of growth proved unsustainable. Today, the legacy of uncontrolled rampant growth is social and environmental problems: governments, urban planners and designers seek to manage traffic congestion, air pollution and social deprivation. Much of the previous rapid fast-track development often neglected the local culture, traditions and the tropical climate. Ambitious rebuilding programmes and upgrading of outdated infrastructure often conflict with the retention of the unique sense of place. Demolition of traditional streetscapes removes the community's ability to connect to its past (Marshall 2003; Vines 2005). Many cities in Southeast Asia are now confronting challenges related to the preservation and revitalisation of their traditional urban fabric (Yuen 2013).

Singapore and Kuala Lumpur developed as part of British Malaya, with Singapore becoming a major trade centre and Kuala Lumpur gradually expanding from a small tin mining settlement (Rimmer & Dick 2009). Although located 400 km from each other in a tropical climate zone, both cities transformed with a similar goal of achieving a global city status but with a different approach to urban management, development and planning.

CASE STUDY **SINGAPORE: CITY-WIDE SCALE REVITALISATION**

Singapore is recognised in the global marketplace as one of the most prominent and leading cities of Asia (Rimmer & Dick 2009). Established by the British as a trading post in the nineteenth century, for the next 150 years, it became the major city of British Malaya. Gaining independence in 1965, the new city-state shifted away from hinterland Malaya. Powerful Chinese business interests sought to capitalise on its location and port facilities. Singapore's economic development tripled per capita income during 1980–1995 (Rimmer & Dick 2009; Yuen 2011). By the 1990s, the city-state firmly established itself as a leading global city (Newman & Thornley 2005). Within half a century, the low-rise British colonial trading post had become a high-rise post-industrial garden city of 5.4 million inhabitants, with an urban area of 700 sq. km (Singapore Department of Statistics 2014). The transformation to a garden city attracted the attention of many urban scholars and professionals (Yuen 2011).

Unlike a traditional city like Hanoi much of Singapore's development is entirely new and built on reclaimed land. However, due to a concern to maximise the development potential of land, rapid development and redevelopment and the lack of preservation policies during the 1960s and the 1970s, only a fraction of the traditional urban fabric remains intact. According to Guillot (2007), the Housing Development Board (HDB), the coordinating body for residential development on the island, actively promoted high-rise residential towers associated with a condominium type of development. As a result, many traditional housing areas were destroyed to pave the way for new modern apartments. Even the establishment of the Preservation of Monuments Board in 1971 did not deter redevelopment given over to systematic large-scale demolition and clearance of parts of the city (URA 1987).

Planning and Redevelopment of Singapore, including the revitalisation of historical and blighted precincts, is the responsibility of the Urban Redevelopment Authority (URA). URA was constituted on 1 April 1974 to take over the functions of renewal and redevelopment from HDB and is of especially critical importance to the developmental city-state because efficient utilisation of land is a paramount requirement in its pursuit of economic growth. The responsibility of the URA is to prepare city-wide long-term planning strategies, which include the Master Plan, the Concept Plan, detailed plans in the form of planning and design guidelines and policies, as well as to coordinate and monitor renewal–revitalisation, conservation and improvement projects (URA 2014a; Yuen 2011).

In 1976, the URA initiated studies involving the preservation and rehabilitation of whole areas, signifying the first steps towards retaining their distinct identity and character (Kong & Yeoh 1994). In the early 1990s, authorities in Singapore realised that building glittering mega-malls and cinema complexes create a high-end 'MacCity' without a distinct identity. As a result, the city focused its attention on the establishment of a specific and unique sense of place by introducing urban conservation of the historical and traditional areas. In the late 1990s, the URA repositioned Singapore as a renaissance city for arts and culture, with preservation of historic precincts. The conservation for private owners' scheme was introduced as early as 1991 (Yuen 2011).

The distinctive feature for the revitalisation of the urban fabric in Singapore includes large-scale redevelopments, city-wide improvements and beautifications along major transport corridors, conservation of historic precincts, promotion of high-intensity development around transit stations acting as catalysts for further revitalisation and small-scale street improvements (URA 2008a). One major large-scale redevelopment coordinated by the URA in 1988 covered 100 hectares of old Singapore, including Chinatown, Emerald Hill, Singapore River, Little India, Kampong Glam, as well as the civic and cultural district (Kong & Yeoh 1994: 250–251). Nevertheless, the rejuvenation of such traditional places does not necessarily lead to the achievement of broader revitalisation aims. As in Kampong Glam, Yeoh and Huan (1996) assert that conservation areas often slice up the organic form and texture of the cultural hearth in an arbitrary fashion. However, a clear flow on effect resulted in contemporary redevelopments in the New Downtown situated along the Bay, a new self-contained city located within a city. The New Downtown called Marina Bay is the size of the current central area. It comprises mixed uses including commercial, residential and entertainment, a 3.5 km waterfront promenade and a 100-hectare recreational green area called Gardens by the Bay. Marina Bay has totally recast the image of the city through urban boosterism and labelled Singapore as the 'world tropical city of excellence' (Marshall 2003: 152; Buck Song 2014; Yuen 2011).

The growing emphasis on protecting the remaining traditional urban fabric, conservation of historical and cultural buildings and national heritage sites became a strong component of the revitalisation strategy for the city (Urban Redevelopment Authority 2008b). The Identity Plan created as part of the 2003 Master Plan for Singapore sets out to conserve historical areas and buildings that are of value to the community. Up to the time of writing, URA has given conservation status to 94 areas involving 6,823 buildings throughout the island.

URA's role in conservation encompasses five areas: planning and research, facilitating and coordinating, regulatory, consulting, and promoting. Planning and research activity includes identifying and recommending buildings of historical, architectural and cultural merits for conservation. Promoting and coordinating adopts a three-point strategy to encourage the private sector to participate in the preservation programme. These strategies include conducting pilot projects, releasing conservation building to the private sector and environmental improvement works to conservation areas.

Since the 1990s the URA strongly supports revitalisation of individual sites by introducing a regulatory framework for conservation including documents and manuals to guide individuals and professionals in their conservation works. Promotion seeks the views of professionals and owners of conservation buildings before deciding on policies and guidelines. The idea is to create a better understanding of conservation with regards to the appropriate restoration methods so as to achieve quality outcomes (Urban Redevelopment Authority 2008b).

A specific strong environmental policy orientated at the beautification of Singapore and creating green zones between settlements as well as along transport routes was one of the foundations of the city's urban design. Intensive tree planting programmes along major road corridors and residential streets are conducted jointly by URA and the Highway Department. Then, systematic streetscape revitalisation projects involve widening of sidewalks, floor-scaping and the provision of quality street furniture. Streetscape revitalisation works in Singapore have focused on tourist, historical and cultural districts such as Orchard Road, Chinatown, the Malay Quarter and waterfront areas along the Singapore River. The special detailing of streetscapes includes promenade railings, paving lighting and street furniture. The Bugis area and the Arab Quarter's streets were converted into pedestrian streets, physically lifting the appeal of the areas (URA 2008b). Newman and Thornley (2005) argue that the beautification of Singapore through creating green zones between settlements and transport corridors was linked to the prime objective of attracting investments in the form of new golf courses and housing estates.

The intensive effort of greening the city and implementing tree planting programmes along the main road corridors and the creation of small parks has labelled Singapore as the Garden City (Ker 1997; Newman & Thornley 2005) (see Figure 7.1).

FIGURE 7.1 Revitalisation of historic precincts in Singapore: (a) (b) the Malay
Quarter and Chinatown; (c) (d) revitalisation of city streets
Sources: The authors (2007–2008).

With efficient public transport, green building strategy, cultural-led urban
revitalisation and compact development, Singapore has advanced a practice
of smart growth and sustainable development (Yuen 2011, 2013).

CASE STUDY **KUALA LUMPUR: ON THE PATH TO A GLOBAL
CITY**

As with Singapore, the government in Malaysia also drove development,
aspiring to transform Kuala Lumpur into a modern globally networked city.
In 1857, Raja Abdullah established Kuala Lumpur as a tin-mining settlement
and trading post at the confluence of the Klang and Gombak Rivers. In the
early stages of the city's development, the rivers served as an important
transportation route. In the twentieth century, urban areas started to expand
away from the rivers' confluence, and their importance as major movement
corridors gradually diminished (Yuen 2011; Shamsuddin *et al.* 2013; Isa and
Kaur 2015). The early twentieth century witnessed the decline of the rivers as
the city's major transport corridors. In the 1920s and 1930s, the city began
its expansion away from the rivers with the rail and road systems gradually

taking over the function of the main transport movement corridors. The new road system lacked systematic planning and led to major traffic congestion (Abdul Latip *et al.* 2009).

In 1957, Kuala Lumpur emerged as a capital of the newly independent Federation of Malayan States. At independence, Kuala Lumpur's population was 316,000 (Isa and Kaur 2015). Since the late 1950s, planners demolished large parts of the existing urban fabric featuring traditional Chinese mansions, shop-houses and Malay kampong houses to pave the way for new international modernist development. Unfortunately, such aggressive fast-track development erased much of Kuala Lumpur's history. Former colonial and other residents returning to the modern city after decades of absence are disorientated by the intensive transformation. Nowadays, contemporary Kuala Lumpur Region is marked with a network of highways, modern buildings lacking tropical design features, mega-malls and complexes. In only a handful of streets can a discerning observer detect traces of previous urban kampongs (urban villages). Sadly, the modern metropolis denies traditional Malaysian tropical interaction with landscapes. Even the two rivers, Gombak and Klang, hitherto the main geographical features and transport routes, are now buried by infrastructure and reduced to concreted drains (King 2008). In short, the very identity of the city is lost.

In Malaysia after the mid-1970s, the accelerated urbanisation was due largely to the rapid expansion of the industrial sector (Macleod and McGee 1996). One significant trend in the process of urbanisation in Malaysia in the period 1960–1990 is the increasing dominance of Kuala Lumpur Metropolitan Region (KLMR) vis-à-vis other cities in Malaysia. Based on this convenient definition, the population of the KL conurbation in 1980 was 2.4 million, or 21.4 per cent of the national population, giving a population density of 286 persons per sq. km. In 1990, this same area had about 3.6 million residents with a density of 439 persons per sq. km, and there are 6.5 million inhabitants today (DBKL 2012). The skewed trend of population agglomeration over the period 1960–1990, especially the specific bias towards the Kuala Lumpur Core Urban Region, has resulted in a marked inability in the recipient areas to cope with traffic congestion, housing and environmental problems. In other words, the quality of the urban environment is deteriorating at a higher speed than either local population growth or local physical expansion. This phenomenon, unless controlled, is bound to affect the quality of life in the region. The situation is compounded by the lack of precise urban development policies to contain population movements. Urban development policies in the 1970s were linked to the exigencies of dealing effectively with, first, the disparities between the rural and urban sectors through better rural–urban linkages and making urban functions more accessible to the rural populations; and, second, disparities between regions and states by stimulating growth in lagging regions. It was only in the mid-1980s that an attempt was made to develop a National

Urbanisation Policy (NUP) to guide urban development (Thong 1996; King 2008).

The KL city administrative area has a population of 1.7 million and the population of the KLMR, covering around 2,700 sq. km, is 6.5 million. The KLMR includes ten local authorities with major centres such as Shah Alam (capital of the state of Selangor), Putrajaya (new federal administrative capital), Petaling Jaya, Ampang, Subang Jaya, Kajang, Selayang, Sepang and Klang. Also, the region is managed by the two state governments, Selangor and Negri Sembilan, and Kuala Lumpur and Putrajaya are declared federal territories controlled by respective local authorities Dewan Bandaraya Kuala Lumpur and Perbadanan Putrajaya (Kuala Lumpur and Putrajaya City Councils) (DBKL 2012; International Urban Development Association 2015).

The KLMR is the fastest growing region in Malaysia and the last decade has witnessed a spate of new residential, institutional and commercial development. Much of this development has been mainly market driven and guided by economic and political reasons. Such rapid property-led development often neglects the local conditions, the natural settings and the local tropical climate and, as a result, has a detrimental impact on the surrounding public space. The current statutory local plans focus mainly on development control addressing issues such as height, bulk and orientation of buildings rather than promoting identity and a tropical sense of place and identity.

Contemporary KLMR provides a planned road-based and low-density urban conurbation, and because of this, with its high car dependency, it is portrayed as the 'Los Angeles of Southeast Asia' (Rimmer and Dick 2009). The urban region is marked with a network of highways, modern buildings lacking tropical design features, mega-malls and complexes. In between the vast concrete jungle and the web of highways and concrete corridors are isolated oases such as Kuala Lumpur Central City (KLCC) or Putrajaya, containing planned and well-designed tropical environments and also some high quality leafy residential precincts including Bangsar and Damansara Heights. Following Idenburg's (2015) classification of two major American cities, Los Angeles and San Francisco, where he refers to the city environment of Los Angeles as a 'dystopia that had gone right' while at the same time San Francisco is labelled as a 'utopia that had gone wrong', the Kuala Lumpur Metropolitan Region can be described as a dystopia with enclaves of utopia. According to King (2008) the KLMR is a juxtaposition of public spaces representing Malay space, Chinese space, Indian space, the space of the Internet, cyberspace and hyperspace, areas of traditions, memory and origins, spaces of the formal and informal economy. Added to this collection should be the comprehensive network of highways cutting across the entire region.

Since the 1970s, redevelopment of the existing urban fabric associated with the destruction of the old urban fabric has been the practice applied and accepted by all levels of government. An example of a major urban transformation was the development of the KLCC complex including the Petronas Towers, large shopping complex and convention centre and a 20-hectare urban park all replacing a former turf horse racing track (King 2008) (see Figure 7.2).

FIGURE 7.2 (a) (b) Kuala Lumpur Central City – major transformation from a horse racing turf club to an urban park and popular public space

Source: The authors (2014).

The decision to redevelop the last remaining peri-urban village within central Kuala Lumpur, called Kampong Bharu, to a high rise commercial and residential precinct was made by the federal government in 2009 (Kampong Bharu Development Corporation 2016) (see Figure 7.3).

FIGURE 7.3 (a) (b) Traditional houses in Kampong Bharu – surrounded by new high rise towers

Source: The authors (2013).

However, in the last decade, there was growing concern regarding the demolition of the traditional urban fabric. In the Ninth Malaysian Plan, the National Heritage Act 2005 (from now on referred to as the NHA 2005) was enacted to give protection and preserve various tangible and intangible cultural heritage and has been promoted for the tourism industry. The Act provides for the conservation and preservation of national heritage, natural heritage, tangible and intangible, cultural heritage, and underwater cultural heritage (Mustafa and Abdullah 2013; Ghafar 2010). Nevertheless, the Act is very general and aimed only at the preservation of specific sites and landmarks and not at conservation and rehabilitation of entire precincts.

There is also an ongoing international discourse on adverse side-effects of fast-track urban redevelopment and its impact on the tropical identity of the metropolitan region (Kozlowski, Ujang & Maulan 2016). The urban design research team based at the Faculty of Architecture and Design, University Putra Malaysia, has conducted research on the selected streets, public spaces and new building complexes of central Kuala Lumpur regarding the climate responsive tropical design and planning. As part of the analysis each street, public space and building complex was assessed against the list of Performance Design Criteria (PDC) for tropical urban environments. The analysis of the selected streets, public spaces and new building complexes in the KLMR identified that a few streets in the remaining older parts of the city retained some form of the sense of place and identity. It was also revealed that older buildings such as traditional Chinese shop-houses performed better against the evaluative criteria than many modern buildings (Kozlowski *et al.* 2015).

In recent years the federal and state governments and local authorities have stepped up initiatives to slow down the diminishing traditional urban fabric. Kampong Bharu Development Corporation was established in 2012 to guide and coordinate the redevelopment of Kampong Bharu but also to protect and retain the existing character of its central part (Kampong Baharu Development Corporation 2016). The other major revitalisation project launched by the federal and state governments and DBKL is the River of Life (ROL) Project including the revitalisation of the Gombak and Klang river corridor in central Kuala Lumpur. DBKL commissioned AECOM, a very large US-based design, planning and project management firm, to produce a master plan for a 10.7 km stretch of the Klang and Gombak river corridors in the central part of Kuala Lumpur. The master plan prepared by AECOM was endorsed by DBKL and the first construction works commenced in 2015 (Kozlowski 2015). In 2015 DBKL endorsed the Urban Design Guidelines for Central Kuala Lumpur which includes an entire section on heritage guidelines. These heritage guidelines for Central Kuala Lumpur identify primary, secondary and specific character zones. The detailed guidelines focus on retrofitting building façades, readapting internal layouts, and on urban infill developments (DBKL 2014). However, the Urban Design Guidelines for Central Kuala Lumpur is a strategic document and therefore its recommendations as yet are not legally binding.

Major findings and conclusions

The analysis revealed that after gaining national independence, both cities embarked on a trajectory of massive physical transformation. The landscape in Singapore was totally changed with large sections of the traditional urban fabric replaced by new housing, institutional and commercial complexes. The relatively small-scale city of Kuala Lumpur expanded into a mega car-dependent urban conurbation, including edge cities and high-tech specialist districts, all linked by a complex system of highways. Both cities witnessed a massive destruction of their traditional built environment. Developers simply erased old historic precincts without reflection on the loss of identity and collective memory. Singapore was first to recognise its mistake. In the late 1970s, Singapore diverted from the one-way fast-track development approach by introducing heritage and conservation protection measures. The revitalisation and greening of all urban streets and converting them into tropical climate responsive and pedestrian-friendly public spaces is nowadays considered enlightened urban development, planning and design.

Only recently has Kuala Lumpur woken up to the damage caused by ill-considered development. Recently, Kuala Lumpur introduced new urban design policies aimed at conservation of old historic buildings and precincts and commenced with the regeneration of the river corridor in the central part of the city. Rhetoric aside, sensitive urban conservation and revitalisation practices are still in their infancy. For one, enlightened revitalisation principles and guidelines are not yet incorporated in any local statutory planning documents. Second, it is only very recently that discourse on urban conservation and heritage has widened from central Kuala Lumpur to the KLMR fringes.

Urban management praxis illustrates sharp differences between the two cities. Singapore has adopted a typical top-down approach with one central agency (the Urban Redevelopment Agency) controlling development, coordinating, managing and implementing all urban development plans and strategies. There is one Master Plan and Concept Plan containing strategies, guidelines and development requirements for the entire city. In KLMR there are ten different local authorities, all with slightly different priorities and agendas. The role of the federal and state governments in urban planning and developments is limited to flagship projects such as KLCC and ROL. Each local authority has prepared a local plan, and there is no regional plan for the entire conurbation or even a regional advisory board that could help in integrating ideas and strategies. Such administrative fragmentation makes it very difficult to initiate a comprehensive city-wide revitalisation strategy aimed at buildings and public spaces.

Policy recommendations and stakeholder implications

Singapore has halted its destruction of the traditional urban fabric and conducted a complete revitalisation of its city streets. Such a move was beneficial for all the key stakeholders including the local community, visitors, local entrepreneurs and decision makers. Singapore has established a smart growth approach set in a well-functioning urban system with urban revitalisation as a key element. The city-wide revitalisation of Singapore's historic precincts and streets profoundly altered the urban environment; however, such an approach can be applied only under specific political conditions supported by a strong and centralised urban management system.

Kuala Lumpur must revitalise its remaining older urban fabric and protect it from further demolition. Also, Malay, Chinese and Indian motifs and traditions should be required in the design of new urban spaces and buildings. Given Kuala Lumpur's urban geographical framework illustrating a growing polycentric structure, a regional approach is critical. Imposing urban revitalisation measures and ensuring the sense of place and identity at the regional level can also ensure a better delivery at the local level by informing the statutory and strategic and statutory local development plans. A regional plan would also ensure that developers apply revitalisation principles and objectives across the urban region beyond central Kuala Lumpur. Apart from promoting regional and city-wide urban revitalisation objectives, smart planning emphasises neighbourhood, streets and individual site scale betterment projects. Micro street-neighbourhood interventions, although piecemeal and selective, can be applied in different urban cultures and most political environments. Introducing a regional approach with urban revitalisation at the regional, city-wide, district and neighbourhood site levels would significantly contribute to a holistic environment and social equilibrium necessary to achieve a smart city status.

Bibliography

Abdul Latip, N. S., Heath, T., & Liew, M. S. (2009) A morphological analysis of the waterfront in city centre, Kuala Lumpur. INTA-SEGA Bridging Innovation, Technology and Tradition Conference Proceedings.

Australian Oxford Dictionary (1989) (Oxford: Clarendon Press).

Beatley, T. (2000) *Learning from European Cities* (Washington, DC: Island Press).

Buck Song, K. (2014) *Perpetual Spring: Singapore's Gardens by the Bay* (Singapore: Marshall Cavendish).

Carmona, M., Heath, T., Oc, T., & Tiesdell, S. (2010) *Public Spaces–Urban Spaces: The Dimensions of Urban Design* (Oxford: Architectural Press).

Dewan Bandaraya Kuala Lumpur (DBKL) (Kuala Lumpur City Hall) (2012) Kuala Lumpur City Plan 2020.

Dewan Bandaraya Kuala Lumpur (2014) Urban Design Guidelines for Kuala Lumpur City Centre.

Fainstein, S (2008) Mega projects in New York, London and Amsterdam. *International Journal of Urban and Regional Research*, 32(4), 768–785.

Gehl, J. (2010) *Cities for People* (Washington, DC: Island Press).

Ghafar, A (2010) Heritage Interpretations of the Built Environment: Experiences from Malaysia retrieved from www.heritage.gov.hk/conference2011/en/pdf/7_Prof%20 Ghafar%20Ahmad.pdf

Gibson, M. S. & Langstaff, M. J. (1982) *An Introduction to Urban Renewal* (London: Hutchinson).

Guillot, X. (2007) Between 'Asianization' and 'New Cosmopolitanism' housing in the twenty first century Singapore, in C. Bull, D. Boontharm, C. Parin, D. Radovic, & G. Tapie (eds) *Cross Cultural Urban Design* (pp. 34–39). (London: Routledge).

Hamnett, S. & Forbes, D. (2011) Risks, resilience and planning in Asian cities, in S. Hamnett & D. Forbes (eds) *Planning Asian Cities: Risks and Resilience* (pp. 1–40). (London: Routledge).

Harvey, D. (2007) *A Brief History of Neo-Liberalism* (Oxford: Oxford University Press).

Idenburg, F. (2015) One Santa Fe housing in Los Angeles by Michael Maltzan. *Architecture Review*, August, retrieved from www.architectural-review.com/today/this-is-the-dirty-magical-realism-future-of-los-angeles/8686180.fullarticle

International Urban Development Association (2015) Kuala Lumpur Metropolitan Malaysia, retrieved from www.Inta-aivn.org

Isa, M. & Kaur, M. (2015) *Kuala Lumpur: Street Names, Guide to their Meaning and Histories* (Singapore: Marshall Cavendish).

Kampong Bharu Development Corporation (KBDC) (2016) Draft Development Master Plan of Kampong Bharu. www.pkb.gov.my/en/kampong-bharu-citizen/download

Ker, L. T. (1997) Towards a tropical city of excellence, in O. G. Ling & K. Kwok (eds) *City and the State: Singapore's Built Environment Revisited* (pp. 44–63). (Singapore: Oxford University Press).

King, R. (2008) *Kuala Lumpur and Putrajaya: Negotiating Urban Space in Malaysia* (Singapore: NUS Press).

Kong, L. & Yeoh, Brenda S. A. (1994) Urban conservation in Singapore: a survey of state policies and popular attitudes. *Urban Studies,* 31(2), 247–265.

Kotin, A. & Szalay, K. (2007) Old Pasadena California: a downtown redevelopment, in R. B. Peiser & A. Schmitz (eds) *Regenerating Older Suburbs* (pp. 70–83). (Washington, DC: Urban Land Institute).

Kozlowski, M. (2015) Kuala Lumpur: transformation towards a World City built environment. *Architecture Malaysia,* 27(2), 70–75.

Kozlowski, M., Ujang, N., & Maulan, S. (2015) Performance of public spaces in Kuala Lumpur Metropolitan Region in terms of tropical climate. *Alam Cipta* (Special Issue, December), 41–51.

Land Transport Authority (2008) Public Transport in Singapore, retrieved from www.ita.gov.sg

Leary, M. & McCarthy, J. (2013) Urban regeneration a global phenomenon, in M. Leary & and J. McCarthy (eds) *The Routledge Companion to Urban Regeneration* (pp. 1–15). (London: Routledge).

Macleod, S. & McGee, T. G (1996) The Singapore–Johore–Riau Growth Triangle, in Fu-chen Lo & Yue-man Yeung (eds) *Emerging World Cities in Pacific Asia* (pp. 417–465). (Tokyo: United Nations University Press).

Marshall, R. (2003) *Emerging Urbanity: Global Urban Projects in the Asia Pacific Rim* (London: Spon Press).

Mustafa, N. B. & Abdullah, N. C. (2013) Preservation of cultural heritage in Malaysia: an insight of National Heritage Act 2005. Proceedings of International Conference on Tourism Development, Penang, Malaysia, February.

Newman, P. & Thornley, A. (2005) *Planning World Cities: Globalization and Urban Politics* (New York: Palgrave Macmillan).

Newman, P., Beatley, T., & Boyer, H. (2009) *Resilient Cities: Responding to Peak Oil and Climate Change* (Washington, DC: Island Press).

Pacione, M. (2005) *Urban Geography: A Global Perspective* (London: Routledge).

Peiser, R. B. (2007) Introduction, in R. B. Peiser & A. Schmitz (eds) *Regenerating Older Suburbs* (pp. 2–35). (Washington, DC, Urban Land Institute).

Rimmer, P. J. & Dick, H. (2009) *The City in Southeast Asia: Patterns, Processes and Policy* (Singapore: NIUS Press).

Shamsuddin, S. (2011) *Townscape Revisited: Unravelling the Character of the Historic Townscape in Malaysia* (Kuala Lumpur: UTM Press).

Shamsuddin, S., Abdul Latip, N., & Ulaiman, A. B. (2013) *Regeneration of the Historic Waterfront: An Urban Design Compendium for Malaysian Waterfront Cities* (Kuala Lumpur: ITBM).

Singapore Department of Statistics (2014) Singapore Statistics, www.singstat.gov.sg

Tallon, A. (2010) *Urban Regeneration in the UK* (London: Routledge).

Thong, L. B. (1996) Emerging urban trends and globalizing economy in Malaysia, in Fu-Chen Lo & Yue-man Yeung (eds) *Emerging World Cities in Pacific Asia* (pp 335–377). (Tokyo: United Nations University Press).

Tiesdell, S., Oc, T., & Heath, T. (1996) *Revitalizing Historic Urban Quarters* (Oxford: Architectural Press).

Urban Redevelopment Authority (URA) (1987) Annual Report 1986/1987 (Singapore: URA).

Urban Redevelopment Authority (2008a) Architecture and Urban Design Excellence, retrieved from www.ura.gov.sg

Urban Redevelopment Authority (2008b) Conservation, retrieved from www.ura.gov.sg

Urban Redevelopment Authority (2014a) Home Page, retrieved from www.ura.gov.sg

Urban Redevelopment Authority (2014b) Master Plan 2014, retrieved from www.ura.gov.sg/uol/master-plan.aspx?p1=View-Master-Plan&p2=Master-Plan-2014

Vines, E. (2005) *Streetwise Asia* (Bangkok: UNESCO).

Watson, D. & Floyd, C. (2003) Community design charrettes, in D. Watson, A. Platus, & R. Shibley (eds) *Time Saver Standards for Urban Design* (pp. 4.10–4.11). (New York: McGraw-Hill).

Wilson, James Q. (1963) Planning and politics: citizens participation in urban renewal. *Journal of the American Institution of Planners*, 29(4), 242–263.

Yeoh, Brenda S. A. & Huan, S. (1996) The conservation–redevelopment dilemma in Singapore: the case of the Kampong Glam Historic District. *Cities*, 13(6), 411–422.

Yuen, B. (2011) Singapore planning for more with less, in S. Hamnett & D. Forbes (eds) *Planning Asian Cities: Risks and Resilience* (pp 201–220). (London: Routledge).

Yuen, B. (2013) Urban regeneration in Asia: mega projects and heritage conservation, in M. Leary & J. McCarthy (eds) *The Routledge Companion to Urban Regeneration* (pp. 127–138). (London: Routledge).

8

CONSTRUCTION FIRMS' COMPETITIVENESS

Puying Li, Simon Huston and Ali Parsa

Abstract

Chapter 8 examines the role of construction firms as horizontal contracting institutions for smart development. With around $108 billion invested in smart city infrastructure worldwide, the sums at stake are staggering. Smart development calls for competitive construction firms. The research blends competitiveness theory with practical analysis of China's urbanization and smart city market and construction firms. From the interplay between theory and practice, the research generated a competitiveness model and isolated key competitiveness indicators (KCIs). Detailed case study analysis helped to confirm KCI relevance. The research explored the competitiveness strategies adopted by a single Chinese intelligent building construction firm and noted the importance of robust corporate strategy, organizational capabilities and a sound financial system.

Key terms

Chinese urbanization, Chinese smart urban development, Construction firms' competitiveness

Learning outcomes

- The smart city is a sustainable method for the solution of urbanization issues.
- The rapid development of the smart city brings huge potential business and improves competition in the market, thus, construction firms must improve their competitiveness in order to survive and grow in this competitive market.
- Construction firms' competitiveness may be divided into three sections: corporate strategy, organizational capability and financial capability.

The construction industry in smart cities

The impact of smart city development on the construction industry has three aspects. First, the preliminary development of a smart city requires integration between industry sectors. In the context of smart cities, the lines between different industry sectors are indistinct and traditional models are shifting (Connock 2015). As IBM's Andy Stanford-Clark said, 'smart cities are an ecosystem of parts played by many different actors. No one company can do the whole thing' (Allidina 2015). Schriener and Doherty (2013) noted that smart cities mandate the convergence of government policy, information communications technology (ICT) and innovative urban design. Therefore, construction firms are required to cooperate with governments, architects and developers in order to incorporate smart initiatives into projects.

The second aspect is the intensification of the fierce competition within the construction industry resulting from the rapid development of the smart city. Competition is already strong in the construction industry because of the large number of construction contractors. In the context of the competitive smart city market, construction companies are required to further improve competitiveness through honing the capability to undertake smart projects. Some could fail if they do not have enough capability to compete. Inadequate capability in smart projects, such as technology innovation, ICT and sustainable construction methods, may cause some construction companies to falter or fail in the increasingly competitive smart city market.

The third aspect is the need for contractors to discover a pathway for the development of business within the context of the smart city market as innovative smart city projects challenge traditional construction firms' means, methods and business models. One example is the French construction firm Vinci which purchased ICT from Imtech for €255 million in 2014. Vinci recognized that constructing smart buildings in a smart city required the application of information technology at every stage of the development (Connock 2015).

As a result, construction firms are required to reconsider their competitiveness in the context of the smart city in order to meet the demands of smart urban development, and thus survive and grow in this competitive market.

Competitiveness theory

The achievement of competitiveness at firm level has received considerable attention since the 1960s (Flanagan et al. 2007). Three main schools are dominant: the competitiveness advantage and competitiveness strategy models (Porter 1980, 1985); the resource-based view and core competence approach (Prahalad and Hamel 1990; Barney 1991); and the strategic management approach (Chandler 1962).

Porter's theory of firm competitiveness is characterized as the industrial organization view of competitive advantage, which proposed that competitive advantage comes from the competitive strategy adopted by a firm in order to neutralize threats or to exploit opportunities obtainable by an industry. Porter (1980) investigated such major factors affecting competitiveness for business internationally versus nationally as cost differences, market differences, regulation differences and resource differences. To provide customers with greater value and satisfaction than their competitors, firms must be operationally efficient, cost-effective and quality conscious. Superior value results through lower prices for equivalent benefits or differentiated benefits that justify a higher price (Porter 1985). But this definition has limitations because it overlooks the fact that business management processes, including human resources, strategic management and operation management, could all affect firms' competitiveness.

In the school of the resource-based view and core competence approach, Prahalad and Hamel (1990) proposed that firms should develop unique resources and so achieve core competence to sustain growth. The main propositions of the resource-based view are:

- A firm can be viewed as a collection of resources.
- Competitive advantage does not depend on market and industry structures, but stems from the resources inside a firm.
- Not all resources are necessarily the source of a firm's competitiveness, it is only the firm's specific resources that meet the criteria of valuable, rare, non-substitutable, imperfect imitability.
- A firm must identify and strengthen those firm's specific resources in developing its core competence.

The resource-based view and core competence approach established the relationship between resources, capability and core competencies. However, it ignored the firm's abilities regarding analysis of the marketing environment and innovation in resources creation.

In the strategic management approach, D'Cruz and Rugman (1992) suggested that competitiveness can be defined as the ability of a firm to design, produce or market products superior to those offered by competitors, considering the price and non-price qualities. Johnson (1992) and Hammer and Champy (1993) considered that the competitive process enhances the ability of an organization to compete more effectively. The strategic management approach pays more attention to the ability of business management processes to achieve greater value or profit but ignores the relationship between a company and its customers. Thus, Feurer and Chaharbaghi (1994) described competitiveness as a value relationship between an organization, its customers and shareholders. They suggested that an organization makes a profit in order to satisfy its shareholders and achieve continuous profit growth, which improves their market position as well as maximizing their potential for making a greater profit to attract the necessary funds provided by their shareholders. It is competitive in the eyes of its customers if it is able to deliver a better value than its competitors.

According to an analysis of key competitiveness theories, some factors affecting firms' competitiveness can be identified, including quality, costs, profits, ability, value, operation and innovation. However, the current definition of competitiveness is not flexible enough to utilize at present. Many scholars provide valuable insight into competitiveness, but they consider a general industry while the construction sector is characterized as heterogeneous (Flanagan *et al.* 2007).

Construction firms' competitiveness

The identification of construction companies' competitiveness has been extensively covered in previous studies. The study by Holt *et al.* (1994) classifies competitiveness under five groups: contractors' organization, financial considerations, management resource, past experience and past performance. Each of these groups also includes various specific indicators. Hatush and Skitmore (1997) classified criteria for assessing contractor competitiveness into five categories including financial soundness, technical ability, management capability, health and safety, and reputation. Rad and Khosrowshahi (1998) considered that quality has been recognized as one of the key factors in the construction industry. However, quality in the construction industry has long been a problem. Great expenditures of time, capital and resources, both human and material, are wasted each year because of inefficient or non-existent quality management procedures (Arditi and Gunaydin 1997). Shen *et al.* (2003) presents a more comprehensive set of contractor competitiveness indicators in the development of a model for calculating a contractor's total competitiveness value (TCV), which divides contractor competitiveness indicators into six categories: social influence, technical ability, financing ability and accounting status, marketing ability, management skills, and organizational structure and operations. Marzouk *et al.* (2013) considered other criteria that should be taken into consideration; the main objectives of the contractors' selection process are to reduce project risk, maximize the quality and maintain a strong relationship between project parties. Fong and Choi (2000) reflected that the significance of three criteria, namely time, cost and quality, should be considered. In a study by Darvish *et al.* (2008), the multi-criteria decision-making method is used, taking into consideration the following criteria which facilitate the successful acquisition of projects by construction companies: technology and equipment, management, experience and knowledge of the technical staff, financial stability, quality, and being familiar with the area or being domestic.

Nevertheless, applications of the competitiveness indicators introduced in previous studies are limited. There has been no investigation of the relevance of competitiveness indicators to different types of project. In fact, project clients have different priorities based on various project objectives, and contractors must have different capabilities for different types of project. Therefore, the project type should also be considered when contractor competitiveness is examined (Shen *et al.* 2006).

Thus, the competitiveness of construction firms cannot be judged in precise theoretical terms due to the variables experienced by different construction firms such as life cycle, the scale of different projects and varying markets. Hence, if construction firms wish to build up consumers' trust, occupy greater market share and explore wider business opportunities in the competitive market, they must consider both existing key competitiveness indicators (KCIs) and those that may still be achieved.

China's urbanization

Since the establishment of the People's Republic of China (PRC) in 1949, urbanization has developed rapidly. The mid-1990s onwards has been a period of rapid expansion in China's urbanization. The PRC's urban population increased from 562.12 million in 2005 to 749.16 million in 2014 (Figure 8.1); from 2008 to 2014, urban areas increased from 178,110.28 square kilometres to 184,098.59 square kilometres; urban construction areas increased from 39,140.46 square kilometres to 49,982.74 square kilometres; urban population density rose from 2,080 people per square kilometre to 2,419 people per square kilometre (National Bureau of Statistics of China 2014).

With the rapid development of China's manufacturing industry, more and more rural people have chosen to live in the city and conurbations. Over the next 20 years, China will see 500 million rural inhabitants move to urban areas in order to get a job and live (Institute of Urban and Environmental Studies of CASS 2012). Thus, China needs to develop vast areas of land and build high residential blocks to meet these demands.

However, Jinglian Wu (2012), a Chinese economist, considered that China wastes an abundance of natural resources and labour in the development of a city,

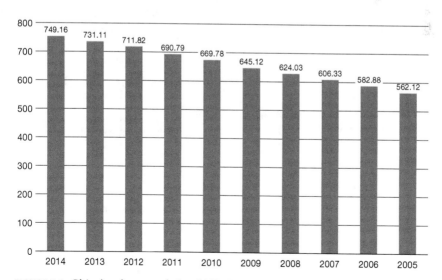

FIGURE 8.1 China's urban population 2005–2014

Source: Adapted from the National Bureau of Statistics of China (2014).

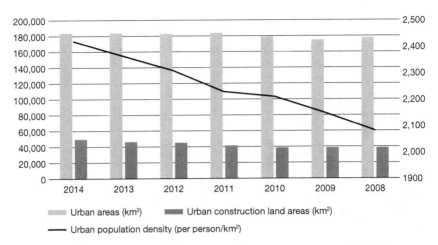

FIGURE 8.2 Statistics for China's urban areas, urban construction areas and urban population density, 2008–2014

Source: Adapted from the National Bureau of Statistics of China (2014).

thus also affecting the sustainable development of that city's environment. First, urban over-expansion causes unsustainable land development. China is a land-scarce country and needs to pay attention to land resource conservation as a part of the process of urbanization. From 2008 to 2014, China's urban construction areas increased 27.70 per cent, but at the same time, the urban population increased 16.30 per cent (National Bureau of Statistics of China 2014); thus the phenomenon of the 'empty city' occurred; modern buildings newly constructed, with wide roads and large squares, yet no people living in them. Moreover, rural immigrants experience difficulty with housing in the city due to its high cost in these newly constructed urban areas. This illustrates the problems of China's serious waste of its land resource – having exploited excessive amounts of land to facilitate urbanization yet without abiding by the demands of the local economy, or considering sustainable development and people's standards of living.

Second, recurrent problems in the nation's major cities, such as overcrowding, traffic congestion, environmental pollution including air, water, noise and waste, and housing problems are not easily solved. For example, more than ¥10 billion is lost to the economy annually in Chinese cities with populations exceeding one million where over 80 per cent of roads are snarled in traffic jams (Lin 2013).

Third, the process of urbanization has involved the demolition of many houses, especially in 2003 when the total residential demolition area was 161 million square metres (Lin 2013). Numerous historical buildings have been destroyed in this unstainable process which has resulted in the loss of traditional built history and local culture in many of China's cities.

Therefore, issues associated with urbanization which China is now facing, such as rapid economic development, overpopulation and environmental damage and so on, make the development of the smart city an inevitable choice in order to solve environmental pollutions, and improve the national economy and the living standards of rural populations.

The development of the smart city has come later to China than developed countries. The PRC's earliest intelligent building was the Beijing Fortune Building, erected in 1990. The smart city started to develop more rapidly with China's publishing of the standards of intelligent building designs, construction and management in 2007 and the 12th Five-Year Plan which pushed the development of the intelligent building. Li (2014), director of the IT and Software Service Division of the PRC's Ministry of Industry and Information Technology, pointed out that the market scale of the smart city in China has a potential $4 trillion of business, which will cover intelligent infrastructure, intelligent traffic systems and public security, among other things. With the support of the 13th Five-Year Plan, the development of the smart city in China is increasing and entering a stage of high-speed development. Chen *et al.* (2015) identified that scales of new intelligent construction in the Chinese market achieved ¥170 billion. Moreover, the Chinese government is encouraging the transformation of existing traditional buildings into intelligent buildings, bringing ¥20–¥30 billion of intelligent building business every year for the construction industry. Thus, the total market scale of intelligent building in China now approaches an annual figure of ¥200 billion. China's current inventory of buildings shows that over 95 per cent are non-intelligent and high consumption buildings and more than ¥1 trillion in expenditure is required to transform these traditional buildings into intelligent buildings. Thus, many potential business opportunities for construction companies have emerged in China's smart city market, due to the numerous existing buildings requiring transformation into intelligent buildings.

China's construction industry

Can the current Chinese construction industry meet the requirements necessary for the development of the smart city? At present, China's traditional construction industry cannot meet the needs of the smart city.

First, the core challenge of China's smart city is that traditional information technology could not meet the smart city's development requirements. For example, the information connection is segregated between region, industry and departments, the opaque information disclosure and the low level of intelligent information services could not implement the overall configuration of urban resources and enhance the operational efficiency of urban systems.

Second, construction activities generate four main types of pollution. Unless pollution is tackled, China cannot realize the smart city.

1   **Water pollution** Inevitably, a great deal of mud and sewage is produced during the urban construction process. Construction sewage, which contains a large volume of sediment, debris and building materials, flows directly into the city

sewers. Construction sewage not only contaminates the urban water environment, but also easily causes siltation which may then cause waterlogging of the city drainage system.

2   **Air pollution** Construction dust has a severe effect on the atmospheric environment. There are two primary aspects to the problem of construction dust: one is the transport of building materials which causes road dust; the other is the handling and stowage of construction materials on site. This dust not only seriously affects the surrounding residential living environment, but also the atmospheric environment. Thus, it is very important to control dust emissions throughout the various stages of the construction process in order to protect the urban ecology and environment.

3   **Noise pollution** Construction noise is mainly produced by related transportation and machinery. It has a serious effect on the quality of the daily lives of the residents surrounding the construction site, and may even give a detrimental impression to visitors to that urban environment. Therefore, reducing construction noise is one of several important tasks in the establishment of a smart city.

4   **Construction waste material pollution** In the process of construction, several solid wastes are inevitably produced which contain numerous broken construction materials. This waste could seriously pollute and erode the urban land environment if not suitably recycled. Such pollution problems could be solved if construction companies paid more attention to environmental protection.

Moreover, there are two considerations affecting construction companies' willingness to improve environmental protection during the construction process. The first is economic: the primary focus is the cost-effectiveness of the project; quality and progress are driven by the need to achieve the greatest economic benefit. Environmental protection issues are rarely considered. In some cases intentional injury to the environment occurs in order to reduce construction costs during the construction process. The second consideration is a frequent lack of environmental consciousness in many construction companies. The majority of construction companies pay attention to the constant improvement of a project's economic and technological aspects, training their construction workers to enhance their technological and management abilities. However, they often neglect to educate workers with regard to their environmental knowledge. This compounds the workers' lack of environmental protection consciousness and their inability to tackle pollution issues as they undertake a project. Thus, construction companies should be required to consider environmental protection issues in their management programmes in order to develop their proficiency in the reduction of construction pollution. For example, such methods as investigating the surrounding environment of a construction site before construction begins in order to formulate guidelines related to reducing environmental pollution; prohibiting the use of high noise level construction equipment during the construction process; keeping construction vehicles clean in order to prevent the carriage of site mud off the construction site; and the recycling of surplus and waste construction materials.

Furthermore, the entry barriers to the smart city market restrict all construction firms from participating in this business. In China, the management system of the intelligent construction industry includes three main sections: the construction licence management, the project management and the technology standard management. The construction licence management includes two qualifications, one is the qualification of engineering professional contractors, at either first class (highest), second class or third class levels; the other is the qualification of intelligent project design and contract, at first class (highest) and second class levels. Construction companies are required to gain these qualifications from the PRC's Ministry of Housing and Urban–Rural Development in order to undertake intelligent construction projects. However, these qualifications are not easy to achieve. Companies must initially apply for the lowest class of qualification and then upgrade to the higher class in order to undertake larger projects, because different classes of qualification can allow the undertaking of different scale projects. But this is a time-consuming journey, generally requiring six years to graduate from the lowest to the highest qualification class, and is dependent on companies' scale, annual revenue, experience, technology, the complexity of projects and so on. At present, a total of around 3,000 Chinese construction companies engage in the intelligent construction business. Of these, only 1,100 companies possess the engineering professional contractor qualification and only 33 companies have both. Thus, the issue of qualification remains a primary entry barrier for construction firms in China (Chen et al. 2015).

Another important barrier is the standard of technology management in building projects. Construction is divided into four main processes: civil engineering, electrical engineering, intelligent system installation and decoration. Intelligent system installation, the third process in building projects, requires specialized contractors who can undertake entire technological tasks including intelligent system design, the provision of advanced equipment, smart device installation and debugging. However, only the larger construction companies have the capability to provide intelligent technology, and this causes many small construction companies the loss of potential business opportunities in the construction of the smart city.

Moreover, huge potential business opportunities exist for construction firms with the widespread introduction of intelligent building in the smart city, but the development of the smart city concept is still in the early stages (Wang 2015). Therefore construction firms' current knowledge of the intelligence required for the smart city is limited, which may restrict project management capabilities in immediately meeting the intelligence needs of the smart city. Thus construction companies must learn and research intelligence knowledge relevant to the smart city, training workers to gain the necessary advanced technology skills and utilizing advanced construction methods to achieve intelligent and sustainable construction.

In summary, the construction industry is very important for the development of the smart city; it is as much about the construction industry promoting a country's economic development which then improves people's living standards, as it is about construction facilitating the realization of ecological and intelligent systems in the smart city. However, the development of the smart city has given rise to a new highly competitive construction market. For example, construction companies do not have an absolute advantage regarding market share in China's smart city market at present. The statistical data showed that the market share of China's top ten construction companies of the intelligent construction market was only 5 per cent (Chen *et al.* 2015). Chinese construction companies are attracted by the intelligent construction market and desirous to participate in order to achieve greater market share. Increasing numbers of contractors are trying to enter this market, giving rise to increased competition. This then requires construction companies to improve their competitiveness in order to survive in this more competitive construction market and meet the development needs of the smart city. Moreover, the construction industry is changing constantly with the innovations and developments of new business methods and technologies. Thus, construction companies have to adopt these applications and develop appropriate strategies to be more competitive in this industry and achieve success in their businesses (Arslan and Kivrak 2008).

Theory

Key competitiveness indicators (KCIs) are the foundation of construction firms' competitiveness. Firms' KCIs could be divided into three sections: corporate strategy, organizational capability and financial capability.

Evidence

The 21 competitiveness indicators (Table 8.1) were identified through analysis of the secondary academic literature reviews and annual reports of 60 international construction firms operating in the UK and 27 large Chinese construction firms.

In order to investigate the importance of these KCIs in the operation of construction firms, a total of 15 Chinese construction industry experts were interviewed by the Modified Delphi method. A Delphi study requires the place of consensus to be established at the start of the study (Crisp *et al.* 1997). Consideration must be given to the level of consensus employed (Hasson *et al.* 2000). To assess consensus, four sets of combined criteria measures are used including:

- Median score of ≥4
- Mean score of ≥3
- The interquartile range (IQR) <1
- The standard deviation (SD) below 1.0 on 5-point scale (Boyd and Smith 2014; Musa *et al.* 2015).

TABLE 8.1 Key competitiveness indicators identified by secondary literature review

KCIs	Cited numbers
1. Quality	120
2. Organization management	112
3. Build safety and healthy environment	95
4. Social responsibility	92
5. Technology management	92
6. Company experience	90
7. Communication and cooperation	89
8. Costs	86
9. Company culture	84
10. High productivity	84
11. Valued-added for stakeholders	80
12. Reputation	79
13. Assets	75
14. Employee management	73
15. Revenue	72
16. Risk management	72
17. Research & Development	72
18. Source finance	54
19. Liability	48
20. Profit	46
21. Information technology	37

For the level of agreement and stability, the stopping criteria followed the Kendall's coefficient of concordance W, which is widely recognized as the best. The value of W ranged from 0 to 1, with 0 indicating no consensus, and 1 indicating perfect consensus.

Delphi first round

In the first round, experts were asked to rate key competitiveness indicators (KCIs) identified from the secondary literature review (Table 8.2) and to provide additional KCIs which they considered contributed to a firm's international competitiveness.

In this round, consensus was reached on 19 out of 21 indicators, but two indicators did not achieve consensus. These include: Research & Development (IQR = 2 > 1; SD = 1.01 > 1) and Liability (Median = 3 < 4). The result shows this round of interviews did not achieve enough satisfactory value, Kendall's $W = 0.287 < 0.5$, suggesting a weak agreement and low confidence among the panellists' rating of the KCIs. This then required a second round of interviews in order to achieve consensus regarding these indicators.

In this interview, some interviewees proposed additional indicators which they considered potentially important to construction firms in international markets, including: the employee's knowledge of/ability in international construction

TABLE 8.2 Key competitiveness indicator rating in the first round interview

KCIs	Maximum	Minimum	Mean	Median	IQR	SD
1. Quality	5	4	4.781	5	0	0.420
2. Social responsibility	5	1	3.906	4	0.75	0.856
3. High productivity	5	3	4.375	4	1	0.609
4. Build safety and healthy environment	5	3	4.563	5	1	0.564
5. Valued-added for stakeholders	5	2	3.719	4	1	0.924
6. Reputation	5	3	4.531	5	1	0.621
7. Research & Development	5	2	4.313	4	1	0.780
8. Organization management	5	3	4.500	5	1	0.622
9. Company's experience	5	3	3.906	4	0	0.588
10. Communication and cooperation	5	2	3.813	4	1	0.821
11. Company's culture	5	3	3.875	4	0	0.554
12. Technology management	5	3	4.438	4	1	0.564
13. Employee management	5	2	4.063	4	1	0.759
14. Risk management	5	3	4.281	4	1	0.581
15. Information technology	5	3	4.063	4	0.75	0.669
16. Assets	5	3	3.938	4	1.75	0.759
17. Costs	5	2	4.125	4	1	0.707
18. Revenue	5	2	4.000	4	0	0.672
19. Source finance	5	2	3.875	4	0.75	0.707
20. Liability	5	2	3.531	4	1	0.718
21. Profit	5	3	4.094	4	1	0.689

Source: The authors (2016).

markets, leadership, managers' decision-making capacity, the welfare and training of expatriate staff and national policy support. Interviewees rated these additional indicators in the second and third round of interviews.

Delphi second round

In the second round of interviews, consensus was reached regarding 17 out of 21 indicators (Table 8.3), while consensus among the expert panel was not achieved regarding four indicators, thus these were removed. The indicators removed included: Valued-added for stakeholders (Median = 3 < 4); Communication and cooperation (Median = 3 < 4); Revenue (Median = 3 < 4) and Liability (Median = 3 < 4), this means that the interviewees considered these four indicators not particularly important to Chinese construction firms. In this round, Kendall's coefficient of concordance $W = 0.55 > 0.50$, confirms a good agreement and confidence among the panellists' rating of the KCIs. Thus, the interviews could be concluded.

In the first round of interviews, participants added five additional indicators which they thought could be important to a construction firms' competitiveness (see Table 8.4). In this round, interviewees rated these indicators on a 5-point scale. All of these indicators achieved the four sets of combined criteria measure, but Kendall's coefficient of concordance $W = 0.290 < 0.50$, thus the third interview is carried on until consensus regarding these five additional indicators is reached.

TABLE 8.3 Second round of interviews

KCIs	Maximum	Minimum	Mean	Median	IQR	SD
1. Quality	5	4	4.93	5	0	0.26
2. Social responsibility	5	4	4.27	4	1	0.46
3. High productivity	5	4	4.47	4	1	0.52
4. Build safety and healthy environment	5	4	4.53	5	1	0.52
5. Valued-added for stakeholders	5	3	3.27	3	0	0.59
6. Reputation	5	4	4.40	4	1	0.51
7. Research & Development	5	3	4.20	4	1	0.56
8. Organization management	5	3	4.47	5	1	0.64
9. Company's experience	5	3	3.80	4	1	0.56
10. Communication and cooperation	4	3	3.27	3	1	0.46
11. Company's culture	4	3	3.73	4	1	0.46
12. Technology management	5	3	4.60	5	1	0.63
13. Employee management	5	4	4.40	4	1	0.51
14. Risk management	4	3	3.87	4	0	0.35
15. Information technology	4	3	3.67	4	1	0.49
16. Assets	4	3	3.87	4	0	0.35
17. Costs	4	3	3.73	4	1	0.45
18. Revenue	4	3	3.33	3	1	0.49
19. Source finance	4	3	3.73	4	1	0.46
20. Liability	4	3	3.27	3	1	0.46
21. Profit	4	3	3.93	4	0	0.26

Source: The authors (2016).

TABLE 8.4 Additional key competitiveness indicator rating by second round interviews

KCIs	Maximum	Minimum	Mean	Median	IQR	SD
1. The employee's knowledge of/ability in international construction market	5	3	4.20	4	1	0.56
2. Leadership	5	4	4.33	4	1	0.49
3. Manager's decision-making capacity	5	4	4.33	4	1	0.49
4. The welfare and training of expatriate staff	5	3	3.80	4	1	0.56
5. National policy support	5	3	3.67	4	1	0.72

Source: The authors (2016).

Delphi third round

In the third round of interviews, the experts were required to rate the additional key competitiveness indicators based on the results of the second round of interviews.

However, consensus was reached on four additional indicators (see Table 8.5), but National policy support's median score = 3 < 4. Thus National policy support was removed from the list of key competitiveness indicators. In this round, Kendall's coefficient of concordance $W = 0.524 > 0.50$, achieved a good agreement and confidence among the panellists' rating of these additional competitiveness indicators. Thus, the interview was finished.

Based on the Chinese interviewees' ratings on the 5-point scale, the key competitiveness indicators (KCIs) can be ranked from the highest mean rating to the lowest mean rating (Table 8.6).

The results show that the most important indicator identified by the experts was quality. In addition to quality, the top ten important indicators were: technology management, build safety and healthy environment, high productivity, organization management, reputation, employee management, social responsibility, leadership and manager's decision-making capacity.

A competitiveness model may be established by the Analytical Hierarchy Process (AHP) (Figure 8.3). The AHP method was first developed by Saaty (1980) and assisted in developing a useful multiple criteria decision-making tool dealing with economic, technical and social issues. One major advantage of AHP is that it can convert a particular subject that is intangible and difficult to quantify into quantified and tangible values by using a systematic approach (Hyun et al. 2008). AHP is only applicable to a hierarchy that assumes a unidirectional relation between decision levels (Cheng and Li 2004). Thus, this competitiveness model with AHP facilitates the construction firms' understanding of which key competitiveness indicators fall under different sections: corporate strategy, organizational capability and financial capability, and assessment of constrction firms' key competitiveness capability.

TABLE 8.5 Additional key competitiveness indicator rating by third round interviews

KCIs	Maximum	Minimum	Mean	Median	IQR	SD
1. The employee's knowledge of/ability in international construction market	5	4	4.20	4	0	0.41
2. Leadership	5	4	4.27	4	1	0.46
3. Manager's decision-making capacity	5	4	4.27	4	1	0.46
4. The welfare and training of expatriate staff	5	4	4.07	4	0	0.26
5. National policy support	4	3	3.27	3	1	0.46

Source: The authors (2016).

TABLE 8.6 Ranking of interview data

Rank	KCIs	Mean	SD
1	Quality	4.93	0.26
2	Technology management	4.60	0.63
3	Build safety and healthy environment	4.53	0.52
4	High productivity	4.47	0.52
5	Organization management	4.47	0.64
6	Reputation	4.40	0.51
7	Employee management	4.40	0.51
8	Social responsibility	4.27	0.46
9	Leadership	4.27	0.46
10	Manager's decision-making capacity	4.27	0.46
11	The employee's knowledge of international construction market	4.20	0.41
12	Research & Development	4.20	0.56
13	The welfare and training of expatriate staff	4.07	0.26
14	Profit	3.93	0.26
15	Risk management	3.87	0.35
16	Assets	3.87	0.35
17	Company's experience	3.80	0.56
18	Costs	3.73	0.45
19	Company's culture	3.73	0.46
20	Source finance	3.73	0.46
21	Information technology	3.67	0.49

Source: The authors (2016).

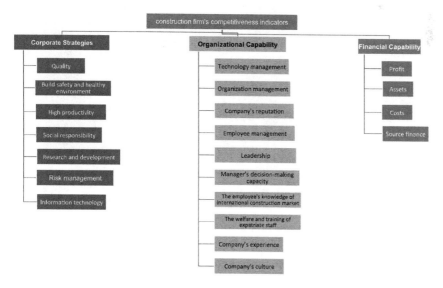

FIGURE 8.3 The Analytical Hierarchy Process (AHP) for construction firms' international competitiveness

Conclusions

This chapter discussed competitiveness theories but noted that conventional business theory had limited traction for construction firms in specific markets. The growing prominence of smart city initiatives worldwide has altered market dynamics. To rise to the challenge, leading construction companies must improve their competitiveness in partnership with government, investors and designers to build up the smart city.

In the context of ongoing Chinese urbanization the smart city guides conditional development to tackle the challenges of overpopulation, lifestyle and environmental protection. Although China has seen the completion of several successful smart city projects to date, the domestic construction industry still requires improvement in many areas. Policymakers need to address three main issues:

First, China must strengthen the domestic construction industry's innovation in information technology ability. Smart urban construction requires the application of advanced information technology in order to implement the overall configuration of urban resources and enhance the operational efficiency of urban systems. Thus, government policies and initiatives are required to optimize the Chinese construction industry's technological innovation ability. For instance, by strengthening IT research and development platforms; strengthening cooperation between government and related institutions to develop new technologies; expanding and developing training programmes for scientific and technological innovation personnel and by providing high welfare incentives for them. A wide range of ongoing public awareness campaigns, staff education and training should improve the industry and workers' awareness of the smart city, thus improving public awareness of smart and sustainable development.

Second, the government should foster a smart city market to guide the traditional industry to transform and pay more attention to intelligent innovation, which may then facilitate construction firms to improve their capability to undertake smart city projects. A more competitive market would encourage construction firms to enter it purely based on competitive ability, so that the most innovative and successful firms will succeed in becoming involved in the process of building a smart city.

Finally, the government must publish and enforce more stringent construction industry guidelines to standardize acceptable levels of implementation and development when firms undertake smart city projects. The current lack of clarity and the perceived high cost of meeting the requirements of the market are two main factors holding many construction firms back from entering the smart city market. Practical guidelines for the operation of construction firms including policy measures, smart city assessment methods and smart city pilot projects would facilitate and strengthen the implementation of smart city policy guidance and supervise construction project parties to enable them to do their best in the building of smart projects.

Government cannot build a smart city without smart construction firms. The chapter found that to be competitive construction firms need a robust corporate strategy, organizational capabilities and a sound financial system.

CASE STUDY

Beijing Tellhow Intelligent Engineering Co., Ltd is one of the PRC's earliest tech enterprises engaging in contract and design of system integration, technical advisory and consultion of intelligent buildings and energy-saving buildings. Tellhow aims to address the energy saving needs of newly built and existing buildings and has independently developed the Tellhow IBMS overall energy saving solution for electrical systems in buildings. Since 2006, Tellhow has provided 'customized' integrated intelligent energy saving solutions for a variety of major sporting venues, luxury hotels, digital parks, government, campus and hospital buildings. Why is Tellhow outstanding in China's intelligent construction industry? The company's success can be analysed via the key competitiveness indicators from the competitiveness model (Figure 8.3).

Quality: Tellhow won the 'Architectural Engineering Luban Awards' three times. It can be seen that Tellhow pays close attention to project quality as the 'Luban Award' is the highest commendation in China awarded annually to the best quality domestic project.

Technology management: Tellhow has its own excellent building technologies. For example Tellhow IBMS integrated intelligent energy saving solutions for electrical systems in its buildings provides 'comprehensive' services for both new and existing buildings, avoiding 'outsized power for petty use', and achieving the connection of subsystems whilst also improving the utilization efficiency of equipment, thus achieving a 30 per cent increase in energy saving and an additional 11 per cent cost savings compared to the conventional model. BEMS Integrated Solution is dedicated to providing a coordinated one-stop service encompassing diagnosis, design, modification, operation and management and provides customized optimum energy saving solutions matching the newly built or existing buildings. The Building Integrated Photovoltaic (BIPV) refers to the installation of solar photovoltaic arrays on the outer surface of the enclosure structure of buildings in order to provide electricity.

Research and development: Tellhow invents many advanced technology products, including Energy Metering, e-Hf Highly-Efficient Lighting Fittings, PV modules, PV plant control system, Intelligent Power Generation, Intelligent Power Distribution Products and Facial Recognition system; in which PV modules has passed more than 30 patent certifications.

Reputation: Tellhow is the 'Most Influential Enterprise of the Year for Chinese Building Energy Saving'. According to the statistics of bill of quantities of intelligent building projects issued by the Ministry of Housing and

Urban–Rural Development of the PRC, the company has ranked among the 'Top Ten in China' for five consecutive years. It has achieved 18 project awards and six technology awards, for example the 2007 ministry of construction 'the construction conserves energy ten big cases', the intelligent construction quality product recommended by the intelligent construction specialized committee and the 2008 outstanding intelligence construction management system.

Social responsibility: Tellhow began energy saving technology R&D and promotion from 2002 and establishes a joint venture with Japan Mitsui and Panasonic. It introduces advanced energy saving technologies from Japan, devises energy saving solutions suitable for China and perfects them through many years' experience in project. From 2007, Tellhow National Action of Energy Saving is launched for popularizing the advanced energy saving ideas and technologies and facilitating the development of China's energy saving and environmental protection. As a responsible enterprise, Tellhow has donated over RMB 2 million to construct eight Hope Primary Schools providing a good learning environment for more than 1,000 children in poverty-stricken areas and supporting the development of education in old revolutionary bases.

Corporate experience: Tellhow has rich project experience in different sectors. From its establishment in 1997, until now, it has completed over 50 projects including government public buildings, venues, airports, buildings for the financial industry and educational institutions, hospitals, hotels, energy-saving and PV plants.

Organization management and company culture: Tellhow illustrates a very clear corporate mission to its employees, facilitating their understanding of the company's aims and what its employess do. For example, the mission statement says, 'creates and leads intelligent technology, the products and the service, improving the quality of human lives'.

Source: Tellhow official website (2016).

Discussion questions

1   Why is the construction industry important to the smart city?
2   Why do construction firms need to improve their competitiveness in order to meet the smart city's development?
3   What is construction firms' competitiveness?

Outline answers

The construction industry is very important to the development of the smart city because it facilitates the realization of a safe, efficient, convenient, energy saving, environmentally sound and liveable environment for the residents. Moreover, construction firms are one of the primary partners, along with government, investors and designers, in the fruition of the smart city.

The smart city's rapid development brings huge potential business and improved competition to the market. Li (2014) predicted that in the future the smart city business will be worth $4 trillion, this projected figure includes intelligent infrastructure, intelligent traffic systems and public security. This will increase the competition in the market. Moreover, the traditional construction industry cannot currently meet the requirements of smart urban development due to its pollution of the environment and non-intelligent technology. Thus, construction firms must improve their competitiveness in order to survive and grow in this increasingly competitive market.

Construction firms' competitiveness cannot be judged in precise terms. Different construction firms have different life cycles, different scales of project and different markets. Thus, identifying the key competitiveness indicators (KCIs) is the core of understanding the meaning of competitiveness. The KCIs were identified in this study by Chinese construction experts with the Modified Delphi method (Figure 8.3). The author proposed that construction firms' competitiveness should be divided into three sections: corporate strategy, organizational capability and financial capability. The construction firms may then assess their competitiveness via these KCIs, in order to identify their existing KCIs and those that may still be achieved.

Bibliography

Allidina, S. (2015). Smart cities mean big business. [Online]. Available at: http://raconteur. net/business/smart-cities-mean-big-business [Data accessed 8 August 2016].

Arditi, D. and Gunaydin, H. M. (1997). Total quality management in the construction process. *International Journal of Project Management* 15(4): 235–243.

Arslan, G. and Kivrak, S. (2008). Critical factors to company success in the construction industry. *World Academy of Science, Engineering and Technology* 2(9): 997–100.

Barney, B. J. (1991). Firm resources and sustained competitiveness advantage. *Journal of Management* 17(1): 99–120.

Barney, J., Wright, M. and Ketchen, D. (2001). The resource-based view of the firm: ten years after 1991. *Journal of Management* 27: 625–641.

Boyd, D. A. A. and Smith, M. (2014). Developing a practice-based body of real estate knowledge: a Delphi study. *Journal of Real Estate Practice and Education* 17(2): 139–167.

Buzzell, R. and Gale, B. (1987). *The PIMS Principles –Linking Strategy to Performance.* New York: Free Press.

Cafiso, S., Graziano, A. D. and Pappalardo, G., (2013). Using the Delphi method to evaluate opinions of public transport managers on bus safety. *Safety Science* 57: 254–263.

Chandler, A. (1962). *Strategy and structure.* Boston, MA: MIT Press.

Chen, J., Ni, X. and Wang, X. (2015). *An analysis of the intelligent city market in China.* Hangzhou: Zheshang Securities.

Cheng, E. and Li, H. (2004). Contractor selection using the analytic network process. *Construction Management and Economics* 22(10): 1021–1032.

Connock, T. (2015). A smarter world: the challenge of building 'smart cities'. [Online]. Available at: www.lawcareers.net/Information/BurningQuestion/Taylor-Wessing-A-smarter-world-the-challenge-of-building-smart-cities [Data accessed 8 August 2016].

Crisp, J., Pelletier, D., Duffield, C., Adams, A. and Nagy, S. (1997). The Delphi method? *Nursing Research* 46(2): 116–118.

D'Cruz, J. and Rugman, A. (1992). *New concepts for Canadian competitiveness.* Toronto: Kodak Canada.

Darvish, M., Yasaei, M. and Saeedi, A. (2008). Application of the graph theory and matrix methods to contractor ranking. *International Journal of Project Managment* 27(6): 610–619.

Feurer, R. and Chaharbaghi, K. (1994). Defining competitiveness: a holistic approach. *Managment Decision* 32(2): 49–58.

Flanagan, R., Lu, W. and Shen, L. A. J. (2007). Competitiveness in construction: a critical review of research. *Construction Management and Economics* 25: 989–1000.

Fong, P. S., and Choi, S. K. (2000). Final contractor selection using the analytical hierarchy process. *Construction Managment and Economics* 18(5): 547–557.

Geist, M. R. (2010). Using the Delphi method to engages stakeholders: a comparison of two studies. *Evaluation and Program Planning* 33(2): 147–154.

Guevarra, L. (2011). Market for smart city technology to reach $16B a year by 2020. [Online]. Available at: www.greenbiz.com/blog/2011/09/29/market-smart-city-technology-reach-16b-year-2020 [Data accessed 1 August 2016].

Hammer, M. and Champy, J. (1993). *Re-engineering the corporation.* New York: Harper Business.

Hasson, H., Keeney, S. and McKenna, H. (2000). *Research guidelines for the Delphi survey technique. Journal of Advanced Nursing.* 32(4): 1008–1015.

Hatush, Z. and Skitmore, M. (1997). Criteria for contractor selection. *Construction Managment and Economics* 15(1): 19–38.

Holt, G. D., Olomolaiye, P. O. and Harris, F. C. (1994). Applying multi-attribute analysis to contractor selection decisions. *European Journal of Purchasing and Supply Management* 1(3): 139–148.

Hyun, C. T., Cho, K., Koo, K. J., Hong, T. H. and Moon, H. S. (2008). Effect of delivery methods on design performance in multifamily housing projects. *Journal of Construction Engineering and Management* 134(7): 468–482.

Institute of Urban and Environmental Studies of Chinese Academy of Social Science. (2012). *Report of China's urban development.*

Johnson, H. T. (1992). *Relevance regained.* New York : Free Press.

Li, Y. (2014). The policy of smart city development in China. [Online]. Available at: http://stock.cnstock.com/stock/smk_qlgg/201411/3246022.htm [Data accessed 19 May 2016].

Lin, J. (2013). Five issues of China's urbanization development. [Online]. Available at: http://theory.people.com.cn/n/2013/0829/c83865-22738584.html [Data accessed 12 April 2015].

Marzouk, M. M., Kherbawy, A. A. and Khalifa, M. (2013). Factors influencing sub-contractors selection in construction projects. *HBRC Journal* 9(2): 150–158.

Musa, H. D., Yacob, M. R., Abdullah, A. M. and Ishak, M. Y. (2015). Delphi method of developing environment well-being indicators for the evaluation of urban sustainability in Malaysia. *Environment Sciences* 30: 244–249.

National Bureau of Statistics of China. (2014). *China Statistics Yearbook.*

Porter, M. (1980). *Competitive strategy.* New York : Free Press.

Porter, M. (1985). *Competitive advantage*. New York: Free Press.

Prahalad, C. K. and Hamel, G. (1990). The core competence of the corporation. *Harvard Business Review* 68(3): 79–91.

Rad, H. N. and Khosrowshahi, F. (1998). Quality measurement in construction projects. 14th Annual ARCOM Conference, Reading.

Saaty, T. L. (1980). *The analytic hierarchy process*. New York : McGrawHill.

Schriener, J. and Doherty, P. (2013). The impact of smart cities on the construction industry. [Online]. Available at: https://enewsletters.constructionexec.com/techtrends/2013/06/the-impact-of-smart-cities-on-the-construction-industry/ [Data accessed 15 May 2016].

Shen, L. Y., Lu, W. S., Shen, Q. P. and Li, H. (2003). A computer-aided decision support system for assessing a contractor's competitiveness. *Automation in Construction* 12(5): 577–587.

Shen, L. Y., Lu, W. S. and Yam, M. (2006). Contractor key competitiveness indicators: a China study. *Journal of Construction Engineering and Management* 132(4): 416–424.

Tellhow (2016). Official website: www.tellhow.com.cn/En/index.php/Aboutus/index/id/54.html

Wang, G. (2015). Project management of intelligent building. [Online]. Available at: www.kanzhun.com/lunwen/165233.html [Data accessed 15 May 2016].

Wu, J. (2012). China's urbanization issues and the political options. [Online]. Available at: http://finance.sina.com.cn/hy/20121013/215913360760.shtml [Data accessed 15 April 2016].

Xiao, Z. and Feng, D. (2015). The current issues and suggestions of sustainable construction. [Online]. Available at: www.chinacem.com.cn/xmgl/2015-03/185383.html [Data accessed 24 November 2016].

PART III

Mechanisms

9

INFRASTRUCTURE-LED REGENERATION

Jaipur, India

Anil Kashyap and Jim Berry

Overview

According to UN projections, by 2030, around 70 per cent of the world's population will live in cities. Even today, cities drive innovation and generate 80 per cent of global GDP (Shah *et al.* 2015). Since 1960, the nominal economic growth rate of India has mirrored the global average, at around 7.5 per cent, but the cumulative figures mask a significant disparity (IMF 2014). In recent decades, growth has accelerated and India now produces 2.8 per cent of global traded output, almost equivalent to that of its former colonial ruler. By 2025, 69 metropolitan cities in India will house 78 per cent of its urban population and will contribute 77 per cent to urban GDP (Planning Commission 2011). However, a lack of transportation infrastructure constrains city growth. Mobility and connectivity remain key challenges for Indian cities. The Delhi Metro Rail Corporation (DMRC) operates the metro system in Delhi, which is an example of progress other Indian cities seek to emulate. However, such multi-modal transportation infrastructure calls for huge planning and consummate project execution. One way to finance such investments is to capture the uplift in real estate prices arising from a growing economy and to hypothecate the increase in value into the construction of well-planned and executed mass transport projects.

Jaipur city provides an interesting case study to investigate the impact of mass transit projects. The city has a population of 3.1 million and is located 260 km from the Indian capital, New Delhi (Census of India 2011). In recent years, Jaipur city has developed rapidly and the present haphazard transport system is under pressure from industrial development, population growth and rapid commercialization. Additionally commuters prefer to use personalized transport which places a further burden on the city infrastructure, particularly on roads and streets. To overcome public resistance, the state government undertook detailed household and traffic surveys to prepare a demand model which incorporated future population projections, employment growth and traffic challenges. Notwithstanding the success of the Delhi Metro Rail Corporation, the Rajasthan government prevaricated on the Jaipur metro project, but finally on 24 February 2011 the foundation stone of the Jaipur Metro was laid. After more than four years of construction work, the Pink Line finally opened on 3 June 2015. From a methodological perspective, the hedonic pricing model has been used in these types of studies to capture the property physical characteristics and indicators of locational attractiveness to explain real estate price increases. Distance to the Central Business District (CBD) and proximity to transport facilities can have a significant impact on property prices which generally decrease in line with increasing distance to/from metro stations. The purpose of this chapter is to take an initial look at the impact of the Jaipur Metro on the immediate area in terms of creating a functional real estate market based on increases in land and property prices linking to an infrastructure-led urban regeneration model.

Learning outcomes

The learning outcome from the chapter will show the significance of the following:

- Effectiveness of metro transit system on property values in respect to magnitude and directions.
- Success of metro rail projects in compact areas.
- Impact of regulatory planning of infrastructure on real estate market.
- Effect of metro station proximity on property prices.
- Systematic discussion and comparison of the key factors affecting the property market.

Stakeholder implications and responsibilities

Whilst consultation is an important aspect of project planning for major transport infrastructure, it is not a panacea. Contention is inevitable as land transformation creates winners and losers (Owens and Cowell 2011). Notwithstanding its limitations, consultation should be well organized and involve a balanced range of stakeholders, including representatives of legislative bodies, the poor and other publicly barred groups (e.g. women) to dispense the information as well as to get the advice about the project plan and its possible impacts. Other key stakeholders such as the important line department, local government legislative body and NGOs will also be consulted. A list of various stakeholders involved in the consultative process associated with large-scale projects is outlined in Table 9.1.

TABLE 9.1 Stakeholders involved in the Jaipur metro

Sr. No.	Stakeholders names	Responsibility and duties
	Feasibility studies bodies	
1	Rajasthan State Urban & Housing Development Authority	Inter-agency coordination, oversee and monitor project implementation as well as the adequacy of overall project funding. Provide necessary policy guidance related to project implementation
2	JDA	Raise funding of amount 150.00 (Rs Cr.) Partially monitoring body
3	Transport and Communications Division of South Asia Department (SATC)	Helping Agency for JMRC Subsidiary of ADB
	Financial bodies	
4	Asian Development Bank	Funding Partner, 969.00 ★ (Rs Cr.) Monitor and review overall implementation of the project in consultation with the EAs
5	Japan International Cooperation Agency (JICA)	Funding for underground portion from Chand Pole to Bari Chopar
6	Other govt. subsidiaries	Raise funding of amount 108.00 (Rs Cr.)
7	Government of India	Raise funding of amount 627.00 ★ (Rs Cr.)
8	World Bank	Funding body, provides financial Aid
9	India Infrastructure Finance Company Ltd. IIFCL	Public sector funding agency, provides financial aid
10	Rajasthan Housing Board	Raise funding of amount 100.00 (Rs Cr.)
11	RIICO	Raise funding of amount 100.00 (Rs Cr.)
12	Government of Rajasthan	Raise funding of amount 1092.00 (Rs Cr.) Executive Agency
	Developmental bodies	
13	DMRC	Executing and monitoring body for construction work
14	E. Sreedharan	Principal Role Advisor of Delhi Metro Rail Corporation for the Jaipur Metro Rail project
15	DMRC employers and workers	Help in implementation and development
16	JMRC	Body established by Govt. of Rajasthan Owner
17	Nihal Chand Goel	Chairman & Managing Director of JMRC
	Construction bodies	
26	ITD Cementation India Ltd	Construction work from Ajmer road flyover to Chand Pole station
27	DSC Ltd	Construction of viaduct consisting of single box by segmental construction, work also includes construction of a ramp and two special spans
	Environment assessment bodies	
28	Archaeological Survey of India (Central/State)	Helps in archaeological researches, maintenance and protection of the cultural heritage of five monuments/sites
29	Central Road Research Institute	Monitoring authority responsible for building roads and pollution impact
30	Expert Appraisal Committee	Responsible for environment clearance

TABLE 9.1 Continued

Sr. No.	Stakeholders names	Responsibility and duties
31	State Expert Appraisal Committee	Subordinating authority of Expert Appraisal Committee
32	Grievance Redress Committee	Partly, responsible for environment clearance
33	Jawaharlal Nehru National Urban Renewal Mission/ BRTS	Data provider during peak hours regarding passengers
34	Indian Meteorological Department	Responsible for survey regarding climate impact
35	Ministry of Road Surface Transport And Highways	Granting authority, Pollution Under Control Certificate
36	Ministry of Environment and Forests	Survey related to status of forest areas, compensatory afforestation norms, etc.
37	National Transportation Planning and Research Centre	Monitoring and supporting authority on transport system
38	Medical Expert Committee	Responsible for survey and impact on health
39	Rajasthan State Road Transport Corporation	Monitoring and supporting body towards RTO
40	Rajasthan State Pollution Control Board	Accountable for the completion of a variety of acts to prevent and calculate pollution to various environmental factors
41	Central Pollution Control Board	Supporting and monitoring body, towards pollution
42	RTO, Jaipur	Surveying authority of transport impact
43	Central Ground Water Resources Board	Accountable for the expansion, dissemination of technology and monitoring of India's groundwater wealth, counting investigation, appraisal, protection, increase, protection from pollution and distribution
44	Rajasthan Department of Archaeology and Museums	Accountable for the discovery, preservation, protection, exhibition and interpretation of the cultural legacy of the state
45	United Nations Education, Scientific, and Cultural Organization (UNESCO)	Concerning the protection of World Cultural and Natural Heritage sites located in city, declared incumbent on the international society
46	NIIT, Jaipur	Helping body in impact and assessment survey
47	Geological Survey of India	Survey related to geology, seismicity, soil and topography
48	Mining department/ MOEFCC	Granting authority, permission for sand mining from river bed
	Power supply	
49	Rajasthan Rajya Vidyut Prasaran Nigam Limited (RVPN)	Ensure high reliability of power supply during metro operation, adequate redundancies in the transmission and distribution were incorporated in the design
50	JVVNL	Responsible of supply and circulation of electricity supply
	Supplier of coaches	
51	Bharat Earth Movers Ltd (BEML), Bangalore	Responsible for manufacturing and supplying the rail coaches for JMRC/ Rs 3.18 billion contract to supply 10 four-car trains for Phase 1

CASE STUDY JAIPUR CITY

Jaipur, the 'sonata in pink', land of colours and festivals, where beauty and ancient art blend fabulously, is the capital of the largest state, Rajasthan, which is located in northern India. It was founded on 18 November 1727 by Maharaja Sawai Jai Singh II, the ruler of Amber, after whom the city was named Jaipur. It was not only planned but its implementation was also organized by Sawai Jai Singh II, in such a manner that a substantial part of the city developed within seven years of its foundation.

The city of Jaipur, surrounded by the Aravali ranges on three sides, each crowned by a fort, is studded with palaces, artistic mansions, landscapes, gardens and parks. Classified as being the first planned city of India and based on the ancient Hindu art 'Shilp Shastra', Jaipur city was built in a rectangular form and is further divided into blocks (Chowkries) with the roads and avenues running analogous to its side. The city was coloured in pink to welcome Prince Albert, consort of Queen Victoria, and from that time this colour has become an integral part of the city which is now known as 'Pink City'. Though it has developed into a modern metropolitan city and commercial centre, the city is a tourist and visitor destination with attractions ranging from cultural, adventure, aesthetic, handicraft, gold, jewellery, precious stones, blue pottery, carving on timber and handmade paper and printing.

The city of Jaipur is located at an altitude of 431 m (1,417 ft) above sea level and at 26.92° N latitude 75.82° longitude (Figure 9.1). The total area of Jaipur district is 326 sq. km and is surrounded by Alwar and Sikar from the north, Ajmer, Tonk and Sawai Madhopur from the south, by Sikar, Nagaur and Ajmer from the west and Dausa district in the east. It has a humid subtropical climate, which receives over 650 mm of rainfall annually.

In 1961, population growth in Jaipur was 4.45 per cent but over the past three decades the population has doubled. In 2001, it reached 2,324,319. Rapid increases in industrial development, population growth and extreme commercialization have put the city's transport system and infrastructure provision under tremendous pressure. Notwithstanding the preference of commuters for personalized transport, traffic gridlock dictated the need for a radical public transport solution in Jaipur. In preparation for this investment, the state government undertook detailed household and traffic surveys to inform a demand model that incorporated future population projections of employment and traffic. The metro commenced commercial operation on 3 June 2015 and thus became the sixth rapid transport system after Delhi, Bangalore, Mumbai, Gurgaon and Kolkata. For effective execution of the project in its initial stage, the work at Phase 1 was divided in two parts: Phase 1A and 1B. Work on Phase 1A stretched from Manasarovar to Chandpole Bazar covering a length of 9.63 km. The remaining Phase 1B is under construction and is expected to be completed by 2018. Additionally, Phase 2 is under active

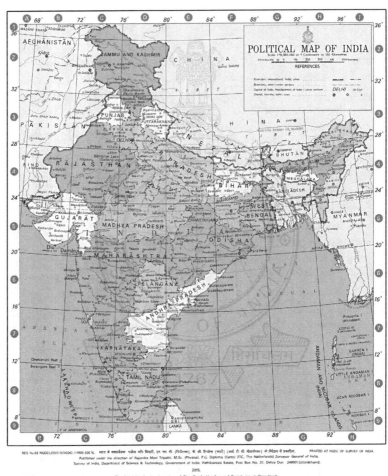

FIGURE 9.1 Location map of Jaipur, India

Source: Survey of India.

attention. Property prices reflect design and physical structure as well as a range of spatial advantages and disadvantages. Metro rail transit systems tend to enhance accessibility and hence inflate property values.

Impact of metro infrastructure on real estate

Connectivity plays a significant role in the growth of a city which is illustrated by the changing face of Jaipur which has been transformed over the course of a few years, boosted by a demand and regulated supply in the real estate market. According to real estate brokers in the city it seems that the property sector in Jaipur showed some growth during the introduction of Jaipur Metro Rail Corporation (JMRC) and then it became stagnant. After a few years the market then showed very good growth in the suburban and outer areas of Jaipur. All of the metro stations including Railway Station, Sindhi Camp and Chand Pole are quite congested and consist mostly of retail and commercial spaces with the result that property prices in these localities are much higher. But as we move towards areas further out, like Sitapura, Pratap Nagar and Sanganer (proposed stations in Phase 2) the overall impact on the property market is less significant. However, in the near future when both phases are fully functional and the outer areas receive increased connectivity with linkages to the inner city it will reduce the travel time and duration to a large extent with the result that people will be attracted and prospective purchasers will wish to buy or rent accommodation in the city.

As for our case study, we have selected the Sindhi Camp metro station, and for now the key question is why? It is the only interconnecting link between the two phases and is therefore a more preferable case study area to investigate and research. Another factor which differentiates it from the other metro stations is the central connecting link of both government as well as the private sector. It is also located near to the CBD. Consequently, our research area consists of the central bus stand of Rajasthan State Road Transport Corporation (RSRTC), a private bus stand of Polo Victory, bus booking offices, hotels, restaurants, retail shops and office spaces. Due to the importance of its location a number of residents live in nearby areas such as Bani Park, Subhash Nagar, Jalupura, Hasan Pura, Civil Lines and Purani Basti. These popular residential areas in the range of 3 km are marked on Figure 9.2 to show the connectivity of the research area.

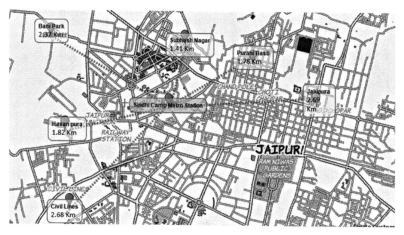

FIGURE 9.2 Residential areas located in 3 km radius of Sindhi Camp

Source: Authors' analysis (2015).

Infrastructure influences the fundamental facilities around a locality including the facilities essential for the economic growth of the locality or area. It typically includes the roads, bridges, water supply, tunnels, sewer, telecommunication and so on, and can be classified in two different categories. First, there are the physical components of interrelated systems which provide and serve societal living conditions like roads, bridges, sewers and so on, and second there is the social component which includes the institutions required to maintain and enhance the economic, health, cultural and social standards of a locality. The core facilities include the financial system, education system, health care system and emergency facilities.

Jaipur city is one of the fastest growing cites in India and is a key destination for investment into the ongoing infrastructure development like the major roads with flyovers and railway over-bridges. Consequently, regularization of residential areas and improvement of overall physical and social infrastructure is contributing to a steep growth curve in the city. Sindhi Camp is a packed and crowded area, because of the nearby CBD, railway junctions and central bus stands. A lot of infrastructure takes place around this locality such as hotels, restaurants, banks and hospitals. It is evident that a good infrastructure provision leads to an active real estate market.

To determine the impact of infrastructure on located properties, we studied an area comprising a 500 m radius from the centre of the metro station and the two stations which are located on either side (Figure 9.3). To identify the distance from Sindhi Camp metro station to both stations we used two approaches: (a) the straight line method, which indicates the distance on the basis of a straight line from our study area to the destination point; and (b) the actual distance method. We found that the actual distance method is slightly higher in comparison to the straight line distance, the reason being that the actual distance method includes the barriers and turns on the pathway whereas the straight line method does not.

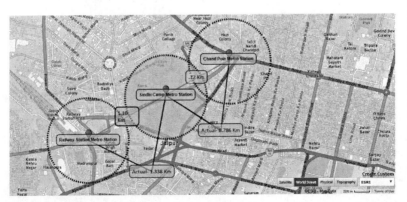

FIGURE 9.3 Catchment area of nearby metro stations

Source: Authors' analysis (2015).

The central Railway Station and headquarters of Northern Western Railway deals with 88 broad gauge, 22 m gauge trains and a daily footfall of 35,000 (approx.) passengers, which indicates that it is one of the busiest metro stations in Jaipur city. From Figure 9.3, we can see the straight line and actual distance from Sindhi Camp metro station is 1.10 km and 1.338 km respectively which is not a big difference between these two stations. Also, the difference in property price will not show a significant difference because of less distance. On the other hand, the distance to Chand Pole metro station indicates a straight and actual distance of .72 km and .786 km respectively; a small segment of Sindhi Camp overlaps with Chand Pole metro station and a significant appreciation in prices here is due to its centrally located position.

As a consequence of the large number of restaurants, lodges and hotels, it is an important factor of price consideration for our data analysis to evaluate the impact on a regular tariff and valuation of these hotels and lodges. There are 20 popular hotels and lodges in a total area of 77.65 ha with a circumference of 3.12 km and radius of 500 m from the centre of the metro station. After the collection of daily tariffs of these hotels from 2010 to 2015, we can see from Table 9.2, the average tariffs of all 20 hotels and lodges continuously escalate every year. Also, the escalation of prices in the year 2013–14 and 2014–15 are 10.71 per cent and 9.72 per cent respectively in comparison of previous years and a total growth of 39.63 per cent from the base year is noticed, which shows impressive results. We can conclude that the introduction of a metro creates a positive impact with price escalation. In addition, we can conclude that in the initial construction stage of a metro the prices are not much different but in moving towards the completion stage the market shows good growth results over the short to medium term.

TABLE 9.2 The average growth of tariff in % from years 2010–15

Years	Tariff appreciation (RS)				
	2010–11	2011–12	2012–13	2013–14	2014–15
Average tariffs	1196	1287	1374.75	1522	1670
Increase %	0.00%	7.61%	6.82%	10.71%	9.72%
Total increase	2010–2015	39.63%			

Source: Primary Survey (2015).

To understand the fluctuation and increase in property prices in our research area we took the average of property prices of Sindhi Camp and compared it with the average prices of Jaipur city. In the initial stages the prices are not particularly high but as the metro moves towards completion the price of properties continuously increases. Also, we can observe in the first and second quarters of 2015, the price curve of Jaipur city goes downwards, one of the

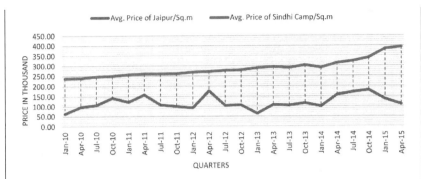

FIGURE 9.4 Comparison of office space prices

Source: Authors' analysis (2015).

reasons being the potential impact of sprawl and expansion of the city on property values.

To find out, we looked at the increase in prices in leased spaces from 2010 to 2015. We took a sample of properties which were approximately similar in size to similar locations within our research area. After that we calculated three components, that is maximum, minimum and average values of these properties for each year over the timeline 2010–15. As we can observe from Figure 9.5, prices do not show significant increases from 2010 to 2013 but in 2014 to 2015 they show significant growth in parallel with the completion of the metro.

In considering the increase in property prices in each direction of every 100 m, we divided our research area into five segments of 100 m, each with a different distance: 1st segment of radius = 100 m, circumference = 0.63 km, total area = 2.92 ha; 2nd with radius = 200 m, circumference = 1.25 km, total area = 12.49 ha; 3rd with radius = 300 m, circumference = 1.91 km, total area = 29.08 ha; 4th with radius = 400 m, circumference = 2.54 km, total area = 51.48 ha; 5th with radius = 500 m, circumference = 3.16 km, total area = 79.55 ha.

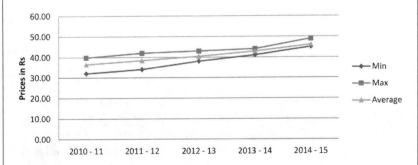

FIGURE 9.5 Increase in lease prices

Source: Authors' analysis (2015).

The first segment shows an average decline of Rs 2 per sq. ft in the leased rent, compared to previous years. In some properties the prices are stagnant or neither increase nor decrease, which can be explained by the overhead construction of a metro which quashes the visibility of those properties and results in the decline in the valuation of such properties. When we move towards the sections 200 to 500 m the property prices show an average growth of Rs 2.5 per sq. ft of lease in all directions except the properties located near the Kasakothi flyover. (It was proposed to reduce the overburden of traffic in this locality but due to political interference the structure got altered in the later stage, with the drawback that this planning resulted in a decrease in valuation.)

Conclusion

The investigation into the Jaipur transport project found that it significantly improved the accessibility to work and to facilities for households and business employees. In addition, it is evident that infrastructure provision and improved accessibility spin-offs will increase adjoining residential and commercial property values. The Jaipur case illustrates that connective infrastructure is now an important aspect of smart city development. The new infrastructure network brought substantive transport benefits and injected commercial vigour into stalled projects, such as the JMRC headquarters, and helped to regenerate brownfield sites. The Jaipur transit project demonstrates that, unless swamped by local negative impacts, infrastructure can result in increased property prices. The taxation of adjacent property price uplifts provides a potential funding mechanism for smart city infrastructure. However, policy makers face several complications in capturing the value. First, it is likely to cause resistance from vested interests. Second, it is likely to generate concerns on timing as, in early metro construction stages, value uplift remains muted. As completion approaches, real estate prices escalate. Finally, it is evident that value uplift can be sometimes be countered by associated negative externalities. For example, in Jaipur, the Kasakothi flyover illustrates how poor infrastructure planning can increase local traffic volumes and undermine rather than lift property values.

Bibliography

Delhi Metro Rail Corporation (2012) Detailed Project Report – Jaipur Metro Phase I Manasarovar to Badi Chaupar, March.

International Monetary Fund (IMF) (2014) World Economic Outlook: Legacies, Clouds, Uncertainties. IMF Policy Paper, Washington, DC: International Monetary Fund.

Owens, S. and Cowell, R. (2011) *Land and Limits: Interpreting Sustainability in the Planning Process*, 2nd edition, Abingdon: Routledge.

Planning Commission (2011) *Investment in Infrastructure during the Eleventh Five-Year Plan.* New Delhi: Government of India.

Press Information Bureau (2015) Launching of Jaipur Metro Is a Significant Step in Making Jaipur a Smart City, 4 June. Government of India, http://pib.nic.in/newsite/efeatures.aspx?relid=122275

Shah, Phoram, Hamilton, Ellen, Armendaris, Fernando, Lee, Heejoo and Armendaris, Fernando (2015) World – Inclusive cities approach paper. Washington, DC: World Bank Group. http://documents.worldbank.org/curated/en/402451468169453117/World-Inclusive-cities-approach-paper

World Bank (2013) *Planning and Financing Low-Carbon, Livable Cities*, 25 Sepember. Available at www.worldbank.org/en/news/feature/2013/09/25/planning-financing-low-carbon-cities

10

UNIVERSITIES, ART AND SCIENCE AND SMART EDUCATION

Negin Minaei

Abstract

Chapter 10 investigates how Higher Education (HE) institutions such as universities can catalyze smart sustainable development. Four common smart education phenomena are identified, including out-branching and outsourcing educational services. The chapter discusses the significant role universities can play in sustainable development by integrating interdisciplinary, multidisciplinary and transdisciplinary research to solve current urban problems. Recent examples and case studies illustrate the university's role and the way this mechanism works. The final section of the chapter illuminates how the HE curriculum can help to promulgate smart disciplines such as the sustainable cities development idea, or Ambient Intelligence and Smart City programs and courses.

Key terms

Higher Education, university, interdisciplinary, multidisciplinary, transdisciplinary, smart city

Learning outcomes

In smart cities, university precincts can inject vibrancy and authentic cultural life or sustain a creative milieu for innovation:

- Instead of attracting foreign students, universities can open overseas branches.
- Transdisciplinary research collaborations and projects have the positive outcome of innovative production and economic growth.
- 'Smart City' programs can propagate smart city practices and advance sustainable futures.

The role of culture for smart city innovation

Social infrastructure is as important as intelligent infrastructure (e.g. a broadband network) for a smart city. Surely social learning and education, and arts and culture can impact social capital, and the role of human capital and education on urban development has been studied (Nam & Pardo 2011: 287). Social infrastructure is about intellectual capital and social capital; it is about a humane city offering a variety of learning opportunities to its inhabitants to increase their knowledge and extend their horizons to live a creative life. Smart people can generate and can benefit from social capital (ibid.: 285). Creating different e-learning platforms, dashboards, e-government services, digital libraries and ICT-enabled networks as user-friendly tools for citizens to access learning opportunities about planning, development, design of cities and engage them in decision making are example attempts by local governments to improve sustainable communities (Allwinkle & Cruickshank 2013). After all smart cities require smart citizens because of the strong link between public engagement and learning and the valuable knowledge and wisdom citizens can bring in to solve the complex issues of a city (Bull & Azennoud 2016: 7, 9), and universities can accelerate the learning process and provide a context for innovative experiments and problem solving, having access to professional human capital.

Smart education and smart practices

IHS Markit has projected by 2025 there will be at least 88 smart cities in the world (Hean 2015). Cities are now effectively engaged in a proxy war or at the very least an invisible global competition to outsmart each other.

One solution to counter the last smart city obstacle is to expand the higher education sector in the city. Identified smart higher education strategy involves four aspects:

1   Distance education and applications of virtual learning (e-learning)
2   Exporting academic services abroad
3   Collaboration of academy with government and industry
4   Emergence of interdisciplinary and multidisciplinary fields of study and projects.

Whilst aspects of these phenomena are long-standing, the rapid evolution of telecommunication technologies has altered the strategic educational landscape. Some national and local governments have reinforced the smart initiative by employing educational technologies such as open data and open access.

Distance education and applications of virtual learning (e-learning)

Distance learning or e-learning has a long history. It has transformed to some degree; it is no longer merely a distribution platform for academic resources but an interactive platform which requires full engagement of the learners. Although most universities in advanced countries have supplied some form of learning management systems and websites for their academics to supplement the face-to-face courses that are taught, none of those courses are fully online in Canada (Patry et al. 2015: 2). Many open universities have started offering courses using e-learning platforms.

Private educational organizations and private companies have been actively engaging with universities to set up websites with free educational materials and courses such as MOOC which stands for Massive Open Online Courses (MOOC 2016). MOOC has been offering open access courses at a large scale to all individuals who can access the Internet. Participants can learn different topics and get a certificate of attendance if they wish but that certificate is not a degree. They may need to pay a small fee for their certificates to be sent to them by the course provider and not the MOOC. Most course providers are universities and academic institutions.

Other examples of widespread open access online courses are Open University and platforms like FutureLearn which offer both courses and programs (FutureLearn 2016a, 2016b). These courses are often self-paced but not self-directed and need participants to interact with the program, with a coordinator and virtual classmates by commenting and sending feedback and submitting assignments. There are, however, online materials, tutorials and training that users can watch without any assessment involved and those could be found on YouTube, Vimeo and other search engines and platforms. A slightly different and limited version which recently has been recommended to use as a professional platform is the Lynda.com from LinkedIn professional social media (Lynda.com 2016). Lynda provides technical training videos to learn skills but it is not open access and it is not totally free. Academic institutions or individuals can subscribe by paying an annual fee to enable their academic staff and students to learn some skills via tutorial videos for free. It is also free to those LinkedIn members who are subscribed to the premium version of LinkedIn. Andergassen et al. (2015) describe an evolution in the e-learning platforms such as Learn@WU which has been happening and has shifted most websites and platforms from content to activity-based learning since the learning assessment is now an important part of the learning process.

Exporting academic services abroad

The common practice for students has been to continue their education in the best universities, often abroad, that is students traveled to another country to study in a known high ranked university to pursue their goals and change their future. The globalization of higher education as a consequence of globalization led to international collaborations in higher education; the growth of international branch campuses was evident. There are suggestions that branching out universities abroad is not a global business (Healey 2015); however, the financial potential of an available market still seems to be among the top motives. Some developed countries in the Far East and the Middle East established links with best western universities and, as a result, new types of academic services have emerged. For example about 14 unaccredited and 32 accredited American universities have been established in parts of Europe, the Middle East and Asia which can be found in the list of 'American Colleges and Universities Abroad' (Wikipedia 2016b). C-BERT reports that, by August 2015, 229 international branch campuses were actively working around the world. The USA and UK are the biggest exporters of education (Healey 2015). American and British universities have expanded their territory and offered their educational services to other countries. These universities have their own permanent academic members of staff staying in the host country and they follow the rules and principles of the central branch of the high ranked university. This can be revolutionary in terms of the possible environmental and economic impacts it can have on students, the host country and the exporting country. Here are some of the benefits:

1   Students can stay in their own country and even city and still have the support of their family.
2   They can spend less money on international trips and flights which contributes to less carbon emission and that benefits the environment.
3   By staying in their own country and paying their tuition there, they financially prevent capital loss.
4   They still get their degrees from internationally known universities with the latest updates and quality lectures which enable them to work in an international market.
5   Host countries can prevent the challenge of the brain drain (emigration of highly educated individuals) and losing their future intelligent human capital which has monetary consequences.
6   The exporter countries can prevent all complications of foreign students coming to their country from the visa process, financial issues, housing affordability, cultural compatibility and cultural planning and implementation costs.

The aforementioned factors can explain the reasons behind the evolution of such common practice in many developing countries, that is instead of students leaving their countries to reside abroad to continue their education in a well-known university, a branch of that university establishes in their country. Surely, governments of the host countries need to be flexible and compatible with the global teaching and learning system and practices and allow the branching model to happen there to prevent human capital flight. There are countries such as Iran which are not open to foreign institutes to provide educational services on a national scale. In 2009 the International Monetary Fund reported that Iran topped the list of countries losing their academic elites: the annual loss of 150,000 to 180,000 specialists equivalent to a capital loss of $50 billion (Wikipedia 2016a).

A more recent version is the outsourcing of academic services such as teaching programs, courses, modules or even research for a short course of time to improve the quality of the programs and promote them to attract higher number of students.

Collaboration of academic institutions with government and industry

While the universal trend of universities' collaboration with industries, communities and governments to collectively transform sustainably has been identified, the missions of universities are still limited to the economic focus rather than educators of sustainable development values (Trencher et al. 2014). One direct result of encouraging start-ups and innovation incubators by governmental bodies is the collaboration between industry, university and even SMEs. This is important as most developing and underdeveloped cities hugely suffer from this separation. Private companies have taken the lead to join different universities with the government bodies and even start-ups to see a real innovative transformation in industry. The business side of being the pioneer has not been neglected; in other words, from the entrepreneurial perspective it is a positive action to help cities develop high quality consulting services, create more job opportunities and produce innovative products designed to solve the problems of our present cities. The scale of such collaborations varies. There are examples of large organizations working on international city projects like Smart City Council, and on national projects such as Future Cities Catapult. In some cases, smaller companies partner up with one another or with a university to work on a particular initiative or idea. Here are some examples.

Emergence of interdisciplinary, multidisciplinary and transdisciplinary fields of study and projects

Debates around the future of higher education and the potential approaches to innovative teaching and learning are fairly new, exemplified in a series of conferences such as 'Reimagining the University' (2014) and 'The Role of Humanities,

---

CASE STUDY **LARGE COMPANIES WORKING ON NATIONAL-SCALE PROJECTS**

Future Cities Catapult, which is a company limited by guarantee, works with 'Intel Collaborative Research Institute (includes Intel Laboratories, UCL and Imperial College), The Royal Parks, the London Borough of Enfield, ScienceScope and City Insights'. They have formed five living laboratories across London at Brixton, Enfield, Elephant and Castle, Hyde Park and Tower Bridge to collect and use real-time city data by employing sensors to measure a range of physical parameters such as human activity and air quality. Furthermore, they have established an Urban Innovation Centre as a 'collaborative hub for businesses, academics, city leaders and entrepreneurs to connect, develop and create smart city solutions' (Future Cities Catapult 2016).

CASE STUDY **SMALL COMPANIES OR INDIVIDUALS LINKED TO A UNIVERSITY**

This project was developed by Carlo Ratti around 2009 while he was working in MIT SENSEable City Lab but now is a bigger project called 'The Copenhagen Wheel'. It is a smart electric wheel with the ability of being attached to any bike. It has an app with a social network to be installed on one's phone. It contains a motor, multiple sensors, batteries, wireless connectivity and a control system. It can save energy while gearing and provide it when it is needed to get up a hill. It can collect data about a trip, the surroundings, physical activities of the user and the calories burnt. It can notify the user if he or she passes or will pass a friend or a colleague (cited in Superpedestrian 2016).

Daan Roosegaarde, a Dutch designer at the State University of New York, could produce a prototype of a little plant which glows when it is dark. He plans to combine luciferin with the DNA of trees. Fire flies and jelly fishes have a compound called 'luciferin' which can light up. Some think that this could be the future of street lights (Hyde 2014), meaning to grow street lights which are absolutely sustainable instead of having lamp-posts and trees separately. While some may be excited about these types of ideas some may consider them futuristic or simplistic. Scott sounds the alarm about simplistic recipes for a 'steady march . . . towards some sort of creative utopia', in that bohemian enclaves are undermined by the reality of 'extensive numbers of insecure low wage jobs' and social marginalization (Scott 2006: 12, cited in Huston, Wadley & Fitzpatrick 2015: 317).

Arts and Transdisciplinary Practice in Higher Education' (2015) organized by the Alanus University (Germany), Crossfield Institute (UK) and the University of Gloucestershire (UK). In the latter, Isis Brook emphasized the importance of distinguishing the idea of 'transdisciplinary' from inter- and cross-disciplinary (as cited in Anderson 2015). The conference called on educators, practitioners and researchers to evaluate the status of universities and teaching–learning methods across different disciplines. That could be a sign of a predicted transformation in higher education in the near future.

Applying transdisciplinary approaches to research is one of the main distinctive factors of smart education and smart development; as it engages stakeholders and subsequently brings multiple expertise and perspectives to a project (Bracken et al. 2015). In this section, the main focus is on interdisciplinary, multidisciplinary and transdisciplinary education and not cross-disciplinary. Definitions of the afore-mentioned terms can be found in this chapter's glossary.

Since sustainable and resilient city concepts have emerged, concerns of protecting the planet have been raised. Many universities have created curriculums to answer to this novel need and prepare our new generation with the knowledge and skills they require to protect the environment and ensure their future. Most of the new interdisciplinary courses that have been added to university curriculums are related to sustainability ranging from sustainable development in the natural and built environment to the psychology of sustainable development and sustainability management.

Creating resilient communities is another mission of a smart sustainable city. Noting the importance of the health, well-being and creativity of people has been one of the key drivers in smart communities. As a result, many new interdisciplinary majors have been defined which have merged art, science and humanities, the combination of arts with psychology leading to new majors such as 'art therapy,' 'theater therapy,' 'music therapy' and 'dance therapy' or environment and psychology resulting in environmental psychology (Minaei & Sattaripour 2015). Some of the new interdisciplinary projects are quite important and have quality outcomes. For example a course called 'medicine, science and Humanities' launched by Johns Hopkins University's Krieger School of Arts and Sciences examines medical and scientific issues but through the lens of humanities studies and aims to promote a functional connection between humanistic cultures and science (HUB 2015). Another example is the 'MA degree in Interdisciplinary Studies' which was launched by the New College of Interdisciplinary Arts and Sciences at the Arizona State University. This program allows students to be innovative and select two or more disciplines and combine them to solve problems with multidisciplinary origins (ASU 2015). There are several examples of various disciplines such as sustainable approaches, community planning, public participatory planning, evidence based design, human based design, ergonomics, environmental psychology, evidence based therapies and art therapies (Minaei & Sattaripour 2015). Some of their discoveries have been life-changing. For instance, scientists have observed that painting proves helpful in depression and improves self-esteem. It is very constructive to children with behavior problems because they are generally unable to communicate their emotions (Dev 2015). Another finding proves that music therapy and singing can be a solution for communication problems of those patients who have experienced shock, stroke or have been diagnosed with PTSD (Post Traumatic Stress Disorder) and lost their ability to talk; they can sing instead and communicate since singing and talking are managed by two different lobes of a human brain (Grimaldo 2015). These are only a few examples of the potential benefits interdisciplinary and multidisciplinary courses can offer and the types of problems they can solve.

Universities and the emergence of 'smart' related disciplines

The main reason for creation of the smart city concept was to ensure that sustainability goals are pursed by governments and cities. Pioneer universities in advanced cities across the world have defined courses and modules in smart cities. See Table 10.1 for information about some of the existing courses, the host institute and the themes that are covered.

TABLE 10.1 Benchmarking smart city courses, programs and modules

University and program or course title	Host department and location	Content
UCL *MSc Smart Cities and Urban Analytics* **(TBCASA, 2016)**	The Bartlett Centre for Advanced Spatial Analysis Multidisciplinary Faculty of the Built Environment London, UK	– City theory – Quantitative methods – Geographic Information Systems and science – Optional module, 15 credits – Smart cities: context, policy and government – Spatial data capture, storage & analysis – Urban simulation
The Open University *Smart Cities* **by Lecturers from University of Birmingham (FutureLearn, 2016b)**	6-week online course for all MK: Smart and Higher Education Funding Council for England (HEFCE) England, UK	– Issues facing smart cities – The role of systems thinking – Living labs, Open data – Crowdsourcing and roadmaps – Challenges of privacy, ethics and security – The value of leadership, standards and metrics – How to co-create a smart city project where you live
Harvard University *Smart Cities: An Introduction to Urban Integrated Networked Solutions* **(HUGSD, 2016)**	Graduate School of Design Department of Urban Planning and Design Cambridge, MA, USA	– A literature review of smart cities – An analytical case study of proposed or practiced smart city solutions – A rigorous cataloguing of urban problems that can be addressed by smart city-inspired solutions – A hands-on approach towards envisioning, proposing, designing, developing and prototypical implementation of ITC-driven, networked and integrated solutions to the catalogued urban problems
MIT *Beyond Smart Cities* **(MITPE, 2016)**	Short program (3 days) Professional Education course for industry and government applicants Cambridge, MA, USA	– Resilient urban cells – New mobility systems – Resilient energy systems – Living space on demand – Shared co-working facilities – Urban food production – Responsive technologies – Trust networks
Arizona State University *Smart Cities*	One semester course Arizona, USA	– *A walk through the smart city* – Turing's legacy – The wired city: computable city:

TABLE 10.1 Continued

University and program or course title	Host department and location	Content
(SpatialComplexity, 2016)		information infrastructure – The smart city as a communications mechanism – Transport and transit: smart systems and Big Data – *Online public networks: communications data* – Big Data and urban information systems – The participatory city: crowd-sourcing, city dash boards, social media – Modeling the city: GIS, 3D and virtual reality representations – Urban simulation and prediction
IEEE Advancing Technology for Humanity *Introduction to Metrics for Smart Cities* **(edx, 2016)**	4-week online course (Professional Education) for students of urbanism, architecture and information technology MIT, MA, USA	– The definition and key elements of a smart city – How to model complex systems of smart cities – How to understand and develop methods to quantify the effectiveness of smart city systems
University of Twente *Smart Cities* **(IGS, 2016)**	Summer School course for students in their third Bachelor's year, entry level Master's students and professionals with an interest in smart cities Enschede, the Netherlands	– Urban metabolism in resilient cities – Future energy: smart grids – Smart ways to make smart cities smarter – Sensing smart cities – Geo-processing and analytics for smart cities – Social innovation – Smart city as inter-professional challenge – Smart rules for smart cities – Governing smart cities
InnoEnergy **MSc** *Energy for Smart Cities* **(InnoEnergy, 2016)**	2-year FT MSc for BSc students in mechanical or electrical engineering, or students with a related science degree Eindhoven, the Netherlands	– Power systems – Power electronics – Nuclear energy: basic aspects – Combustion engines – Numerical methods in energy sciences – Electrical drives: energy efficiency of electrical machines – Power system calculations – Regulatory affairs

TABLE 10.1 Continued

University and program or course title	Host department and location	Content
		- Energy economics - Thermal systems and energy management
IHS + TSCPL *Smart Cities* **Institute for Housing and Urban Development Studies (IHS) and Erasmus University Rotterdam, Total Synergy Consulting (TSCPL) (CDIA, 2015)**	2-week Professional Education course for government officials, policy makers, administrators and practitioners and professionals New Delhi, India and Rotterdam, the Netherlands	- Definition of smart cities which is then applied to solve emerging challenges of rapid urbanization - Various components of smart cities - Identification and application of a systematic, strategic planning approach for smart cities design and implementation - Role of various stakeholders for the development of smart cities - Identification of available public–private partnerships and building partnerships as mechanisms for mobilizing private finance
Singapore Polytechnic *Digital Technologies for a Smart City* **(SP, 2016)**	A Specialist Diploma for 285 hours (1 year) for engineers and technical support officers from electronic engineering, computer engineering and information technology Singapore, Singapore	- Smart systems and cloud computing - Sensors and communication technologies - Smart city system design - Cloud computing & analytics - Mobile application development
University of Windsor, Faculty of Engineering Sustainable Smart Cities (uwindsor.ca, 2017)	A course for Master's students of three Engineering programs of Civil, Environmental and Mechanical Engineering Windsor, Ontario, Canada	- Sustainable cities versus resilient cities - Urban metabolism in urban cells - Smart city definitions, key elements and case studies - Sustainable and smart transport - Urban food agriculture - Sustainable and smart waste management - Resilient Energy Systems and STEEP - Information infrastructure, trust and security - Responsive technologies, sensors and communication technologies - Social innovation, crowdsourcing and the participatory city

Source: The author.

TABLE 10.2 Different sensing modes and their applications

Sensing type	Common uses
Strain and pressure	Floors, doors, beds, sofas, scales
Position, direction, distance and motion	Security, locator, tracking, falls detection
Light, radiation and temperature	Security, location, tracking, health safety,
Solids, liquids and gases	energy efficiency
	Security and health, monitoring, pool
	maintenance, sprinkler efficiency
iButton	Used to identify people and objects
Sound	Security, volume control, speech recognition
Image	Security, identification, context understanding

Source: Cook et al. (2009: 281).

A new interdisciplinary field of study in the smart age is 'Ambient Intelligence' which links environmental design ranging from architectural design, interior design and mechanical engineering to 'Ubiquitous Computing' or 'pervasive computing.'

Ambient Intelligence (AI) as an emerging field in computer science aims to equip our environments with intelligence to make them responsive, sensitive, adaptive, ubiquitous and transparent to humans and human interaction to support their daily life; AI technologies use sensor technologies and sensor networks relying on sensory data from the environment. Sensors can vary based on their functionality and type. See Table 10.2 for their variety (Cook et al. 2009: 277–280).

O'Grady and O'Hare (2012) bring three good examples of the effects of ambient intelligence on our quality of life. They explain that art galleries and museums can use a digital information space to enable personalization so the needs of visitors are met. Visitors could be guided via their smart phones to their favorite exhibit and have a satisfactory experience rather than having to visit all exhibits to discover the ones they are interested in.

Another example is the application of GPS on smart phones or wrist watches for people who suffer from dementia. The GPS tracker can document patterns and model patients' behaviors from which they can activate alarms on the patient's smart phone or that of a family member. These two examples clearly illustrate the way computer science is combined with our daily life and the methods in which it can save us and governments the costs of medical and social services.

University, smart university and the resulting changes to cities

The emerging mission and the fundamental change in social function and role of universities are clear. Universities are no longer providing a simple contribution to the economy but instead they lead a culture of transformation and co-creation

(Trencher et al. 2014: 169). They can be leaders in achieving sustainable development by employing diverse research and incorporating governments, academics and various stakeholders and decision makers.

Economic status and some degree of affordability can affect the health and well-being of individuals, particularly the younger generation. Smart technologies have become the driving force in teenagers' success. Studies showed having a laptop could make an unexpected difference in teenagers' life including an increase in their grades, college applications and job applications and a decrease in missed school days. In addition, psychologically speaking, their self-esteem was improved and depression and suicidal thoughts were diminished (Compassionate Cities 2016). This is a considerable impact that having access just to one technology can have on a teenager's success and quality of life. This is the reason that academics such as Suzanne Mettler are concerned with social inequality in higher education and specifically the university crisis in the USA (cited in Christopherson et al. 2014: 209). Academic institutions like the European Universities Association (EUA) mind the matter and continue to support universities as economic and social drivers of a society (cited in Christopherson et al. 2014). In Canada, each province has its own way of supporting students. For instance, Ontario Student Assistance Program (OSAP) aims to support students with financial needs to pay their tuition (OSAP 2016). HEFCE is another good example of such support in UK universities although most universities in the UK have become more of a private sector nowadays (Christopherson et al. 2014: 212). Many countries have a tuition-free system for their citizens in higher education such as Denmark, Sweden and Iran; Hotson states countries like Germany have recently provided tuition-free education for all domestic students (cited in Christopherson et al. 2014: 210) while even 15 years ago there were many German universities offering tuition-free courses to foreign students as long as those students were able to learn in German; even that requirement has changed with the emergence of globalized higher education and many international courses are offered in English nowadays.

Universities are the driving force of this change by employing five channels defined by Trencher et al. (2014: 160) to carry out their new missions that include:

1 Knowledge management
2 Technical demonstration projects and experiments
3 Technology transfer and economic development
4 Reform of built and natural environment
5 Socio-technical experiments.

There are some indicators to measure the 'smartness' of a learning ecosystem such as a university, listed by ASLERD, including: infrastructure, food services, environment, info/admin services, mobility, safety, support to social interactions, satisfaction, challenge and self-fulfillment. These indicators have been applied to the Aveiro University in Portugal to measure its smartness (Galego et al. 2016: 147–149).

The 'Smart University-Ambient Intelligence Project' by Fontys University of Applied Sciences has applied sensing and incorporation of new types of technology such as face recognition, motion detection and so on, to allow a person to measure workload and efficiency both in classrooms and in offices across the university right from his or her smart phone (Krüger 2013).

Some universities have already started integrating smart technologies and smart education in their curriculum and educational system. Spaces such as innovation centers and districts, living labs and information marketplaces are all pointing to the fact that smart technologies are gradually changing the composition of spaces even in cities.

Some believe to enable cities to implement smart cities, living labs (urban environments as premises for real-world data, testing new ideas and becoming innovation areas) and innovation districts (small plots of land within a town or city composed of creative, artistic industries or start-up companies) need to collaborate. Apart from those living labs owned and managed by private firms, most of the living labs are in universities (Cosgrave et al. 2013: 671–672) since universities have professional human resources and technical support and they potentially can receive funding from their governments. The interesting point about the geographical location of innovation districts is that they form in parts of a town or city where rents are cheap and similar types of start-ups have occupied the space where good transport access is available and overall it is a good quality place to live (ibid.: 672). Preservation Green Lab (2014) reported that older and smaller buildings provide affordable flexible spaces for new businesses within a district rather than new big buildings; they measured character scores of several districts within American cities and revealed the important contribution of older smaller blocks to livability, diversity, robustness of the local economy and entrepreneurial activities. This reminds us of Florida's theory (2014) as he talks about the creative class, the places that attract smart talented people, and regeneration projects in areas with historic or cultural potential to grow and transform to innovation districts. In his terminology the creative class falls under the categories of engineering and design; technically university graduates and people with artistic or computer skills. The types of places that attract smart talented people are areas where like-minded people prefer to live. Florida talks about the new motors of economic development and most of that starts from university. Marlet and van Woerkens use Dutch data and conclude that Florida's creative class contributes more to explaining urban employment growth than to indicators of education because it matters what people actually do at work (2007, cited in Huston et al. 2015: 316). Florida (2014) says the number of jobs a place provides is not as important as the quality of those jobs, or the tolerance and welcoming culture of people who live in that place. The fact to consider is that some places are welcoming at a time but not always; to clarify, there are two conditions:

1 Some small towns are open to foreign university students because they see students as funding sources helping the local economy (paying rent and tuition,

transit and other living costs) but simultaneously these towns are not open to graduates to stay there and work there as conversely people see graduates as threats to their local economy who win jobs; thus, the welcoming culture may exist during studentship but not after graduation.

2   Some big cities are open to university students and prefer to keep them after graduation to continue with their principles and standards of working; they see no reason to invest in training new human resources, so those universities are the ones who retain talents in the place and affect the welcoming culture by providing job opportunities for their graduates.

Rethinking the concept of places with a welcoming culture may raise such questions as to what extent Brexit and its consequences can affect smart cities of the UK? Can programs such as 'Welcome to Toronto, We Have Been Expecting You' by the City of Toronto to educate people, promote the culture of welcoming and create memorable experiences for visitors (Toronto 2016) affect the smartness of Toronto?

Smart infrastructure such as Internet networks and the connectedness they bring to the world in a virtual dimension have provided spaces for creativity. These spaces are not necessarily dependent on a university or a government but merely rely on people to create and be creative. Kickstarter is one example of such a platform that provides funding for start-ups, artists, designers and creators and helps them bring their creative projects to life. It somehow is a living lab because it works with some social media such as Facebook and measures the possible demands of a creation by the number of times the link to that creation has been shared. They have other websites as partners for the promotion of the innovative prototypes; for example, Insider Design which broadcasts the news around technological innovations mentions where that project can be found such as Kickstarter.com; it also mentions whether that creation needs supportive funding to make a prototype. Therefore, interested investors or individuals can donate and raise money for the project. The number of likes and the comments one idea gets in social media is an indicator as to whether that invention will have a future market and is worth investing in.

Conclusion

The indicative link between smart cities and university ranks, and extensive econometric evidence suggest that universities are central to smart cities. As discussed, social capital is as important as intelligent infrastructure for a city. Studies have shown that a mutual direct association between a university and its urban context exist; both can impact one another and influence their success. Universities play multiple roles in local economic development and resilient sustainable futures. Inventing new fields of studies to approach a city's challenges by applying multi-disciplinary and transdisciplinary approaches, or using mixed-methodology and engaging stakeholders, collaborating with government and industry are a few examples of universities' missions in our era. From one side they attract students and capital to the city which helps the local economy (e.g. rental housing) and creates a diverse cultural environment; from the other side, due to the universal problem of unemployment, they train students to become future entrepreneurs by offering skilled courses in their innovation centers. Graduates can start as entrepreneurs and establish their start-ups in older and smaller buildings within the city which are affordable and livable and characterful. This can prevent dereliction and abandonment of older districts of a city and will promote adaptive reuse and regeneration in older and historical areas of a city which is a sustainable and a smart move for a city to attract talents and preserve historical spaces.

Glossary of the key terms

AI Ambient Intelligence.
ASLERD Association for Smart Learning Ecosystem and Regional Development.
C-BERT Cross Border Education Research Team.
Cross-disciplinary Viewing one discipline from the perspective of another (Jensenius 2012) and explaining its knowledge (Chandar 2014).
Interdisciplinary Combining knowledge and methods of two or more different disciplines, using a synthesis of approaches (Jensenius 2012) to address a common challenge (Chandar 2014).
Intradisciplinary Working within a single discipline (Jensenius 2012).
MOOC Massive Open Online Courses.
Multidisciplinary People from different disciplines working together, each drawing on their disciplinary knowledge.
OSAP Ontario Student Assistance Program.
SME Small and medium-sized enterprises.
Transdisciplinary Creating a unity of intellectual framework or a research strategy crosses many disciplinary boundaries with a holistic view (Jensenius 2012; Chandar 2014) which is linked to the industry and stakeholders and has nine distinguishing factors from other forms mentioned above (Brook 2015, cited in Anderson 2015).

Bibliography

Allwinkle, S. & Cruickshank, P. 2013. Creating Smart-er Cities: An Overview. In M. Deakin (Ed.), *Creating Smart-er Cities*. New York: Routledge.
Andergassen, M., Ernst, G., Guerra, V., Mödritscher, F., Moser, M., Neumann, G. & Renner, T. 2015. The Evolution of E-learning Platforms from Content to Activity Based Learning: The Case of Learn@WU. Interactive Collaborative Learning (ICL), 2015 International Conference on (pp. 779–784). IEEE.
Anderson, F. 2015. Review on the Role of Humanities, Arts and Transdisciplinary Practice in Higher Education. International Conference at Alanus University, Germany, 29–30 May. www.alanus.edu/fileadmin/downloads/fachbereiche_und_studienangebote/fb_bildungswissenschaft/Review_The_role_of_humanities.pdf
ASU. 2015. Arizona State University, Interdisciplinary Studies (MA). Phoenix, AZ https://newcollege.asu.edu/interdisciplinary-studies-ma

Bracken, L. J., Bulkeley, H. A. & Whiteman, G. 2015. Transdisciplinary Research: Understanding the Stakeholder Perspective. *Journal of Environmental Planning and Management* 58(7): 1291–1308, http://dx.doi.org/10.1080/09640568.2014.921596

Bull, R. & Azennoud, M. 2016. Smart Citizens for Smart Cities: Participating in the Future. *Proceedings of the Institution of Civil Engineers – Energy* 169(3): 93–101, http://dx.doi.org/10.1680/jener.15.00030

Chandar, S. 2014. Is There a Difference between Cross-disciplinary and Multi-disciplinary? *Quora*, 6 July, viewed 28 November 2016, www.quora.com/Is-there-a-difference-between-cross-disciplinary-and-inter-disciplinary

Christopherson, S., Gertler, M. & Gray, M. 2014. Universities in Crisis. *Cambridge Journal of Regions, Economy and Society* 7: 209–215, DOI: 10.1093/cjres/rsu006

Cities Development Initiatives for Asia (CDIA). 2015. Metro Manila, viewed September 2016, http://cdia.asia/2015/08/08/training-events-international-course-on-smart-cities-5-16-oct-2015/

Compassionate Cities. 2016. What a Difference a Laptop Can Make in a Foster Teen's Life. Smart Cities Council, viewed 14 September 2016, http://smartcitiescouncil.com/article/what-difference-laptop-can-make-foster-teens-life

Cook, D. J., Augusto, J. C. & Jakkula, V. R. 2009. Ambient Intelligence: Technologies, Applications and Opportunities. *Pervasive and Mobile Computing* 5: 277–298. doi:10.1016/j.pmcj.2009.04.001

Cosgrave, E., Arbuthnot, K. & Tryfonas, T. 2013. Living Labs, Innovation Districts and Information Marketplaces: A Systems Approach for Smart Cities. Conference on Systems Engineering Research (CSER 13) Eds.: C. J. J. Paredis, C. Bishop, D. Bodner, Georgia Institute of Technology, Atlanta, GA, 19–22 March.

Deakin, M. (Ed.). 2013. *Creating Smart-er Cities*. New York: Routledge.

Dev, D. 2015. Art Therapy – A New Help for Kids with Behavior Problems. *Health Aim*, 22 January, viewed 2015, www.healthaim.com/art-therapy-a-new-help-for-kids-with-behavior-problems/10828

EDX. 2016. IEEEx, viewed 20 November 2016, www.edx.org/course/introduction-metrics-smart-cities-ieeex-scmtx-1x

Florida, R. 2014. The Creative Class and Economic Development, *Economic Development Quarterly* 28(3): 196–205, DOI: 10.1177/0891242414541693

Future Cities Catapult. 2016. *Sensing London*. London, viewed 21 November 2016, http://futurecities.catapult.org.uk/project/sensing-london/

FutureLearn. 2016a. The Open University, England, viewed 25 November 2016, https://www.futurelearn.com/

FutureLearn. 2016b. The Open University, England, viewed 26 September 2016, https://www.futurelearn.com/courses/smart-cities

Galego, D., Giovannella, C. & Mealha, O. 2016. Determination of the Smartness of a University Campus: The Case Study of Aveiro. *Procedia, Social and Behavioral Sciences* 223: 147–152. DOI: 10.1016/J.SBSPRO.2016.05.336

Giovannella, C. 2014. Where's the Smartness of Learning in Smart Territories? *Interaction Design and Architecture(S) Journal – IxD&A* 22: 60–68.

Goddard, J., Coombes, M., Kempton, L. & Vallance, P. 2014. Universities as Anchor Institution in Cities in a Turbulent Funding Environment: Vulnerable Institutions and Vulnerable Places in England. *Cambridge Journal of Regions, Economy and Society* 7: 307–325.

Grimaldo, M. 2015. Lake Charles Woman Learns to Speak through Music Therapy after Suffering Stroke. *KPLC*, 29 January, viewed 2015, www.kplctv.com/story/27977949/lake-charles-woman-learns-to-speak-through-music-therapy-after-suffering-stroke

Harvard University Graduate School of Design. 2016. HUGSD, Cambridge, MA, viewed 20 November 2016, www.gsd.harvard.edu/course/smart-cities-an-introduction-to-urban-integrated-networked-solutions-fall-2012/

Healey, N. 2015. Universities that Set Up Branch Campuses in Other Countries Are Not Colonisers, *The Conversation*, viewed 25 November 2016, http://theconversation.com/universities-that-set-up-branch-campuses-in-other-countries-are-not-colonisers-46289

Hean, C. K. 2015. How We Design and Build a Smart City and Nation, TEDxSingapore, viewed 14 October 2016, www.youtube.com/watch?v=m45SshJqOP4

HUB. 2015. Johns Hopkins Adds New Interdisciplinary Major: Medicine, Science and Humanities. Johns Hopkins University, http://hub.jhu.edu/2015/01/22/major-medicine-science-humanities/

Huston, S., Wadley, D. & Fitzpatrick, R. 2015. Bohemianism and Urban Regeneration: A Structured Literature Review and Compte Rendu. *Space and Culture* 18(3): 311–323. DOI: 10.1177/1206331215579751

Hyde, R. 2014. The Sci-fi Future of Lamp-posts. *The Guardian*, 13 November, viewed October 12, 2016, www.theguardian.com/cities/2014/nov/13/sci-fi-future-lamp-posts-street-lighting

InnoEnergy: The European Company for Innovation. 2016. *Business Creation and Education in Sustainable Energy*, Eindhoven, viewed 20 November 2016, www.innoenergy.com/education/master-school/msc-energy-for-smart-cities/

Institute for Innovation and Governance Studies. 2016. IGS, University of Twente, Enschede, Netherlands, www.utwente.nl/igs/smartcities/

Jensenius, A. R. 2012. Disciplinarities: Intra, Cross, Multi, Inter, Trans. *ARJ, Alexander Refsum Jensenius*, Oslo, 12 March, viewed 26 November 2016, www.arj.no/2012/03/12/disciplinarities-2/

Krüger, S. 2013. *Smart University –Ambient Intelligence Project*, online video, 27 June, viewed November 2016, www.youtube.com/watch?v=ubRLwzv8CtI

Lynda.com, From LinkedIn. USA, viewed 25 November 2016, https://www.lynda.com/

Minaei, N. & Sattaripour, A. 2015. The Vital Role of Humanities, Arts and Design of Built Environments on Human Well-being and Health. International Conference on The Role of Humanities, Arts and Transdisciplinary Practice in Higher Education, Alanus University of Arts and Social Sciences, Germany, www.researchgate.net/publication/277278787_The_Vital_Role_of_Humanities_Arts_and_Design_of_Built_Environments_on_Human_Well-being_and_Health

MIT, Professional Education. 2016. MITPE, Cambridge, MA, viewed 20 November 2016, http://professional.mit.edu/programs/short-programs/beyond-smart-cities

MOOC, MOOC List, viewed 25 November 2016, www.mooc-list.com/

Nam, T. & Pardo, T. A. 2011. Conceptualizing Smart City with Dimensions of Technology, People, and Institutions. In *Proceedings of the 12th Annual International Digital Government Research Conference: Digital Government Innovation in Challenging Times* (pp. 282–291). ACM.

O'Grady, M. & O'Hare, G. 2012. How Smart Is Your City? *Science* 335(6076): 1581–1582, DOI: 10.1126/science.1217637

OSAP. 2016. Ontario Student Assistance Program, Ministry of Advanced Education and Skills Development, viewed 25 November 2016, https://osap.gov.on.ca/OSAPPortal/index.htm

Patry, A., Brown, E., C., Rousseau, R. & Caron, J. 2015. Evolution of the Instructional Design in a Series of Online Workshops, *Canadian Journal of Learning and Technology* 41(3): 1.

Preservation Green Lab. 2014. Older, Smaller, Better: Measuring How the Character of Buildings and Blocks Influences Urban Vitality. National Trust for Historic Preservation,

Preservation Leadership Forum, http://forum.savingplaces.org/connect/community-home/librarydocuments/viewdocument?DocumentKey=83ebde9b-8a23-458c-a70f-c66b46b6f714

Reimagining the University. 2014. Reimagining the University: New Approaches to Teaching and Learning in Higher Education, in online conference programme organized by Crossfield Institute, Alanus Hochschule and University of Gloucestershire, Cheltenham, UK, viewed 26 November 2016, www.academia.edu/8846996/Reimagining_the_University_New_Approaches_in_Teaching_and_Learning_in_Higher_Education

Singapore Polytechnic Academy. 2016. SP, Singapore, viewed September 2016, www.sp.edu.sg/wps/portal/vp-spws/pace.courses.catalogue.details?WCM_GLOBAL_CONTEXT=/lib-pace/internet/part-time+courses/specialist+diploma+in+digital+technologies+for+a+smart+city

SpatialComplexity. 2016. Michael Batty, Arizona State University, viewed 20 November 2016, www.spatialcomplexity.info/outline-of-the-lecture-course

Superpedestrian. 2016. The Copenhagen Wheel, viewed 21 November 2016, www.superpedestrian.com/

The Bartlett Centre for Advanced Spatial Analysis. 2016. TBCASA, UCL, London, viewed 20 November 2016, www.bartlett.ucl.ac.uk/casa/programmes/postgraduate/msc-smart-cities-and-urban-analytics

Toronto. 2016. Get Involved in the We've Been Expecting You Program. City of Toronto, viewed 29 November 2016, www1.toronto.ca/wps/portal/contentonly?vgnextoid=12111c8ab77da310VgnVCM10000071d60f89RCRD&vgnextchannel=fcb77ce2eb32e310VgnVCM10000071d60f89RCRD

Trencher, G., Yarime, M., McCormick, K. B., Doll, C. N. H. & Karines, S. B. 2014. Beyond the Third Mission: Exploring the Emerging University Function of Co-creation for Sustainability. *Science and Public Policy* 41: 151–179. DOI:10.1093/scipol/sct044

UWindsor. 2017. Winter Course Offerings, University of Windsor, Ontario, viewed 1 January 2017, www.uwindsor.ca/engineering/civil/114/2017-winter-graduate-course-offerings

Wikipedia. 2016a. Human Capital Flight from Iran, 20 November 2016, *Wikipedia,* viewed 29 November 2016, https://en.wikipedia.org/wiki/Human_capital_flight_from_Iran

Wikipedia. 2016b. List of American Colleges and Universities Abroad, 2 November 2016, *Wikipedia,* viewed 25 November 2016, https://en.wikipedia.org/wiki/List_of_American_colleges_and_universities_abroad

11

GREEN CORRIDORS AND WATERCOURSES

Matthew Axe

Abstract

Smart regeneration should incorporate some consideration of green spaces for Eco-system Services and human health benefits. Green corridors provide a connective network of variable habitats and perform various Ecosystem Services. Their multi-purpose nature creates design and implementation issues, necessitating a balance of local human and ecological needs to achieve long-term inclusive benefits. Chapter 11 investigates some of the critical issues associated with green corridors using a case study. Green corridor designs are getting smarter, as they incorporate more human requirements and Ecosystem Services than classical wildlife corridors. Local community engagement is essential. It is recommended that small-scale green corridors are included in developments, where they can form part of a landscape-scale vision, and should be provisioned as an opportunity, rather than from planning ordinances or ecological restoration projects.

Key terms

Green corridor, greenway, riparian corridor, Ecosystem Services, community engagement

Introduction

Planning officials face significant pressures to accommodate fast-growing urban populations (Owens & Cowell 2011; Alexander, Fealy & Mills 2016). Most modern planning regimes at least attempt notions of 'sustainability' or consider the environmental impacts from new developments. Notwithstanding these concerns, human, and especially economic, interests dominate utilitarian planning processes. In short, the planning process is essentially anthropogenic (Nilsson & Florgård 2009; Austin 2014). All too often, urban development leads to habitat destruction, with local extinctions and long-lasting effects (Soulé 1991; Czech, Krausman & Devers 2000; McKinney 2002). From an alternate bio-centric view, no matter how profitable or desperate the social need, some human activities or urban development projects simply should never go ahead. As well as negative site-specific repercussions, ill-considered or excessive development has diffuse impacts such as on climate or the pattern of water flows (Sukopp & Starfinger 1999; Kinzig & Grove 2001).

To mitigate some of the negative externalities, urban development can be conducted with a greater awareness of the human dependence on the natural infrastructure; that is by incorporating the provision of ecosystems functions that directly benefit humans into the development plan. These functions, such as net primary production of organic compounds, carbon cycling and water purification, are commonly described as Ecosystem Services (ES) (Millennium Ecosystem Assessment 2005). In addition to these biochemical functions, most people visiting green spaces gain benefits for their physical and psychological health (Matsuoka & Sullivan 2011; Tilt 2011).

With 'smart' landscape design, developed land incorporates green spaces for positive ES benefits. The green infrastructure outlined in Chapter 2 has both open spaces and a connective network of green corridors with variable habitats and various functions. Smart green corridors assist the movement of wildlife, provide a linear habitat of their own, and deliver other ES, but importantly also incorporate societal needs.

Design and implementation issues accompany green corridors because of their multi-purpose nature, and the variable scales on which they can be implemented. Smart urban regeneration objectives require pragmatic, responsive enterprises – balancing local human and ecological needs to secure long-term inclusive benefits. Here we examine the multiple objectives of green corridors in urban settings and the means to implement them.

Ecological and anthropological aspects of green corridors

Biodiversity

Without connections between each other plant and animal populations become isolated, and lose genetic diversity (Merriam 1991). The island biogeography model of a reduction in species numbers where habitat area is reduced has been applied to the terrestrial urban environment (Adams & Dove 1989). In Europe, recognition came about in the 1970s that strips of existing industrial/post-industrial land in urban areas afforded a beneficial movement corridor for wildlife, free from public interference (Kelcey 1975; Teagle 1978). These were inadvertent urban wildlife corridors, then chiefly of interest only to naturalists. However in the USA, Leedy (1978) made a step change by recognising the need to integrate wildlife management needs into urban planning. In particular, Leedy (1978) attempted to show how the requirements for wildlife movement corridors could be deliberately incorporated within new urban landscape designs.

Despite the good intentions of Leedy (1978) and others, the evidence for the necessity of conduit corridors for wildlife movement between isolated habitat patches in towns and cites has been vigorously challenged (Adams & Dove 1989; Dawson 1994). Many animal species do not need green corridors for dispersal through the urban environment, although barriers (bridges, roads and large sections of hard landscaping) do restrict some. Short corridors, such as lengths of hedgerow, with a suitable habitat for a group of species, can be just as effective, creating 'stepping stones' in an otherwise adverse landscape (Dawson 1994). Dispersal of higher plants has been shown to be relatively unaffected by the presence, or not, of green corridors (Dawson 1994; Angold et al. 2006). Methodological concerns with wildlife corridor studies arise because most compare wildlife movement data between two different landscapes either with, or without, corridors. This approach risks misinterpreting and confounding effects from other landscape factors. However the principle two bio-centric arguments for retaining wildlife corridors in an urban setting are that: (a) they maintain habitat connectivity where prior to development all habitats were interconnected (Adams & Dove 1989), and (b) they focus human attention on local habitat conservation needs, without causing any detrimental impacts to wildlife (Beier & Noss 1998). Specific habitats are important for dispersal routes of some mammals and the modern consensus on wildlife corridors is that they are best developed with a focal species in mind (Dawson 1994; Barker 1997; Beier & Noss 1998; Angold et al. 2006). This focal species should be tied into the local area ecological strategy.

By focusing on a target wildlife species a consideration of food sources and protection from predators will determine planting selection. The wildlife corridor is then best considered as a linear habitat. Despite the importance of native plants for natural habitats, exotic plants can also support wildlife (Goddard, Dougill & Benton 2010). Urban areas can never substitute for natural ecosystems, but with thoughtful design green spaces can contribute to biodiversity conservation (Kowarik 2011). There may be a strong case for accepting novel existing urban ecosystems containing mixtures of alien and native species, since these have already adapted to the urban environmental conditions (Kowarik 2011). Such adaptive transformations that help deliver a higher objective is an essence of a smart urban design.

Climate, carbon and water regulating eco-services

Urban areas contain dark and impervious surfaces, when compared to vegetated areas. Consequently raised air and surface temperatures produce a well-known phenomenon, the urban heat island (Grimmond 2007). Compared to neighbouring rural areas, research in the USA has shown increases across all seasons in daily minima and mean temperatures, even for towns with populations of less than 10,000 (Karl, Diaz & Kukla 1988). Such raised temperatures present a risk to human health

(Patz *et al.* 2005). Plants provide cooling effects by evapotranspiration (the evaporation of water through leaf pores) and by shading, particularly from shrubs and trees. Green spaces in urban areas can thus provide cooling effects to potentially help cities adapt to climate change (Gill *et al.* 2007), and green corridors provide particularly good opportunities for this, since tree and shrub canopy layers frequently form along their boundary with adjoining land.

Indeed shrub or tree layers have further benefits for regulating atmospheric carbon, the most prevalent greenhouse gas (GHG). Carbon is both fixed into plants by photosynthesis and emitted by respiration; the decay of plant residues on or under the soil surface returns further carbon to the atmosphere (Falkowski *et al.* 2000; Ciais *et al.* 2013). In managed systems the use of fossil fuel for maintenance activities needs also to be taken into account. The net effect of this determines if the land is a carbon *sink* or *source*. Green corridors have good potential to be carbon sinks, but few empirical studies on the plant and soil organic carbon (SOC) stocks in urban settings have been carried out.

Research conducted in urban Chicago found the carbon emitted by lawn maintenance was sufficient to make grass areas alone carbon sources and not sinks, but when the presence of the woody shrub layer was also taken into account the areas became net carbon sinks (Jo & McPherson 1995). Lengths of narrow woodland may form part of green corridors. Shrubs and trees can also be introduced as individuals or small groups, increasing the carbon stock, while providing aesthetic or habitat benefits. Developing hedges or linear rows of trees present interesting opportunities for carbon sequestration in green corridors. Carbon stocks of agricultural hedges have only recently been quantified (Axe, Grange & Conway 2017). In urban settings hedges may be present as a remnant agricultural or an ornamental feature; however they can also be newly introduced for habitat creation (see Case Study). Unlike trees they are typically a more managed feature, usually being trimmed regularly for aesthetic reasons, and allowing more light into an area; but this does not degrade their capacity to provide a greater carbon store than herbaceous plants. The varying management activities on trees, shrubs and hedges can make a difference to their carbon sequestration rate and carbon stocks, and so these activities need to be considered in the complete system. For example, application of mulch, compared to grass or bare soil, increased density of root biomass on urban trees (Watson 1988).

Artificial hard surfaces in the urban environment alter the movement of surface water from that established over time in the natural environment, so that unaltered watercourses consequently cannot cope with high rainfall events in urban areas, leading to flooding (Austin 2014). The risk of flooding restricts building development near watercourses, so there are frequent occurrences of wetland or riparian corridors in urban areas. Vegetation along riparian corridors can significantly reduce surface run-off in urban catchments, particularly from low-intensity, short duration rainfall events (Ellis 2013). In the USA, riparian vegetated buffer zones are protected by planning controls, which allow them to be incorporated into restoration works or development plans as green corridors (Austin 2014).

Human use

Urban green spaces are valuable for human physical and physiological health, as well as for wildlife (Matsuoka & Sullivan 2011; Tilt 2011). Few studies have examined physical health benefits of green corridors specifically, but the presence of walkable green spaces in urban areas has been found to increase the longevity of senior citizens, independent of their socio-economic status (Takano, Nakamura & Watanabe 2002). The length of green corridors can have a positive effect on exercise regimes; a walk or run that makes use of linear routes is more likely to be repeated regularly than one requiring repetitive circuits of a park (Austin 2014). As well as providing a route for exercise, green corridors can provide access for pedestrians or cyclists to larger green spaces for recreation, separated from noisy and potentially dangerous vehicular routes. Ready access to nature for residents also has psychological benefits for improving concentration (Kuo & Sullivan 2001a), and a sense of relaxation or well-being (Day 2008; Kaplin 2001). A green corridor-specific study in Berlin found regular users of a canal towpath corridor had significantly lower cortisol levels, combined with higher life satisfaction, than less frequent users (Honold et al. 2016). Humans benefit from social and community interaction with each other and with nature. Green corridors in the local neighbourhood are also advantageous by offering opportunities for environmental education and involvement in conservation (Bryant 2006); this may be informal and spontaneous, or controlled through local conservation groups.

It is important to appreciate that vegetation in urban green space can elicit the fear of attack by criminals hiding in undergrowth; but despite these preconceptions, greener urban surroundings have reportedly fewer crimes (Kuo & Sullivan 2001b), and greenways are relatively safe places (Flink, Olka & Searns 2001). The individual's sense of threat can be alleviated by opening views along paths; Harrison et al. (1995) specified a minimum 4-metre width of path, with graded vegetation at edges to reduce fear from potential entrapment or concealed attackers. A further threat is that posed by scavenging animals; although less prevalent in Europe, corridors in North America potentially can increase bear and human conflicts.

Scale and implementation

Wildlife corridors, and river catchments that regulate surface water flows operate on landscape scales (Selman 2006), and it is on this scale that they are usually identified. The classical wildlife-specific corridors advocated by Leedy (1978) need now to be more nuanced, with a convergence to other ecological and human requirements as identified in this chapter. Smart green corridors should, to varying degrees, include a target species for habitat improvement, maximise on the woody vegetation for multiple ES, enhance the vegetation near the edge of watercourses, foster local community interest in the corridor and promote the corridor for non-motorised human transport. In addition these smart green corridors stand not in isolation but form part of a landscape-scale green infrastructure network.

In practice development professionals are usually faced with design or restoration plans on a land-parcel scale (Austin 2014). This can make identification of suitable green corridors difficult. Further more the economic motives for development tend to dissuade developers from designing in social/environmental benefits that do not generate profit at the point of sale (Adams 2012); from this viewpoint a green corridor would be seen as a loss of further development land. Without an obvious return on investment it is difficult to engage with developers on green infrastructure projects (Merk *et al.* 2012); hence, unless the developer has an altruistic motive, green corridors are likely to be instigated via a public finance or planning instrument. Thus green corridors are typically both identified and implemented via government planning controls, river catchment plans or other landscape-scale public regeneration projects. However this need not be exclusive, and given the prevailing availability of modern detailed geographical and biological information, an ecological consultant at the site scale should be able to identify opportunities where a green corridor might be applicable to a development.

The implementation of a landscape-scale green corridor project in Birmingham, UK, is given as a case study for comparison with smart urban regeneration concepts.

Analysis of case study and conclusion

In the case study there were individual undertakings frequently on small parcels of land, but these were coordinated as a landscape-scale improvement to urban green and wetland corridors. Taken together they should have a sizeable impact on biodiversity. A target species for habitat improvements was not always explicitly given, and it would seem a more general approach was taken, since hedgerows, for example, would improve the habitat for several taxa; where endangered species were identified, such as for the Wandering Water Vole project, then clear aims for the habitat were set. Woody vegetation did not seem to have been identified for ES benefits other than habitat but, in contrast, improvements to vegetation and banks near the edge of watercourses were clearly carried out to improve water quality as well as habitat. Perhaps the most interesting aspect was the local community engagement, mainly via the partnership of a diverse range of NGOs. There was no promotion of the corridors for walking or cycling per se; however, because of the activity invested in some of the canal corridors, there was a follow-on improvement programme of the towpath surfaces for cyclists (Birmingham City Council 2016).

Clearly the case study was a public, rather than private, investment for improving mainly the biodiversity aspects of a green corridor network in a major urban conurbation. On the landscape scale, and short temporal scale, it delivered its objectives amply. Many of these objectives met the smart urban regeneration needs. A key feature was that links were forged between residents, nature conservation groups and the local green corridors. When residents invest time and interest into green space, it is less likely to become neglected and subject to criminal activity. Another positive feature was that improvements to the canal corridor for a cycle route then followed the initial project, showing that momentum

CASE STUDY **BIRMINGHAM NATURE IMPROVEMENT AREA**

The Birmingham & Black Country Nature Improvement Area (NIA) covered Birmingham, Dudley, Sandwell, Walsall and Wolverhampton urban conurbation, with a population of over 2 million people. It was the only urban NIA established in a UK government-instigated competitive funding process. Funding ran from 1 April 2012 to 31 March 2015. Objective setting and delivery was achieved by a partnership of existing local agencies, as a deliberate step change from site-focused to landscape-scale nature conservation (BARS 2016a).

In the project planning phase, both green and wetland corridors were identified as one target theme. Canals, rivers and streams extending throughout the conurbation were recognised as often providing the primary opportunity for wildlife to enter and cross the urban space. Wetland corridors often had an associated green corridor. Larger-scale corridors were also identified from remnant agricultural land and reclaimed post-industrial sites. Many of these corridors were considered to have been disrupted by developed land and/or were of degraded habitat value (BARS 2016b).

The corridors delivery theme was established to enhance ecological links and engage public involvement. There were 124 undertakings in this theme alone, of these, 44 restored or created new hedgerow corridors, 42 improved floodplain corridors and 7 naturalised watercourses. These were achieved through the engagement of individual volunteers and local groups. Two notable examples of these works were the Birmingham and Fazeley Canal towpath enhancement, where 940 m of canal towpath were planted with orchard trees and pollinator-friendly plants (BARS 2016c, 2016d); and the Wandering Water Vole project which increased the habitat available to the three remaining populations of water voles along the canal network – the bioengineering techniques employed protected banks, while creating habitats for several species and improving water quality (BARS 2016e).

on green infrastructure can be can be maintained once public interest is raised. However it is worth noting the absence of private investment, especially considering the social benefits from improved green spaces would likely raise neighbourhood real estate values. A developer-instigated smart green corridor project, without public finance or planning ordinances, still seems likely to be a rarity – but relevant inclusion of a multifaceted urban green corridor, integrating wildlife management, other ecosystem services and local socio-economic needs, should be an aspiration of any smart urban development.

In conclusion green corridor designs are getting smarter, by incorporating more human requirements and ES than classical wildlife corridors. They are an essential part of any urban green infrastructure, and need local community engagement to succeed.

Recommendations

Small developments should include smart green corridors, with or without planning ordinances, when they can form part of a wider green network. For wildlife, this does not mean a direct connection to other green space is required, since a chain of habitats can create a stepping-stone effect. When developing these habitats, it is advisable to focus on a target species that has local relevance.

Corridors also provide other ES, particularly by including trees, shrubs and hedges. Ideally, green corridors should improve walkable links within cities, but a relatively small corridor is sufficient to meet the human need for green space. Fostering local community interest in the corridor through environmental education should help prevent neglect of the land and reduce crime.

References

Adams, D. (2012). *Urban planning and the development process.* Abingdon: Routledge.

Adams, L. W. & Dove, L. E. (1989). *Wildlife reserves and corridors in the urban environment: a guide to ecological landscape planning and resource conservation.* Columbia, MD: National Institute for Urban Wildlife.

Alexander, P. J., Fealy, R., & Mills, G. M. (2016). Simulating the impact of urban development pathways on the local climate: a scenario-based analysis in the greater Dublin region, Ireland. *Landscape and Urban Planning, 152,* 72–89.

Angold, P. G., Sadler, J. P., Hill, M. O., Pullin, A., Rushton, S., Austin, K., *et al.* (2006). Biodiversity in urban habitat patches. *Science of the Total Environment, 360,* 196–204.

Austin, G. (2014). *Green infrastructure for landscape planning: integrating human and natural systems.* Abingdon: Routledge.

Axe, M. S., Grange, I. G., & Conway, J. S. (2017, article under review). Carbon storage in hedge biomass – a case study of actively managed hedges in England. *Agriculture Ecosystems and Environment.*

Barker, G. (1997). *A framework for the future: green networks with multiple uses in and around towns and cities.* Peterborough: English Nature.

BARS (2016a). *Birmingham & Black Country nature improvement area projects.* Retrieved from: http://ukbars.defra.gov.uk/project/show/36537

BARS (2016b). *Corridors – improving quality, linkage & bridging gaps.* Retrieved from: http://ukbars.defra.gov.uk/project/show/36565

BARS (2016c). *Birmingham and Fazeley Canal – Towpath enhancement (TP1).* Retrieved from: http://ukbars.defra.gov.uk/action/show/243317

BARS (2016d). *Birmingham and Fazeley Canal – Towpath enhancement (TP2).* Retrieved from: http://ukbars.defra.gov.uk/action/show/112749

BARS (2016e). *Wandering Water Voles project.* Retrieved from: http://ukbars.defra.gov.uk/project/show/41359

Beier, P. & Noss, R. F. (1998). Do habitat corridors provide connectivity? *Conservation Biology, 12*(6), 1241–1252.

Birmingham City Council (2016). *Birmingham cycle revolution.* Retrieved from: www.birmingham.gov.uk/cs/Satellite/BirminghamCycleRevolution?packedargs=website%3D4&rendermode=live

Bryant, M. M. (2006). Urban landscape conservation and the role of ecological greenways at local and metropolitan scales. *Landscape and Urban Planning, 76*(1), 23–44.

Ciais, P., Sabine, C., Bala, G., Bopp, L., Brovkin, V., Canadell, J., *et al.* (2013). Carbon and other biogeochemical cycles. In Stocker, T. F., Qin, D., Plattner, G.-K., Tignor, M., Allen, S.K., *et al.* (Eds), *Climate change 2013: the physical science basis. Contribution of working group I to the fifth assessment report of the intergovernmental panel on climate change* (pp. 465–570). New York: Cambridge University Press.

Czech, B., Krausman, P. R., & Devers, P. K. (2000). Economic associations among causes of species endangerment in the United States. *Bioscience, 50,* 593–601.

Dawson, D. (1994). *Are habitat corridors conduits for animals and plants in a fragmented landscape? A review of the scientific evidence.* English Nature Research Report No. 94. Peterborough: English Nature.

Day, R. (2008). Local environments and older people's health: dimensions from a comparative qualitative study in Scotland. *Health & Place, 14*(2), 299–312.

Ellis, J. B. (2013). Sustainable surface water management and green infrastructure in UK urban catchment planning. *Journal of Environmental Planning and Management, 56*(1), 24–41.

Falkowski, P., Scholes, R. J., Boyle, E. E. A., Canadell, J., Canfield, D., Elser, J., *et al.* (2000). The global carbon cycle: a test of our knowledge of earth as a system. *Science, 290*(5490), 291–296.

Flink, C., Olka, K., & Searns, R. (2001). *Trails for the twenty-first century: planning, design, and management manual for multi-use trails.* Washington, DC: Island Press.

Gill, S. E., Handley, J. F., Ennos, A. R., & Pauleit, S. (2007). Adapting cities for climate change: the role of green infrastructure. *Built Environment, 33,* 115–133.

Goddard, M. A., Dougill, A. J., & Benton, T. G. (2010). Scaling up from gardens: biodiversity conservation in urban environments. *Trends in Ecology & Evolution, 25*(2), 90–98.

Grimmond, S. (2007). Urbanization and global environmental change: local effects of urban warming. *Geographical Journal, 173,* 83–88.

Harrison, C., Burgess, J., Millward, A., & Dawe, G. (1995). *Accessible natural greenspace in towns and cities: a review of appropriate size and distance criteria – guidance for the preparation of strategies for local sustainability.* English Nature Research Reports No. 153. Peterborough: English Nature.

Honold, J., Lakes, T., Beyer, R., & van der Meer, E. (2016). Restoration in urban spaces nature views from home, greenways, and public parks. *Environment and Behaviour, 48*(6), 796–825.

Jo, H. K., & McPherson, G. E. (1995). Carbon storage and flux in urban residential greenspace. *Journal of Environmental Management, 45*(2), 109–133.

Kaplan, R. (2001). The nature of the view from home psychological benefits. *Environment and Behavior, 33*(4), 507–542.

Karl, T. R., Diaz, H. F., & Kukla, G. (1988). Urbanization: its detection and effect in the United States climate record. *Journal of Climate, 1*(11), 1099–1123.

Kelcey, J. G. (1975). Industrial development and wildlife conservation. *Environmental Conservation, 2*(2), 99–108.

Kinzig, A. P., & Grove, J. M. (2001). Urban–suburban ecology. In Levin, S. A. (Ed.), *Encyclopaedia of Biodiversity Vol. 5.* (pp. 733–745). Cambridge, MA: Academic Press.

Kowarik, I. (2011). Novel urban ecosystems, biodiversity, and conservation. *Environmental Pollution, 159*(8), 1974–1983.

Kuo, F. E., & Sullivan, W. C. (2001a). Environment and crime in the inner city: does vegetation reduce crime? *Environment and Behaviour, 33(3),* 343–367.

Kuo, F. E., & Sullivan, W. C. (2001b). Aggression and violence in the inner city: effects of environment via mental fatigue. *Environment and Behaviour, 33(4),* 543–571.

Leedy, D. L. (1978). *Planning for wildlife in cities and suburbs*. Fish and Wildlife Service, U.S. Dept. of the Interior.

McKinney, M. L. (2002). Urbanization, biodiversity, and conservation: the impacts of urbanisation on native species are poorly studied, but educating a highly urbanised human population about these impacts can greatly improve species conservation in all ecosystems. *Bioscience, 52*(10), 883–890.

Matsuoka, R., & Sullivan, W. C. (2011). Urban nature: Human psychological and community health. In Douglas, I., Goode, D., Houck, M. & Wang, R. (Eds), *The Routledge Handbook of Urban Ecology* (pp. 408–23). Routledge.

Merk, O., Saussier, S. Staropoli, C., Slack, E., & Kim, J-H. (2012). *Financing green urban infrastructure*. OECD Regional Development Working Papers 2012/10. OECD publishing.

Merriam, G. (1991). Corridors and connectivity: animal populations in heterogeneous environments. In Saunders, D. A., & Hobbs, R. J. (Eds), *Nature conservation 2: the role of corridors* (pp. 133–142). Chipping Norton: Surrey Beatty & Sons.

Millennium Ecosystem Assessment (2005). *Ecosystems and human well-being: synthesis*. Washington DC: Island Press.

Nilsson, K. L., & Florgård, C. (2009). Ecological scientific knowledge in urban and land-use planning. In McDonnell, M. J., Hahs, A. K., & Breuste, J. H. (Eds), *Ecology of cities and towns: a comparative approach* (pp. 549–556). Cambridge: Cambridge University Press.

Owens, S. & Cowell, R. (2011). *Land and limits: interpreting sustainability in the planning process* (2nd ed.). Abingdon: Routledge.

Patz, J. A., Campbell-Lendrum, D., Holloway, T., & Foley, J. A. (2005). Impact of regional climate change on human health. *Nature, 438*(7066), 310–317.

Selman, P. (2006). *Planning at the landscape scale*. Abingdon: Routledge.

Soulé, M. E. (1991). Conservation: tactics for a constant crisis. *Science, 253*, 744–750.

Sukopp, H. & Starfinger, U. (1999). Disturbance in urban ecosystems. In Walker, L. R. (Ed.), *Ecosystems of the world 16. Ecosystems of disturbed ground* (pp. 397–412). Oxford: Elsevier.

Takano, T., Nakamura, K., & Watanabe, M. (2002). Urban residential environments and senior citizens' longevity in megacity areas: the importance of walkable green spaces. *Journal of Epidemiology and Community Health, 56*(12), 913–918.

Teagle, W. G. (1978). *The endless village. The wildlife of Birmingham, Dudley, Sandwell, Walsall and Wolverhampton*. Nature Conservancy Council, West Midlands Region.

Tilt, J. H. (2011). Urban nature and human physical health. In Douglas, I., Goode, D., Houck, M., & Wang, R. (Eds), *The Routledge Handbook of Urban Ecology* (pp. 394–407). Abingdon: Routledge.

Watson, G. W. (1988). Organic mulch and grass competition influence tree root development. *Journal of Arboriculture, 14*(8), 200–203.

12

THE DEVELOPMENT OF SMART CITIES IN CHINA

Andreas Oberheitmann

Abstract

In China, about 800 million inhabitants of cities are increasingly looking for intelligent and resource saving living conditions. In order to fulfil these needs, the Chinese government is promoting the concept of Smart Cities in almost 300 areas in China including Beijing, Shanghai, Tianjin, Wuhan. Generally, these concepts share four different layers building up on each other: (a) sensors (camera, RFID, detectors, sensors, smartphones and reception devices); (b) networks (telecommunication, Internet, TV, power grid and private networks); (c) platforms (service support, network management, information processing, information security); and (d) applications (smart agriculture, industrial monitoring, public security, smart governance, smart health care, smart transportation and smart energy). However, the jury is still out on whether China's 'digital' Smart City emphasis is enough to tackle underlying ecological challenges and structural imbalances to realize Smart Urban Development.

Key terms

*   *Made in China 2025*: China's national initiative to comprehensively upgrade Chinese industry. The initiative was inspired by Germany's 'Industry 4.0' plan, which was first discussed in 2011 and later adopted in 2013. The heart of the 'Industry 4.0' idea is intelligent manufacturing, that is applying the tools of information technology to production. Made in China 2025 highlights ten priority sectors: (1) new advanced information technology; (2) automated machine tools & robotics; (3) aerospace and aeronautical equipment; (4) maritime equipment and high-tech shipping; (5) modern rail transport equipment; (6) new-energy vehicles and equipment; (7) power equipment; (8) agricultural equipment; (9) new materials; and (10) biopharma and advanced

medical products (CSIS 2015).

- *Didi Kuaidi*: China's most prominent smartphone app to call a taxi. In 2016, Didi Kuaidi acquired Uber's China operations.
- *WeChat (Wei Xin)*: China's most widely used text message system similar to WhatsApp. WeChat has a variety of different other functions such as electronic payments, hold-to-talk voice messaging, broadcast messaging, video conferencing, video games, sharing of photographs and videos or location sharing.

The first chapter of the book noted various interpretations of 'smart' which, although all broadly compatible insomuch as they integrate environmental, social, economic, technological and governance considerations, stress different development aspects. China has mainly adopted a 'digital' Smart City view with somewhat less consideration of 'green infrastructure' and local deliberation. Chinese government Smart City instruments include a range of guidance and pilot projects as mentioned above. The Chinese population has adopted Smart City appliances such as WeChat for communication or everyday payments or Didi Kuaidi to call a taxi. The majority of the population do not appear to share Western privacy and civil society concerns and have enthusiastically adopted Smart City appliances such as surveillance cameras to improve public security.

In China, Smart City development is rather technically driven, the paradigm being largely utilitarian. On the contrary, low carbon city approaches are more environmentally and energy security motivated. However, in terms of sustainability planning (Neves 2009), there are overlaps between the two concepts, such as household energy saving with smart meters, the dilemma of increased electricity consumption through the growth of data centers necessary for the digitalization of the society or the reduction of traffic jam related emissions through autonomous driving of automobiles. The social purpose in China is more collective, but only to a small extent participatory rather than libertarian. The scale of intervention is on the city level; governance however is on the central government level. The implementation of technological Smart City agendas in China mandates three prerequisites:

- Put in place appropriate government actor constellation for planning a Smart City, especially on the central, provincial and on the local level. It is important to ensure the city mayor's support.
- Provide appropriate governmental implementation tools, that is build up a Smart City Planning Task Force (智慧城市规划领导小组 – Zhihui chengshi guihua lingdao xiaozu). It should include the Office of the Mayor, being the head of the commission; the Municipal Development and Reform Commission (DRC); the Municipal Science and Technology Commission (STC); the Economic and Trade Commission (ETC); the Municipal Construction Commission (MCC); the Urban Planning Bureau (UPB); the Municipal Transport Bureau (MTB); the Municipal Environmental Protection Bureau (EPB), as well as university research institutes and associations and so on.

- Include a range of relevant stakeholders, especially industry and commerce as well as the civil society. International technology companies involved with urban-sensor networks should be included given that over 500 Chinese cities are considering building them.

Introduction

Since the Industrial Revolution in the eighteenth century, cities have become the main driving force for economic development (Oberheitmann and Ruan 2012). Owing to growing urban populations and increasing city sizes around the world, cities have been the dominant agents of land use transformation, consumption and pollution. According to statistics, in 1800, urban residents accounted for only 3 per cent of the total world population. Today, more than half of the world's population is living in cities (UN-HABITAT 2008). It is estimated by the UN that by 2030 more than 60 per cent of the world's population will live in cities, among which the urban population in developing countries is likely to increase from 1.9 billion in 2000 to 3.9 billion in 2030 (UN-HABITAT 2008). Hence, cities are the crystallization point of different issues and solutions. Inter alia, cities highlight a variety of environmental problems such as large consumption of resources and impacts on ecological systems (Stern 2006).

Currently, with 1.4 bn., China has the world's largest population with an urbanization rate of 56 per cent in 2015. By 2030, with 1 billion inhabitants, the rate is expected to grow up to 70 per cent (McKinsey Global Institute 2009). Ren (2011) charts China's rapid transition from a command economy to a capitalist one, which has unsettled traditional society and radically transformed landscapes. Chinese cities have multiplied but now need to manage growth more sustainably to help mitigate its worst impacts such as climate change and its associated and extreme weather events (IPCC 2007). Other urbanization challenges include energy and water scarcity, traffic congestion, waste disposal issues or other infrastructure risks. Ren (2013) attributes the radical transformation of Chinese landscapes on an 'urban bias in policymaking' (p. xv), at both central and local levels. Ren (2011) agrees that central and local pro-development policies playing to the gallery of international recognition fueled commercially driven place-making initiatives. Developers assisted local and transnational real estate or architectural firms. However, ill-considered fixation with GDP growth left a legacy of uneven urban–rural development and social disparities. Smart City policies could help re-balance the Chinese economy for sustainable growth. The technical aspects involve digitalization to make a city more efficient, resource saving and to increase livability (EU–China Smart and Green City Cooperation 2014).

This chapter reviews the background to Smart City planning in China, looks at Smart City implementation prerequisites, paying particular attention to:

- Establishment of an appropriate Smart City government actor constellation.
- Provision of appropriate governmental implementation tools.
- Inclusive consultation with relevant stakeholders.

Beijing's Haidian District piloting for a Smart City District provides a useful case study which allows some tentative conclusions and policy recommendations.

Background

Strategic foresight, citizens' needs and access to technology mediate smart urban development pathways mode, for example a low carbon city development is result of the citizens' need for climate change mitigation and adaptation and available climate-friendly technology (Oberheitmann 2012a, 2012b). A Smart City development is the result of the citizens' needs for efficiently utilizing the opportunities of digitalization. A Smart City integrates different information and communication technologies (ICT) and the Internet of Things (IoT) in order to manage a city's assets. These assets – inter alia – include private and public buildings (schools, libraries, transportation systems, hospitals, power plants), local departments' information systems, water supply networks, public and private waste management and other community services.

The concept of a Smart City has been developed since around 2008. By then, technologies such as radio-frequency identification (RFID) sensors, wireless connectivity, electronic payments and cloud-based software services started to enable new approaches to collaborative urban solutions based on extensive data collection (bid data). The concept was extensively developed by global technology companies such as IBM, which promoted the concept of 'Smarter Planet'.

The concept of Smart Cities has been widely accepted by all levels of government in many countries around the world, including China. The development of urban infrastructure projects, including significant Smart City elements in their construction, has been implemented in China since around 2010 and the Smart City technology market is booming. In January 2013, China's Ministry of Housing and Urban–Rural Development (MOHURD) formally announced the first list of 90 national pilot Smart Cities. By April 2015, there were over 285 pilot Smart Cities in China, as well as 41 special pilot projects (EU SME Centre 2015). Currently, there are no regulations directly governing Smart Cities in China. However, in order to foster this process, the Chinese government has introduced a number of guidance documents:

- The 'Notice to Speed up the Project Implementation of Smart Cities' issued by the National Development and Reform Commission (NDRC) and the Ministry of Industry and Information Technology (MIIT) on 15 January 2014 identified different sectors where digitalization may be able to improve societal development such as health care, education, social insurance, employment and elderly care.
- The 'National New Urbanization Plan (2014–2020)' issued by the State Council on 16 March 2014 included a chapter on Advanced Smart Cities Development utilizing the Internet of Things, cloud computing and big data and emphasized cross-regional and inter-departmental coordination and cooperation.

- The 'Guidance on Promoting the Sustainable Development of Smart Cities' issued by NDRC, the Ministry of Science and Technology (MOST), the Ministry of Finance (MOF), MOHURD and the Ministry of Transport (MOT). The document from 27 August 2014 pointed out that by 2020, a number of Smart Cities will be developed as role models for other cities in China (EU SME Centre 2015).

Key targets of the Chinese government include the enhancement of information, the development of leading technologies and strengthening network security management in order to make public services more efficient and to build a livable environment. In this sense, the aim to develop leading technologies is also embedded in China's recent innovation strategy 'Made in China 2025', which includes 'new advanced information technology' and 'automated machine tools & robotics' as the first two (and maybe most important) fields of technologies to be promoted in this government initiative (CSIS 2015).

According to Li, Liu and Geertman (2015), the Chinese concept of a Smart City has four dimensions or layers building up on each other:

1   Sensors (camera, RFID, detectors, sensors, smartphones and reception devices).
2   Networks (telecommunication, Internet, TV, power grid and private networks).
3   Platforms (service support, network management, information processing, information security).
4   Applications (smart agriculture, industrial monitoring, public security, smart governance, smart health care, smart transportation and smart energy).

For example Wuhan, capital of Central China's Hubei province, very early announced that it aims to become a leader in terms of applying Smart City technology. According to Wuhan's city mayor, the city planned to finish the information technology infrastructure construction by the year 2015 and by the year 2020 become a Smart City. The city has already launched several programs to improve citizens' lives and offer more convenience. A municipal administrative service center which provides 24-hour self-service public service has been set up. Wuhan also launched a two-dimensional code food tracing system to ensure food safety and water resource monitoring system to protect the environment (*China Daily* 2012). Another example for a Smart City development in China which will be described in more detail below is Beijing Haidian District.

Smart City prerequisites

Appropriate government actor constellation for planning a Smart City

This section discusses the governance structure for implementing Smart Cities in China and addresses appropriate actor constellations. The actor constellation is important to understand who plays a role in planning and implementing Smart Cities in China.

As for China, different administrative levels have to be taken into account for the implementation of Smart Cities. The first levels are the national and provincial levels. Basic decisions on politics and measures relating to Smart Cities, such as the guideline documents mentioned previously, have been taken at the national level and are transformed into provincial policy. Especially in the current Chinese government under President Xi Jinping, Chinese policy decisions are being centralized.

The second level and, in this context most importantly, is the municipal policy level. Here, the most important actors include city mayors and senior officials, as shown in Table 12.1.

The most important driver of building a low carbon city is the political will of the local government. If the mayor of a city is convinced and willing to implement such a project, it can be undertaken more easily in China. A directive from the central government only seems insufficient to provide enough power for a quick and successful implementation. Table 12.1 shows the main tasks and responsibilities of municipal departments to plan and implement Smart Cities in China.

Smart Cities strive to engage citizens in collaborative and transparent decision-making, while being mindful of social equity concerns. To achieve this, different government implementation tools are applied as the next section elaborates.

Appropriate governmental implementation tools

The most important governmental implementation tool for the planning and implementation of low carbon cities is the establishment of a designated Smart City Planning Task Force (智慧城市规划领导小组 – Zhihui chengshi guihua lingdao xiaozu). This commission, in many cases, includes the following actors:

- Office of the Mayor, being the head of the commission;
- Municipal Development and Reform Commission (DRC);
- Municipal Science and Technology Commission (STC);
- Economic and Trade Commission (ETC);
- Municipal Construction Commission (MCC);
- Urban Planning Bureau (UPB);
- Municipal Transport Bureau (MTB);
- Municipal Environmental Protection Bureau (EPB);
- University research institutes and associations, and so on. (Zhang et al. 2009).

For a bottom-up and top-down approach, the main planning and implementation issues should be discussed and decided in a consensual manner. This approach ensures that every major political stakeholder is involved in the decision-making process and, after a consensus is reached, the policies and actions can be implemented without large resistance from a single political stakeholder.

TABLE 12.1 Main stakeholder implications and tasks of municipal government departments for Smart City planning and implementation

Department	Main stakeholder implications and tasks
Municipal Development and Reform Commission (DRC)	• Developing economic and social development strategies, including ecological and environmental protection measures • Approving and submitting for approval fixed asset investment projects
Municipal Science and Technology Commission (STC)	• Organizing the formulation and implementation of plans, including annual plans of advanced technology development and industrialization • Municipal earthquake prevention and disaster risk reduction
Municipal Economic and Trade Commission (ETC)	• Involvement of foreign investment
Municipal Construction Commission (MCC)	• Formulating urban and rural development strategies of construction • Fundraising and use of urban construction funds • Organizing, directing, coordinating and supervising the implementation of important construction projects of the municipality
Urban Planning Bureau (UPB)	• Carrying out and enforcing state urban planning • Providing recommendations for location selection of construction projects
Municipal Environmental Protection Bureau (EPB)	• Supervising the exploitation of natural resources • Overseeing ecological environment construction, ecological conservation and recovery of ecological damage
Municipal Transport Bureau (MTB)	• Organizing and formulating medium- and long-term plans on the development of roads, auxiliary facilities and communication industry • Working out annual expenditure plans for funds earmarked for communication

Source: Zhang et al. (2009).

The following indicates the key issues that need to be addressed by the Smart City task force.

- Provide smart and resource saving access to safe housing, energy, water, sanitation for all citizens and with priority to the poor in an ecologically sound manner to improve the quality of life and human health.
- Build cities for people. Roll back urban sprawl. Minimize loss of rural land by all effective measures, including regional urban and peri-urban smart and ecological planning.

- With Smart City mapping identifying sensitive areas, define carrying capacity of regional life support systems. Also identify those areas where more dense and diverse development should be focused in centers of social and economic vitality.
- Build cities for safe pedestrian and non-motorized transport use with smart, efficient, convenient and low-cost public transportation. Increase taxation on vehicle fuels and private cars and spend the revenue on Smart City projects and public transportation.
- Provide strong economic incentives to businesses for smart buildings with tax reductions for smart city solutions.
- Provide adequate, accessible education and training programs, capacity-building and local skills development to increase community participation and awareness of Smart City design and management.
- Support community initiatives in Smart Cities.

According to the EU SME Centre (2015), the following barriers have to be overcome, for example by such a task force, in order to be able to put appropriate governmental implementation tools in place:

- Legal and regulatory issues (policy inconsistencies, if there is a change in policy direction in one or more departments involved; lack of clear working mechanisms among all government departments as the single ministries have their own agendas).
- Market barriers (lack of information openness, data and privacy protection, technology exclusiveness and subsequent lack of standardization, unclear public–private partnership models).
- Operational barriers: lack of understanding of local culture and local awareness.

Inclusion of relevant stakeholders

A Smart City, however, is not only implemented by the government. Other relevant stakeholders have to be addressed, namely industry, commerce and civil society.

Smart City development in China is an opportunity for technology companies such as Huawei, China Mobile, Wei Xin, Alibaba and so on, and also foreign companies such as Siemens or General Electric to expand their businesses. These companies have to develop Smart City technologies such as online household electricity monitoring and saving devices and mobile phone Smart City solutions such as the taxi app Didi Dache. The competition in China on the Smart City market is fierce. In 2015, Didi Dache merged with competitor Kuai Di Dache which generated a new US$6 billion worth company, 'Didi Kuaidi' (Reuters 2015). In July 2016, Didi Kuaidi acquired Uber's China operations (*Economic Times* 2016).

The Smart City market in China is also very attractive for foreign investors, for example in May 2016, Cisco Systems and General Electric announced a joint venture with a spin-off from China's Academy of Sciences to help build new-wave data

CASE STUDY BEIJING HAIDIAN DISTRICT

Haidian is one of Beijing's most modern districts and well known for its high-tech area Zhonguancun. With an area of 431 sq. km and 3.7 million inhabitants, it consists of 22 sub-districts and seven townships. In 2014, Haidian District had a per-capita GDP of 116,600 RMB or about 19,000 US$, higher than that of the Slovak Republic (18,479 US$) or Latvia (16,144 US$) (IMF 2014). Haidian District government issued different documents regarding the development of a Smart City District, including the 'Smart Haidian Top-Level Design', 'Smart Haidian Development Programme' and 'Smart Haidian Construction Programme' (EU–China Smart and Green City Cooperation 2014). The following are the main strategy elements and policies of the Haidian Smart City District plans.

Main elements:

- *Smart administration*: Government service innovation, deep integration and sharing of resources, efficient operational systems and intelligent decision support.
- *Smart parks*: Building eco-friendly smart parks with complete information infrastructure, efficient interaction between businesses and government, active industrial services and smart park management for efficient business operations.
- *Smart urban areas*: Building smart urban areas to fully detect urban components and events and manage the district in a lean way.
- *Smart homes*: Creating smart homes that deliver services such as social security at the community level and benefit people with public services by building integrated community information service system to raise the quality of life.
- *IT industry highland*: Building internationally influential information industry clusters by strengthening demonstrating applications of products and technologies in the parks and actively nurturing the new generation of information technology industry.

Main technologies and applications:

- Wireless and fiber optic and other basic networks and data centers in Smart Haidian are shared by different projects.
- ICT infrastructure is shared by different projects via cloud platform built with cloud computing technology, Haidian administration network, data center, public network and other communication technologies.
- Haidian spatial data sharing is built for the Haidian GIS technology and other business applications.

- Haidian has achieved deep resource integration and sharing of government departments, efficient operational synergies and smart decision support through smart administration.
- It explores and shares resources between sectors and levels, accurately masters economic operation, public opinion and other economic and social developments and trends.
- It strengthens the integration of portals and service hotlines according to the needs of residents to build a one-section and one-stop service system for enterprises and residents.

(EU–China Smart and Green City Cooperation 2014)

Some of these applications are utilized in the low carbon and smart building of Tsinghua University's School of Environment (Figure 12.1).

Smart Haidian is built under the responsibility of Haidian District government in cooperation with the Haidian Smart City Industry Alliance as an intermediary organization. In order to ensure the construction and implementation, a work leading group has been formed. The leading group is led by the district governor. It also contains deputy district governors, leaders of district bureaus and industry experts. Other stakeholders are participating in accordance with the government management processes undertaken (EU–China Smart and Green City Cooperation 2014).

FIGURE 12.1 The low carbon and smart building of Tsinghua University's School of Environment

Source: Andreas Oberheitmann.

networks with such features as video surveillance and sensors to monitor traffic and air quality (*Wall Street Journal* 2016). In China, more than 500 Chinese cities are considering building urban-sensor networks. These efforts, however, may also impose a potential threat to privacy as the data collected by city officials for legitimate-sounding purposes could theoretically be shared with other agencies with more repressive intentions, according to the *Wall Street Journal* (2016).

Against this background, the inclusion of civil society is an important success factor for the development of a Smart City. In China, the majority of people, however, are in favor of the government's Smart City policy. In 2016, more than 800 million Chinese are already using smart phone apps such as WeChat (Wei Xin) (Statista 2016), for example to pay for public parking services, or use Didi Kuaidi to order a taxi. Many Chinese also do not share privacy concerns around electronic surveillance, for example on streets or in public buildings, which is already part of their daily life and seen more as an improvement in security rather than a threat to privacy.

Conclusions and policy recommendations

China's rapid transformation is unprecedented in human history but has unsettled traditional society and radically transformed landscapes, via urbanization and intensification. China now needs to manage growth more sustainably to help mitigate its worst impacts. As an interim measure, China has adopted a 'digital' Smart City view to attenuate the worst aspects of rampant construction. The jury is still out as to whether this approach can resolve structural imbalances and tensions around uneven urban–rural development, ecological impairment and widening inequality. In 2015, 56 per cent or almost 800 million people in China live in cities. By 2030, the Chinese urban population is projected to rise by 70 per cent to 1 billion. Increasing urbanization is likely to accentuate energy and water scarcity, traffic congestion and other infrastructure risks. Increasing digitalization and the development of a Smart City may help to make a city more efficient, resource saving and livable. By 2015, the Chinese government had already listed 285 pilot Smart Cities in China, as well as 41 special pilot projects. In order to foster this development, the Chinese government introduced different guidance notices, inter alia, namely:

- Notice to Speed up the Project Implementation of Smart Cities;
- National New Urbanization Plan (2014–2020);
- Guidance on Promoting the Sustainable Development of Smart Cities.

These policies adopt a technological, scientific and people-centric approach to developing Smart Cities in China. Data openness and the development of leading technologies is also a key target for these guidance notices (EU SME Centre 2015). The research investigated the Beijing Haidian District pilot Smart City program whose constituents were (a) Smart administration, (b) Smart parks, (c) Smart urban areas, (d) Smart homes and (e) IT industry highland. The main technologies and applications adopted were wireless and fiber optic, other basic networks, data centers, cloud computing technology and so on.

This chapter concludes its preliminary review of Smart City development in China, with some tentative policy suggestions:

- Establish an appropriate government actor constellation for planning a Smart City, especially on the central, provincial and on the local level. It is important to ensure the city mayor's support.
- Provide appropriate governmental implementation tools, that is build up a Smart City Planning Task Force (智慧城市规划领导小组 – Zhihui chengshi guihua lingdao xiaozu). It should include the Office of the Mayor, being the head of the commission; the Municipal Development and Reform Commission (DRC); the Municipal Science and Technology Commission (STC); the Economic and Trade Commission (ETC); the Municipal Construction Commission (MCC); the Urban Planning Bureau (UPB); the Municipal Transport Bureau (MTB); the Municipal Environmental Protection Bureau (EPB), as well as university research institutes and associations and so on.
- Engage authentically with a range of relevant stakeholders; including civil society, planners, academic experts, industry and commerce as well as international companies with critical technology plays.

However, China's 'digital' Smart City emphasis leaves underlying ecological challenges and structural imbalances that leave Smart Urban Development unresolved.

References

China Daily (2012). Wuhan to become smart city by 2020. Retrieved November 4, 2016, from www.chinadaily.com.cn/business/2012-11/16/content_15935840.htm

CSIS (2015). *Made in China 2025.* Center for Strategic and International Studies, June 1. Retrieved November 7, 2016, from www.csis.org/analysis/made-china-2025

Economic Times (2016). Didi's takeover of Uber sparks monopoly, fare concerns in China. August 2. Retrieved November 13, 2016, from http://economictimes.indiatimes.com/news/international/business/didis-takeover-of-uber-sparks-monopolyfare-concerns-in-china/articleshow/53505285.cms

EU–China Smart and Green City Cooperation (2014). *Comparative Study of Smart Cities in China and Europe.* Study prepared for Ministry of Industry and Information Technology (MIIT) and DG CNECT, EU Commission. March. Mimeo.

EU SME Centre (2015). Smart Cities in China. Mimeo.

IMF (2014). *World Economic Outlook database.* Washington.

IPCC (2007). *Climate change 2007. Synthesis report.* Contribution of Working Groups I, II and III to the Fourth Assessment Report of the Intergovernmental Panel on Climate Change. Core writing team, Pachauri, R. K. and Reisinger, A. (Eds.). Geneva: IPCC.

Li, Y., Lin, Y. and Geertman, S. (2015). The development of smart cities in China. Mimeo.

McKinsey Global Institute (2009). Preparing for China's urban billion. Mimeo.

Neves, M. F. (2009). 4 P's for sustainability planning. *China Daily,* November 27. Retrieved December 6, 2016, from www.chinadaily.com.cn/thinktank/2009-11/27/content_9067182.htm

Oberheitmann, A. (2012a). CO_2-emission reduction in China's residential building sector and contribution to the national climate change mitigation targets in 2020. *Mitigation and Adaptation Strategies for Global Change*, 17(7), 769–791.

Oberheitmann, A. (2012b). Development of a low carbon economy in Wuxi City. *American Journal of Climate Change*, 1, 64–103. doi:10.4236/ajcc.2012.12007

Oberheitmann, A. and Ruan, X. (2012). Low carbon city planning in China. In: Frauke Urban and Johan Nordensvard (Eds.). *Low Carbon Development: Key Issues.* Abingdon: Routledge, 270–283.

Ren, Xuefei. (2011). *Building Globalization: Transnational Architecture Production in Urban China.* Chicago, IL: University of Chicago Press.

Ren, Xuefei. (2013). *Urban China.* Cambridge: Polity Press.

Reuters (2015). *China taxi apps Didi Dache and Kuaidi Dache announce $6 billion tie-up.* February 14. Retrieved November 13, 2016, from www.reuters.com/article/us-china-taxi-merger-idUSKBN0LI04420150214

Statista (2016). *Number of monthly active WeChat users from 2nd quarter 2010 to 2nd quarter 2016 (in millions).* Retrieved November 13, 2016, from www.statista.com/statistics/255778/number-of-active-wechat-messenger-accounts/

Stern, N. S. (2006). *Review of the economics of climate change.* Retrieved November 19, 2016, from http://webarchive.nationalarchives.gov.uk/

UN-HABITAT (2008). *Urbanization: facts and figures.* Retrieved November 19, 2016, from www.unhabitat.org/mediacentre/documents/backgrounder5.doc

Wall Street Journal (2016). U.S. and Chinese tech firms team up on sensor networks for 'smart cities', May 12. Retrieved November 13, 2016, from www.wsj.com/articles/u-s-and-chinese-tech-firms-team-up-on-sensor-networks-for-smart-cities-1463081921

Zhang, K., Oberheitmann, A. and Cui, D. (2009). *Low carbon economy in the cities in China.* Study for InWent. Mimeo.

13

VALUATION SYSTEMS

The case of Dubai

*Simon Huston, Ebraheim Ali Lahbash
and Ali Parsa*

Abstract

To manage development pressures, smart cities rely on reliable information systems. In turbulent times, a robust valuation system helps to regulate transactions and guide property investment. Reliable valuations are particularly important for Gulf cities like Dubai whose property and capital markets absorb regional oil and increasing Asian liquidity. The transformation of the Emirate from an inconsequential fishing village to a glittering global trade, tourism, financial and logistics hub stimulated property markets. However, the 2008 Global Financial Crisis (GFC), continued regional instability and sustainably issues raise concerns. The research investigates whether the Emirati residential valuation system (RVS) is 'fit for purpose'. Valuation system criteria included valuation output reasonableness, stakeholder information transparency, administrative capability, end-user trust and valuation standards salience. The research collected evidence, using embedded observations, transactions analysis, expert interviews and discussions. Initial results suggest that key UAE residential valuation system improvements should focus on information dissemination, institutional capabilities and dissemination of valuation standards.

Key terms

Residential valuation system, market value, information systems, capabilities, trust, valuation standards salience

Purpose

The research investigates whether the Emirati residential valuation system (RVS) as currently constituted is fit for purpose. Evaluation criteria include:

- Reasonableness of valuations
- Information systems data transparency and accessibility
- Institutional administrative capabilities
- End-user trust
- Valuation standards salience.

Introduction

Smart development optimises resource allocation in reasonably stable markets. A high calibre bureaucracy impartially interacts with private sector market players and the public, informed by a robust Residential Valuation System to:

- Use foresight to set long-term sustainable objectives
- Prevent ill-considered urban sprawl and set environmental limits
- Impartially navigate land-use contention
- Facilitate deliberation but enforce the rule of law
- Capture unearned land value increments to invest in place enhancement
- Stimulate local innovation and culture.

Robust geospatial, title, transaction and planning data help avoid valuation inaccuracy or bias for appraisal, improve modelling fit and inform impartial decision-making by officials (Crosby 2000; Crosby and Hughes 2012; McGreal and Taltavull 2012; McParland et al. 2002). The RVS facilitates property market functioning and includes system boundaries, elements (stakeholders, institutions), processes (valuation practice and standards), outputs (registration certificates, title transfers, valuations, market reports) and feedback mechanisms (vacancies, prices, court cases, complaints). The Emirati property market is relatively immature and, anecdotally, the valuation system has data governance and information issues, compounded by region instability, oil price uncertainty and dependence on regional financial inflows. The research undertook a systematic investigation to make a measured RVS assessment. We employed mixed methods (Creswell et al. 2008; Creswell 2009) to evaluate the Emirati residential valuation system to assess if it is 'fit for purpose'. Formally, the main research question is: Is the current Emirati residential valuation system appropriate? [RQ$_{main}$]

The Red Book (RICS 2014a: PS2 3.1) suggests key professional competence features for a RVS:

- Appropriate academic/professional qualifications which demonstrate technical competence (*capabilities*).
- Membership of a professional body (RICS) which demonstrates a commitment to ethical standards.
- Sufficient knowledge about asset class, demonstrated by skills and local, national or international market experience (*trust*).
- Compliance with legal regulations governing valuation practice (*salience*).

Five aspects appear central to the proper functioning of a RVS:

- Reliable output
- Information systems which disseminate reliable data to stakeholders
- Institutional capabilities around governance, human resources, administration, (professional people), technologies (DSS, GIS, AIS) and meta-cognition
- Stakeholders should trust system
- Standards are salient (prominent) and widely implemented.

Table 13.1 illustrates indicative RVS assessment criteria based on critical features.

RVS institutional capacity involves collecting, registering, categorising and synthesising diverse intelligence sources on property markets and values. Relevant operational indicators of intelligence capability could include, for example, the application of spatial decision technologies such as GIS mapping for property registration, taxation, utilities charging and boundary dispute resolution. A modern RVS should make use of remote technologies to conduct desktop research for registration, due diligence, mass appraisals or locales quality criteria for planning. To supplement mass appraisal, qualified professionals (surveyors) should undertake systematic site visits to measure subject properties, ascertain condition and record

TABLE 13.1 Critical RVS features or robustness assessment criteria

Area	Details
Output	Reasonable valuations and regular market reports
Information	Data securely and systematically captured to provide necessary valuation details including: valuation base, date, location, legal matters (title, lease and encumbrances), property characteristics and conditions
Capabilities	• Governance, transparency (auditability) • Competent people foster completeness, accuracy, relevance, comparability, clarity, neutrality • Administration: valuation purpose (base) and date (e.g. MV, IV, RV, MGV or fair value). If not market-based, valuation and price can differ. If fair value, specify one of two recognised definitions (IASB in IFRS 13 or IVSC in IVS Framework paragraph 38) • Technological adaptation • Meta-cognition
Trust	Stakeholders can run queries or data analytics (descriptive stats, performance indicators, valuations)
Standards	Valuation standards are disseminated and used, professionals conduct valuations who are knowledgeable about scope, (period), materiality, measurement, prudence

Source: Adapted from Lamberton (2005).

key features or encumbrances. For market value (MV) determination, grounded evidence supports the selection of appropriate comparable properties (similar recently sold or leased premises). Practical indicators of RVS institutional capacity could involve:

- Documenting the technologies utilised
- Recording surveyor qualifications
- Observing or questioning practice.

Indicators of RVS intelligence capabilities regarding market dynamics could include use of advanced modelling techniques to inform cyclical determination (Hepsen and Vatansever 2011).

RVS principles

Having reviewed the real estate literature and the systems and valuation sources, the research distils five principles for a balanced RVS evaluation (Figure 13.1). The principles help generate the operational explanatory framework (see later Figure 13.3).

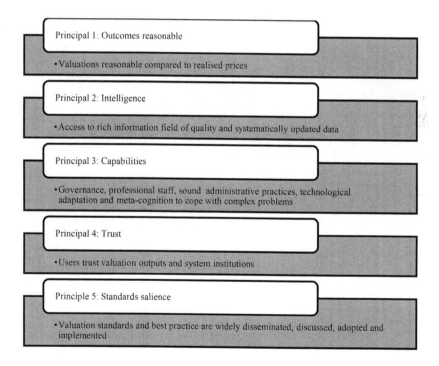

Principal 1: Outcomes reasonable

• Valuations reasonable compared to realised prices

Principal 2: Intelligence

• Access to rich information field of quality and systematically updated data

Principal 3: Capabilities

• Governance, professional staff, sound administrative practices, technological adaptation and meta-cognition to cope with complex problems

Principal 4: Trust

• Users trust valuation outputs and system institutions

Principle 5: Standards salience

• Valuation standards and best practice are widely disseminated, discussed, adopted and implemented

FIGURE 13.1 RVS principles underpinning RVS draft explanatory framework (toolkit)
Source: The authors (2016).

Methodology

The research uses explanatory sequential mixed methods, involving the development of an explanatory framework, explored and investigated in four phases:

- Conceptual
- Exploratory (qualitative and statistical analysis)
- Operational (archival, observational and qualitative interviews)
- Reflective (confirmatory panel discussion).

Exploratory Phase

The Exploratory Phase of the research involved site visits to Dubai, desktop analytical reviews and statistical analysis. In practice, the exploratory examinations had two stages, split into various examinations:

Stage 1

- Preliminary site visit [practical] to establish general market conditions, make contacts, observe and discuss issues and secure data.

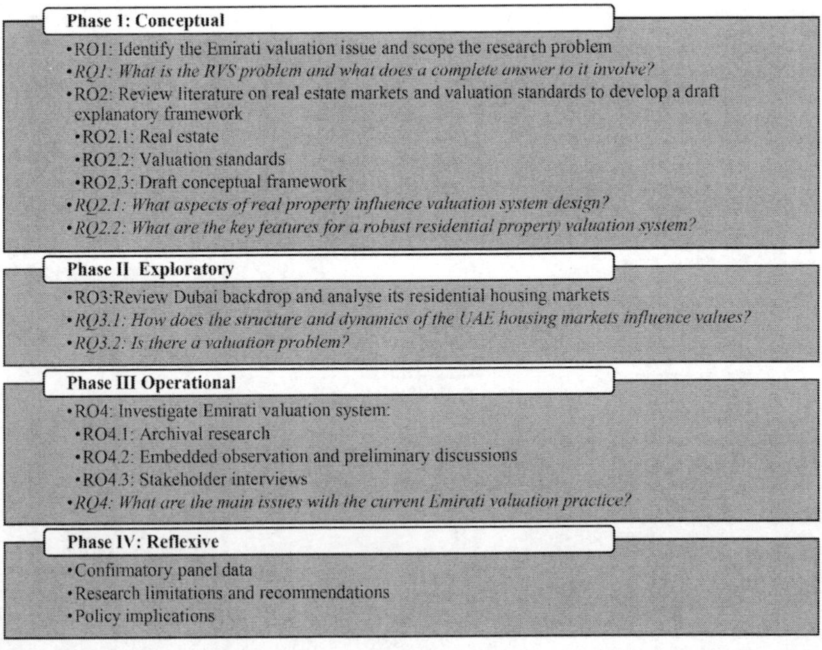

Phase I: Conceptual
- RO1: Identify the Emirati valuation issue and scope the research problem
- *RQ1: What is the RVS problem and what does a complete answer to it involve?*
- RO2: Review literature on real estate markets and valuation standards to develop a draft explanatory framework
- RO2.1: Real estate
- RO2.2: Valuation standards
- RO2.3: Draft conceptual framework
- *RQ2.1: What aspects of real property influence valuation system design?*
- *RQ2.2: What are the key features for a robust residential property valuation system?*

Phase II Exploratory
- RO3:Review Dubai backdrop and analyse its residential housing markets
- *RQ3.1: How does the structure and dynamics of the UAE housing markets influence values?*
- *RQ3.2: Is there a valuation problem?*

Phase III Operational
- RO4: Investigate Emirati valuation system:
- RO4.1: Archival research
- RO4.2: Embedded observation and preliminary discussions
- RO4.3: Stakeholder interviews
- *RQ4: What are the main issues with the current Emirati valuation practice?*

Phase IV: Reflexive
- Confirmatory panel data
- Research limitations and recommendations
- Policy implications

FIGURE 13.2 Overview of the Emirati RVS research phases with objective and research questions

Source: The authors (2016).

- Secondary examination of the Emirates' economic backdrop and qualitative examination of its property market to establish *how the structure and dynamics of the UAE housing markets influence values?* [RQ3.1]
- Primary qualitative investigations into a sample of Emirati submarkets (photos, site visits and expert discussions).

Stage II

- Quantitative analysis of transaction prices and market values (Examinations E1–E5), looking to:
  - o   Understand market structure and dynamics
  - o   Establish the reliability of DLD valuations.

Operational Phase

The Operational Phase looked at the extent to which the current system reflects 'ideal' valuation system principles (Figure 13.1). Operational Phase research involved two stages.

Stage I

Embedded research (2015) to investigate Dubai Land Department (DLD) institutional environment via:

- Scrutiny of archival material to evaluate record keeping (valuation paperwork)
- Tracing sample of transactions
- Attendance at meetings
- Observation of valuation administrative practices such as valuation procedures and standards implementation
- Discussions with Emirati practitioners about valuation issues.

Stage II

Semi-structured expert stakeholder interviews (2016) to ascertain system operations and evaluate expert practitioner views on matters linked to RVS framework:

- Information systems
- Institutional capabilities (governance, qualifications, practices, technologies, cognition)
- Trust between various stakeholders (banks, developers, planning authorities and real estate agents)
- Valuation standards, discussion, dissemination and implementation.

Reflexive Phase

Once data were analysed, the final Reflexive Phase of the study involved a stake-holder confirmatory focus panel to corroborate initial interview findings. The experts corroborated findings, but also flagged institutional issues pertinent to the UAE valuation systems and helped formulate key policy recommendations.

Exploratory Stage I secondary data analysis

Macroeconomic backdrop

Throughout its history, the Emirates traded in silk, spices, ivory, gold, guns and slaves (Davidson 2008a, 2008b). Other activities included piracy, pastoralism or pearling and fishing. Prosperity waxed and waned in line with strategic shifts in regional power as the Caliphate replaced the Sasanids (636 CE) and Seljuk Turks replaced the Umayyad dynasty. Portuguese colonialists, interested in maintaining naval bases to secure trade with India, were supplanted by first the Dutch then the English. Eager to secure the route to its Raj in India, Great Britain signed treaties with the Trucial States. Before the First World War, the British worried about the Russians occupying Bandar Abbas but later Imperial strategic concerns centred on securing Persian oil concessions. In 1975, the UAE formally gained its independence from the UK but its strategic importance in the Gulf remains.

Nowadays, the UAE economy has several pillars:

- Trade, re-export and logistics related activities (aviation and shipping)
- Abu Dhabi's hydrocarbon output (oil and gas)
- Dubai's service, tourism and hospitality sectors
- Construction (real estate, infrastructure, public works and megaprojects)

Controversially, Davidson (2008) highlights its black economy (human trafficking, money laundering, arms dealing). In the volatile Middle East, investors perceive the UAE as a safe haven and it benefits enormously from regional growth, especially in the Indian subcontinent. However, GDP fluctuations cloud judgement about long-term sustainable growth prospects.

Oil and gas still accounts for about one-third of the Emirates exports but 74 per cent of government revenue. Since the collapse of oil prices after the 2008 Global Financial Crisis, the Emirates dependence on hydrocarbons has somewhat moderated. However, the long-term situation remains precarious with multiple challenges. Structural reforms need to address long-term challenges (environmental, water and energy security) in a tight and prudent fiscal regime. Regional security and inter-fiscal policy coordination would help policy reform. The 2008 GFC showed how vulnerable the Emirates are to global sentiment, credit conditions and oil price fluctuations. In 2008, the GFC hit overpriced markets badly. Dubai's housing-exposed Government Related Entities (GREs) struggled. The property developer Nakheel had racked up AED 26bn in infrastructure costs for Palm Jumairah alone. Abu Dhabi provided US$20 billion funding to restructure Dubai's debt and put its GREs on a sustainable footing. Defaulting entities like Nakheel and Dubai World resumed interest repayments and continued reclamation work on Palm Jebel Ali (although only 4 per cent of its infrastructure works are finished).

In summary, over the past 30 years, the Emirates has developed very fast as a progressive, trading and business hub, fuelled by regional, Indian and global growth and cheap, migrant labour. Prosperity is linked to global growth rates but vulnerable to longer-term substantive challenges. Consequently, the RVS needs to monitor regional instability, global sentiment and oil prices and scan risks:

- Food, water and energy security
- Environmental degradation and climate change
- Demographic pressures
- Regional instability and insecurity
- Oil markets volatility
- Real estate market collapse
- Banking system default
- Fiscal short-termism.

In Dubai, house prices fluctuate. Efforts to stabilise the market, such as the new 4 per cent real estate registration fee, are distractions from the real volatility issues discussed above. Respondent 4, for example, stated that it, 'decreased transactions by 60 per cent', (RVS Interview Responses 2016 Q1, p3).

Dubai submarkets

In Dubai, multiple positive and negative factors segment property markets. Spatial differentiators include status, socio-economic profile, climate risk exposure, waste dumps, access to jobs and facilities, air quality, views and cultural suitability to buyer segments. Dwelling stock is also diverse due to design, structure and energy efficiency.

Dubai Marina (DM)

Located on the site of a former public beach beyond Burj Al Arab, the first phase of Dubai Marina finished in 2003. DM is a westernised, high-density district which comprises around 200 modern high-rise towers (freehold), commercial and hotel complexes. Landmarks include its prominent Yacht Club, Dubai Marina Mall, luxury hotels and Dubai Media city. Beachfront properties command a premium and cater to more upmarket tourists but a vibrant, if superficial, social lifestyle and pretentions of luxury also appeals to multicultural, young professional, expatriates who work in its range of commercial and retail outlets. There is adequate parking and public transport links (metro and bus stations) enable access to a range of entertainment and facilities such as schools, clinics and parks. Entertainment amenities include cinema complexes and a diverse range of restaurants. GRE developers involved with DM include Emaar (E), Dubai Properties (DP), Meraas (M). Projects completed include Jumeirah Beach Residence (DP), Dubai Marina (E); Blue Waters (M). Construction continues.

Burj Khalifa (BK)

A new, iconic, upmarket central district is centred on the world's tallest building, the Burj Khalifa. The 1.5 sq. km district consists of freehold modern and stylish towers surrounding Dubai's main luxury landmark. Completed in 2010, the Burj Khalifa rises 828 ms (2,717 ft) and contains 156 occupied floors. The iconic building is the focal point for celebrations and firework displays. Close by is Dubai Mall, the world's biggest retail floor space, which offers a large choice of restaurants, cinema complex, ostentatious water features (including an aquarium). BK has all the usual amenities to cater for ambitious global elites, including a Canadian university, branded private schools, cosmetic surgery clinics, international banks and a range of entertainment facilities, including an ice rink and theme parks. Infrastructure incudes good metro facilities, bus stations, large underground parking. Population growth is around 10 per cent annually but increasing as new towers emerge from the desert. Its diverse, aspirational residents are largely young Arab and Asian professionals who work in the banking, real estate, retail and hotel sectors. Emaar built Dubai Mall and the Burj Khalifa tower and Emaar Square (completed 2011).

Warsan First (W1)

Warsan is one of the most affordable locations in Dubai comprising mainly of Asian community, established since 2008. Segregated into country clusters (China, UAE, England, Italy, Russia, Morocco, Spain, France, Persia, Greece), it is a highly dense built-up area, consisting of three–four storey affordable residential and commercial blocks, and with strong links to the Chinese commercial sector. This is a multicultural community with a very high percentage of Asian business entrepreneurs. Approximate population growth here is around 7–8 per cent, with a potential growth an extra 2–3 per cent per year. Employment opportunities are generally in the commercial trading sector and hotels, restaurants. Facilities such as Dragon Mall, a wide range of warehouses, business facilities, offices at affordable prices, small medical centres and the Desert Palm Polo Club are very attractive to local and expatriate residents. The major developer is Nakheel (N), which has already completed such projects as International City, Dubai Textile city. Infrastructure offers medium access to major roads, bus services, limited school and university opportunities, average architectural design and no high-rise blocks. Total area size of this location: 10 sq. km.

Thananaya Third (T3)

Thananaya Third is a multinational cultural vibrant community established in 2008. It has a significant iconic landmark of the UAE, the oldest Golf Club. The population is mainly young professionals and families, as well as foreign expatriates living in gated communities. The average growth of this population is approximately 7–8 per cent a year. The major employment opportunities are in such sectors as telecommunications and media, hotels, business facilities, Golf Club and other leisure centres. Range of facilities of this area offer easy access to metro and bus services, close proximity to international schools, medical clinics, shopping malls and restaurants. The main developer for this area is Emaar, which completed the Greens and Lakes. Well-developed infrastructure enables residents to have good access to international airport, metro and bus stations; easy access to the beach and the Palm Island. Urban morphology comprises mainly of high quality architecture, spacious comfortable buildings and business offices. The total built-up area including park zone is 12 sq. km.

Thananaya Fifth (T5)

Established in 2009, T5 has a multicultural, westernised community. It is a vibrant location, offering a luxury lifestyle, free zone area, lots of business opportunities, with close proximity to Dubai Marina. The population has a steady growth of approximately 7–8 per cent comprising mainly of young mainly western professionals, families. The area offers a wide range of job opportunities in Dubai Multi Commodities Centre, government offices, hotels, retail, banks, UAE free trade zone centre. Facilities offer good range of schools, hospitals, parks, hotels, entertainment centres, shopping malls, supermarkets. The major developer is Nakheel, which completed Jumeirah Islands, Jumeirah Park, Jumeirah Heights, Jumeirah Lakes. Good infrastructure includes metro and bus stations, good access to the Jumeirah Beach, proximity to Golf Club and major highways. The area consists of high quality buildings and towers covering approximately 12 sq. km.

Having described each of the five locales, Table 13.2 summarises their key characteristics.

Conclusion of Exploratory Phase: Stage I

The exploration into the structure and dynamics of the UAE housing markets found that Emirati property markets are prone to exogenous risks and spatially polarised or segmented. The implication is that a robust Emirati RVS requires substantial macro and micro information to adjust valuations, according to location. A full understanding of submarket structure and dynamics now calls for statistical examinations (E3.1–E3.4) of housing transactions.

TABLE 13.2 Key characteristics of five residential housing markets selected for examination

Criteria	DM	BK	W1	T3	T5
Major projects	Dubai Marina (E); Jumeirah Beach Residence (DP), Blue Waters (M)	Dubai Mall/ Burj Khalifa Emaar Square (E)	International City, Dragon Mall, Dubai Textile city	Greens, Lakes	Jumeirah Islands, Jumeirah Park, Jumeirah Heights, Jumeirah Lake Towers
Established	2003	2009–11	2008	2008	2009
Place character	High density tourist and expatriate	Prestige central area	Downmarket, Asian	High-end multinational	Western, vibrant
Major developers	E, DP, M	E	N	E	N
Built forms	Freehold towers	Iconic luxury freehold towers	Low-end, low-rise residential or commercial blocks (3-4 storey)	Premium commercial and villas	High-end towers
Demographics	Mainly young western professionals	Multicultural young professionals	Multicultural, mainly Asian entrepreneurs	Young professionals and families (gated expatriate communities)	Young mainly western professionals, families
Population growth	Average 10% per year	Average 10% per year	Approx. 7–8%	Approx. 7–8%	Approx. 7–8%
Employment	Mainly retail commercial (Dubai Marina Mall, Dubai Media city)	Diverse (banking, property agents, real estate developments, retail, hotels)	Commercial trading sector, hotels, restaurants	Telecom-munications, media, hotels, business facilities	Commodities Centre, government offices, hotels, retail, banks, UAE free trade zone centre
Facilities	Range, yacht, marina	Dubai Mall	Dragon Mall	Golf Club, beach and the Palm Island	Range, beach
Transport	Metro, tram	Metro, bus underground parking	Roads, bus services	Metro, bus	Metro, bus
Area size	4 sq. km	1.5 sq. km	10 sq. km	12 sq. km	12 sq. km

Source: The authors (2016).

Exploratory Stage II: statistical market analysis

Having looked at the general Emirati economic backdrop and qualitatively examined the housing submarkets, the research statistically investigated several thousand residential market sales transactions (2007 to 2014) in five Emirati locales with two goals, looking to, first, strengthen initial qualitative submarket analysis and, second, assess the reasonableness of valuations. A range of techniques were employed to gain market insights:

- Descriptive statistics (E3.1)
- Preliminary regressions (E3.2)
- Non-Parametric Testing (E3.3)
- Nearest Neighbour Analysis (E3.4)
- Automatic Linear Modelling (E3.5).

Quantitative analysis confirmed the importance of the geospatial submarkets. Regression models suggest *Locales* is a significant price predictor. The research checked a sample of valuations using OLS and ALM techniques and found them broadly reasonable compared to realised prices [RVS Principal 1] and [RQ3.2]. As expected, complications arise with upper-floor Burj Khalifa and other 'trophy' properties outliers. However, a complete answer to the main research question [RQ$_{main}$] requires further scrutiny of DLD operations via embedded research and stakeholder interviews.

Operational Phase: embedded research

Overview

A complete answer to [RQ$_{main}$] requires further scrutiny via DLD embedded research of current Emirati valuation practices [RQ4]. Embedded Operational Phase fieldwork was conducted over 2015–2016 in DLD and via development and valuation practice in Fujairah. The privileged embedded position facilitated:

- *Embedded Test_1* (EB1) Archival research (granting access to DLD and Emirati valuation documentation)
- *Embedded Test_2* (EB2) Observation of DLD committees, administration procedures and valuation practices
- *Embedded Test_3* (EB3) Preliminary discussions with valuers, and the broader RVS community in DLD as well as wider discussions with a range of RVS players.

The RVS toolkit or explanatory framework (Figure 13.3) underpinned operational investigations. Specifically, the research needs to establish the extent to which the RVS system complies with the four remaining RVS principles (Figure 13.1):

- *Principal 2* [Intelligence]: RVS information system is reliable with rich information field (quality and systematically updated data).
- *Principal 3* [Capabilities]: RVS institutions are properly configured and governed, administration is competent and supported by appropriate technologies, staff are professional and some have the skills and experience to cope with complex valuations.
- *Principal 4* [Trust]: RVS players trust system and its valuation outputs.
- *Principle 5* [Standards salience]: valuation standards are widely disseminated, discussed and best practices implemented.

Fieldwork practicalities

Whilst working as a developer in Fujairah for a decade from 2006, the author experienced the full spectrum of UAE administrative documentation relating to approvals and valuations. Over the period 2015–2016 inclusive, the author made several fieldwork trips to Dubai to shadow the DLD Valuation Department over several months but also to engage more broadly with the Emirati system. DLD archival investigations involved:

- Access documentation
- Audit of sample archival valuation data to document operational procedures
- Assess valuation standards compliance
- Judge institutional valuation capability.

Valuation process

If they have multiple properties, applicants must complete an additional VR1 form. Applicants submit the completed forms to the Appraisal Department with supporting documentation. The Appraisal Department sends the application to the Dubai Municipality Surveyor to obtain up-to-date information (photos, maps, legal documents and site reports for completed units) and GIS Section (for external and internal photos). The survey and GIS sections in Dubai Municipality upload the information and disseminate it to the Appraisal Department, supported by all the relevant documentation and site visits reports. Exceptionally, in case of vacant land, after filling the request for valuation with the supporting documents, the Appraisal Department Committee reviews the paperwork without any site visits.

After creating a folder for each valuation request, the Committee considers the request for valuation twice a week. The Committee appraisal panel usually consists of eight members (two from Dubai Municipality, three independent developers and three members from Dubai Land Department). Prior to the meeting of all the panel members, DLD's Appraisal Department first Administrative Officer compiles all the relevant documentation for valuation. Appraisal Committee valuations call for a minimum of five panel members of which at least two members are from Dubai Municipality (DM). DM surveyors check building quality, measured areas and estimated construction costs. The AC assigns each plot a municipality number. The first Administrative Officer uploads all the relevant documents for digital display and panel sequential review. Once the panel reaches a valuation consensus about an application, they authorise the issue of the Valuation Certificates (VC). The Administrative Officer collates reviewed applications and enters data into the *Al Tabu* system (Land Registry System) then issues VCs. Appraisal Department staff directly contact clients, via text message or via EM Dubai (an internal messaging system) to notify them that VCs are available.

The embedded fieldwork generated five test results in three areas (with alternate system or institutional foci):

- Embedded Test_1 (EB1) DLD archival research, looking at governance for system [EB1_IS(g)] and separate institutions [EB1_C(g)]
- Embedded Test_2 (EB2) DLD observation, looking at institutional level professionalism [EB2_C(p)] and technology [EB2_C(t)]
- Embedded Test_3 (EB3) system-wide preliminary discussions, looking at IS technologies [EB3_IS(t)].

Embedded Test_1

The DLD archival reviews (EB1) also supplemented by observation and discussions discovered no evidence of independent valuation audits. From the point of view of both information systems [RVS Principal 2: Intelligence] and [RVS Principal 3: Capabilities (Governance)], the separation of powers and institutional checks and balances is extremely critical to prevent conflicts of interest and to maintain confidence in the system's impartiality and reliability. The conclusion for EB1 is that governance (g) is weak at both system level (IS) or [EB1_IS(g)] and at the institutional level within it (capabilities) or [EB1_C(g)]. Further investigation into institutional capability (governance) or independent valuation oversight is called for during operational interviews.

Embedded Test_2

In terms of professionalism, the embedded observation (EB2) discovered that most valuers have formal business management qualifications or engineering certificates and some relevant experience in the industry. For built-up areas, normally professionally certified valuers conduct sites visits but for vacant land no visits are required by surveyors and GIS staff. However, the second embedded test (observation) on professionalism [EB2_C(p)] concludes that, compared to developed real estate markets, overall Emirati professional certification is 'weak'.

In terms of institutional capability technological support [EB2_C(t)], first, GIS reports are integral to normal valuation procedures. Second, the Appraisal Committee (panel) can access robust decision GIS reports (internal and external images). The embedded observation conclusion re technology [EB2_C(t)] is 'OK'.

Embedded Test_3

Wider appraiser discussions (EB3) discovered that the system (unit of analysis) is actually quite fragmented. There are important differences in the IS available or in-house technical capabilities between the government (DLD or Dubai Municipality), government quangos (GRE such as NK, DP, Emaar) and the private sector. The latter is very diverse with large, multi-functional real estate consultants (JLL, CBRE, Savills, etc.) and smaller regional entities. Multinationals can access multiple external and internal databases to inform their valuation or sales price judgements, notwithstanding issues of confidentiality and conflicts of interest. The conclusion for the third embedded test on system-wide IS technologies [EB3_IS(t)] is weak.

Operational Phase: interviews

Operational Phase Embedded research found reasonably qualified staff and technological support but some concerns about the separation of powers (governance) during DLD valuation process. To further probe potential system weaknesses, the thesis conducted interviews with 29 Emirati RVS experts and stakeholders during 2016. Its preparation involved operationalising the RVS draft framework (Figure 13.1) into the RVS Operational Test Kit (Figure 13.3). Principals 2–5 (Figure 13.1) guide system evaluation:

- P2: *Intelligence systems* (IS) – data governance, source transparency, rich information field, systematic updates, analytics (statistics/forecasts/modelling).
- P3: Institutional *capabilities* (C) – governance, professionals, administrative competence, supportive technologies (adaptive) and meta-cognition.
- P4: *Trust* (T) – collaboration, users trust valuations, users trust system.
- P5: Standards *salience* (S) – valuation standards disseminated, discussed, implemented.

After ethical review, the research piloted the interviews on three experienced UK valuers to fine-tune the instrument. Researchers conducted the main interviews during the period May 2016–June 2016 in Dubai and other Emirates. To ensure interviews were properly conducted (protocols, no bias, leading questions), the first dozen were conducted jointly with Dr Simon Huston in April 2016. Respondents included a broad range of valuation system players, including: bankers, developers, real estate agents, government officials.

Framework and indicators (questions)

The RVS Operational Test Kit (Figure 13.3) provides the explanatory framework to structure interview questions and semi-structured discussion over a range of indicators; particularly Q8 (about the use of valuation standards) which tests IS sources, administrative competence, standards dissemination *and* implementation.

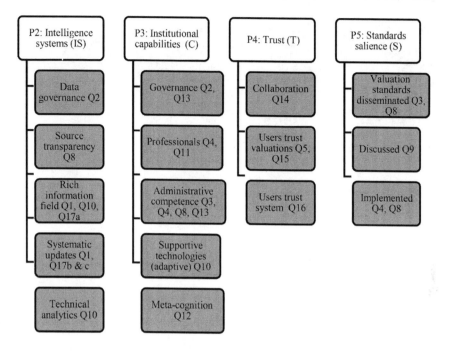

P2: Intelligence systems (IS)	P3: Institutional capabilities (C)	P4: Trust (T)	P5: Standards salience (S)
Data governance Q2	Governance Q2, Q13	Collaboration Q14	Valuation standards disseminated Q3, Q8
Source transparency Q8	Professionals Q4, Q11	Users trust valuations Q5, Q15	Discussed Q9
Rich information field Q1, Q10, Q17a	Administrative competence Q3, Q4, Q8, Q13	Users trust system Q16	Implemented Q4, Q8
Systematic updates Q1, Q17b & c	Supportive technologies (adaptive) Q10		
Technical analytics Q10	Meta-cognition Q12		

Key

Clear rectangles: RVS principles/concepts/notions or 'latent variables'.

Shaded rectangles: represent measureable indicators (variables or KPIs).

FIGURE 13.3 RVS Operational Test Kit, based on the five RVS principles linked to instrument questions (indicators)

Source: Adapted from Streiner (2006).

P2 Intelligence

The second RVS Principal [Intelligence] or 'P2' requires an adequate and reliable Information System (IS). Access to quality information underpins a robust RVS and it involves proper governance around data input, a rich and regularly updated information field (Table 13.1: Critical RVS features). Operationally, a robust RVS should systematically collect, register, categorise and synthesise [*ibid.*] diverse property data sources for legal clarity, dispute resolution and for valuations. The IS should also disseminate property or capital market information so players can make informed judgements, for example about cycles. Indicators of intelligence capability include, for example, the availability of decision support systems (DSS) with spatial technologies (GIS mapping). The IS includes title details, tenure, boundaries, liens, encumbrances, tax band, utilities charges and so on. In a modern RVS, real estate agents, brokers, bankers, planners, tax collectors and other officials can access the IS for desktop research for various purposes, including registration, due diligence, mass appraisals or locales quality criteria for planning. As well as descriptive statistics, advanced IS technical analytics informs users about market dynamics and cycle stage via univariate or multivariate forecasts.

P2 Indicators

The RVS Explanatory Framework breaks the IS notion down into its multiple constituents (indicators), reflected in instrument questions:

- P2 (g) Data governance Q2
- P2 (st) Source transparency Q8
- P2 (f) Rich information field Q1, Q10, Q17a
- P2 (u) Systematic updates Q1, Q17b & 17c
- P2 (t) Technical analytics Q10.

P2 Results

P2(g) Data governance (Q2)

- Q2 [IS-gov] *What do you understand by the term Residential Valuation System in Dubai (RVS)?*

Governance is about structuring institutions to avoid the Agency Problem (Fama 1980; Kaplan 2012) or officials diverting resources to further their own private ends rather than performing their public duties. To this end, it is important that RVS officials/players have a conceptual overview of the interactions of all RVS elements and its ultimate function (i.e. reliable records and valuations). However, the interviews revealed a wide diversity of views and, some, confusion. For Q2, Respondent 3, gives a flavour of some of the muddle:

> [R3] – *Don't have familiarity. Maybe something to do with RERA. I read recently document about implementation of a valuation system*

However, for Q2 other respondents provided clearer definitions of the valuation system:

> [R15] – *Part of the whole valuation in general that constitutes the property residential price*
> [R16] – *Residential system in Dubai is the criteria and standards to value residential units. We have an appraisal committee who values properties and also looks at property prices and market situation. On the other hand, we have external international companies who perform valuations according to international standards.*

Conclusion: WEAK.

P2 (st) Source transparency (Q8)

• Q8 [IS-source] *In your work, how relevant are valuation standards?*

Mainstream international investors require transparent sources of information that complies with international standards (RICS; IVSC; FRS), indicated by, for example, the Jones Lang LaSalle RE Transparency Index. In the Emirates, relatively poor market transparency has led firms to prepare their own databases but, because methodologies can vary, so can interpretations, leading to market distortions and mispricing. Typical is where an in-house database valuation fails to clarify the basis of value. Whilst market value (MV) is usually most appropriate, in specific circumstances, clients can request and appraisers adopt alternative bases (RICS, VPS 4, *Bases of value*). For Q8 [IS-source], Respondent 4 highlighted the importance of valuation standards in sourcing data:

> [R4] *Critical. It is very important to keep up the standards of valuation due to the huge volume of mortgage transactions and to avoid any risks associated with them.*

However, 9/27 (a third) considered standards either 'irrelevant' or 'largely irrelevant'. The issue of sources of information also surfaced in Q6 [OPEN] about *most critical factors influencing the system reliability*:

> [R28] – *Clients do not include all the information required for valuation or wrong information is provided.*

> [R29] – *We don't have available information for new land and new projects which hugely impacts the valuation price.*

Overall, it is fair to say that the Emirati RVS has a data transparency issue. Q8, transparency assessment = WEAK.

P2 (f) Rich information field (Q1, Q10, Q17a)

• Q1 [IS-updates] *Could you please tell me something about the current state of the Dubai residential market?*

Sophisticated real estate markets such as in the United States of America are 'awash with data providers' such as 'CoStar, Real Capital Analytics, CoreLogic as well as Multiple Listing Services' and a 'proliferation of real estate indices (NCREIF, Moody's, or Case Shiller House Price Index, FTSE NAREIT)' (RICS 2014b: 54–55). Mechanistic data-driven research won't solve the Emirati valuation issue but a rich property information field is necessary to inform sound judgements (meta-cognition). To this end, Q1, investigates how well-informed RVS players are about market conditions. One issue with IS information field assessment is the unit of analysis. Fragmentation in the market between global real estate consultants

(Jones Lang, Cushman Wakefield, Savills, CBRE, etc.) who have dual appraisal and brokerage functions and can avail of multiple external (and internal) databases to inform their valuation or sales price judgements, notwithstanding issues of confidentiality and conflicts of interest. Even for agents operating in well-resourced global or GRE firms with decades of market experience expressed (either in response to Q1 or Q6[OPEN]) frustrations about data quality (poverty of information field).

> [R3] – *DLD data issues (date logged variability and even logged same property twice), inadequate fine grain property characteristics details. I use comparisons with which we are familiar or rely on brokers who are active in the market.*

> [R1] – *Lack of access to reliable data (transparency) for comparables. RERA data from DLD goes to REDIN. Also issue of relationship based selective data release such as to Cavendish Maxwell.*

> [R2] – *Transparent sales and rental records as well as transparency and standardisation with detailed breakdown of service charges.*

Clearly, the Emirati RVS has data issues, particularly regarding the reliability and/or selective access to databases help by Cavendish Maxwell or REDIN. Q1 conclusion for P2 (IS system information field) = WEAK.

- Q10 [IS-tech & C-tech] *In your work, to what extent is technology such as Geographic Information Systems or AVMs statistical mass appraisal software important for valuation?*

Q10 evaluates whether respondents made use of advanced technologies or mass appraisal techniques. Almost half of the respondents (14) stated that they considered these advanced technology 'critical', whilst 4 respondents considered it 'very important', and 11 respondents indicated that it was 'largely irrelevant', due to the nature of their work. Therefore, the Q10 conclusion is a mixed one with some players considering it critical but others irrelevant.

When the results are standardised using a Linked Analysis Excel spreadsheet, the thesis generates a representative weighted average score of 3.3.

- Q17a [IS-field+updates] *In your view, how will does the valuation system handle a) Difference in place & building quality?*

Concerning entity specific IS, most respondents considered data handling reasonable with only one respondent complaining:

> [R3] *Generally, BQ is not good (lifespan 40 yrs). Only with master planned projects like the Greens is much attention payed to urban design.*

However, system-wide data issues are another matter. The interviews elicited 21 separate mentions of data quality issues. Clearly, a major concern for a wide range of Emirati valuation system experts is data quality. A few examples of the sorts of data field deficiencies illustrate the point which were raised in OPEN questions such as Q6 (concerns) or Q7 (critical issues).

[R1] – *Lack of access to reliable data (transparency) for comparables. RERA data from DLD goes to REDIN. Also, issue of relationship based selective data release such as to Cavendish Maxwell.*

[R2] – *Transparent sales and rental records as well as transparency and standardisation with detailed breakdown of service charges.*

[R3] – *Transparency and lack of reliable data. Deficiency in DLD data. Brokers not always reliable but Cavendish Maxwell produce a Property Monitor database.*

[R5] – *Access to the information for the relevant and delegated professionals. However, the government nowadays trying to ease access to knowledge and information, but this matter has a conflict of the tradition keeping confidentiality and privacy for landlords.*

[R25] – *The accuracy and availability of information at the start and the end of the whole project, which covers all the relevant data such as area size, building quality, rental income and location.*

[R26] – *The lack of information, relative to particular buildings such as rental income and building age and other required data and information.*

The conclusion for Q17a + [OPEN] is that the information field in the available system information (IS) is WEAK.

P2 (u) Systematic updates (Q1, Q17b &17c)

- Q1 [IS–updates]

Looking at answers to Q1 from the point of view of information updates, a handful of responses suggest some concerns with the system:

[R1] *Some RERA data is out of data since transactions are dated (2 years post transaction).*

[R6] *–We don't have measurement to evaluate and monitor supply and demand in the market nowadays.*

[R14] – *There is a mutual trust in daily transactions nowadays. We do protect all the parties (sellers, buyers, broker) when we are dealing with the title deed transfer.*

However, during the interviews respondents raised the issue of information 51 times. So for example, R1 again for Q6 (Critical issues) is relevant here.

> [R1] – *Lack of access to reliable data (transparency) for comparables. RERA data from DLD goes to REDIN. Some RERA data is out of data since transactions are dated (2 years post transaction). Also issue of relationship based selective data release such as to Cavendish Maxwell.*

Therefore, overall, given the range of concerns and the fragmented access, the Q1 assessment for information updates is 'WEAK'.

- Q17b [IS-updates] *In your view how well does the valuation system handle b) fast pace of change in locales?*

For the second IS update element, *the fast pace of change in locales*, 17 respondents considered it reasonably developing; very well and quite well – 12 respondents, which suggests that the majority were quite satisfied with the pace of change in locales.

- Q17c *In your view how well does the valuation system handle c) fluctuating sentiment in Dubai?*

For the third IS update element – *fluctuating sentiment in Dubai*, 25 respondents considered IS monitoring of fluctuating capital markets as 'reasonable' while 4 respondents considered it 'quite well' or 'very well'. Therefore, overall, respondents were satisfied with the data provided by the IS on capital market sentiment (cycles).

P2 (t) Technical analytics (Q10)

- Q10 [IS-tech] is dealt with earlier (Rich information field) When the results are standardised it generates a representative weighted average score of 3.3.

P2 Conclusion

For P2, IS the multi-criteria analysis generates a score of 2.30, or, in simple terms, the IS of the Emirati RVS needs improvement. In terms of Principal 2: Intelligence: Access to rich information field (quality and systematically updated data), the thesis finds, on balance, the Emirati RVS-IS is inadequate and unreliable.

P3 Capabilities

Principal 3: [Capabilities] RVS institutions are properly configured and governed, with professional staff and sound administration, supported by appropriate adaptive technologies, and the skills and experience to cope with complex valuations.

P3 Indicators

- P3 (g) Governance Q2, Q13
- P3 (p) Professionals Q4, Q11
- P3 (c) Administrative competence Q3, Q4, Q8, Q13
- P3 (t) Supportive technologies (adaptive) Q10
- P3 (mc) Meta-cognition Q12.

P3 Results

Governance (Q2 and Q13)

- Q2 [C-gov] *What do you understand by the term Residential Valuation System in Dubai (RVS)?*
- Q13 [C-gov+admin] *Could you please explain how your department monitors and checks valuation accuracy?*

Q13 deals with both governance capabilities and administrative competence. For organisational governance to avoid the Agency Problem, the basic principal is separation of duties. In practice, this means that independent staff should oversee a valuation to monitor the process and check its accuracy (reasonableness). Worryingly, Q13 results indicate that 6 respondents considered monitoring and checking 'non-applicable'. However, further questioning made clear that this was due to the nature of their work; 2 respondents mentioned quality and assurance department; 3 respondents – independent and internal reviews; 2 respondents – technical and information support; 12 respondents mentioned the importance of appraisal committees and departments; 1 respondent stated the importance of senior staff; 4 respondents referred to financial auditing office; 12 respondents – Dubai Appraisal Department and committee; 2 respondents – technical software.

Translating these results into an ordinal ranking is somewhat tricky but the thesis assumes a graduation from weakest to strongest. The responses varied, indicating a range of approaches and variable governance competence:

[R1] – *Quality management (QM) involves 1) independent review and Quality Assurance (QA) checklist signed off by PA.*

[R19] – *We check the transactions based on daily average price transactions using specialist software which is designed for appraisal department.*

Overall assessment for Q13 = 2.59 OR 'WEAK' – 'OK'.

Professionals (Q4, Q11)

- Q4 [C–prof+C–admin & S–imp] *For each of the valuation purposes you just mentioned, could you please tell me something about the method(s) of valuation practised?*

Discussion of valuation methods was not applicable for 10 respondents but appraisers mentioned a broad range of approaches with, as expected, comparison (13) and income (10) the most common.

Some respondents were obviously very knowledgeable about nuances in the markets and adopted suitable approaches (comparison within submarket not DR cost) to capture the spatial variations in values due to differences in perceptions about construction quality.

> [R1] *Generally, comparison is the favoured method for residential units. The depreciated cost basis is unreliable due to differences in market perception. For example, Arabian ranches on one side of Emirates road is perceived favourable by the market as EMAAR's reputation is strong and people are willing ot pay a 25 per cent premium compared to the opposite development by Dubai Properties which suffers from adverse sentiment notwithstanding that in actual fact often the same subcontractors were used, land prices are identical and build quality is similar. Another example involves DAMAC which overstretched itself but overcame these liquidity issues and this challenge had some adverse impact of a while on the market.*

Other respondents outsourced the valuation tasks.

> [R29] – *We have a third party to perform valuation and discuss the results with EMAAR committee and other property valuers.*

So overall, a mixed range of responses, some reflecting very high degree of professionalism but other responses were less convincing. The conclusion, given that almost 25 per cent thought the question was 'not applicable' is a score of 3.93.

- Q11 [C–prof] *For valuations, what professional qualifications do you consider important?*

Q11 is also linked to the first concept of Institutional Capabilities which is associated with qualified staff in UAE RVS valuation. Whilst several respondents mentioned RICS qualifications or international standards, many other system agents simply relied on local university or professional qualifications (property training) and local valuation work experience. Other qualifications mentioned included knowledge of TGOVA and Emirates Valuation Book, and technical engineering, the use of new technology, valuation certificate from Dubai Real Estate Institute. Examples include:

[R26] – *Experience in property residential market, training and qualification certificate.*

[R 17] – *University qualification certificate and Dubai Real Estate Institute qualification, experience in the field are very important.*

[R 21] – *Experience and practice in this field.*

[R 27] – *Experience, up-to-date knowledge, ongoing practice.*

[R 13] – *Experience and adoption of the best international practices. The use of new technology and joint venture with other relevant entities such as Municipality to know more about the building quality.*

[R14] – *Market and construction experience. University qualification certificate.*

Using the reordering ordinal Likert scoring approach generates Q11 score of 2.25 or 'WEAK'.

Administrative competence (Q3, Q4, Q8, Q13)

• Q3 [C-admin] *In your organisation, when are valuations required and for what purposes?*

Respondents gave several reasons for valuation but three main valuation purposes emerged. First is for *commercial* purposes [registration (6); loans (2); mortgage (2); banks (9); auditing (7); inheritance (10); donation (12); zakat (2)]. Second is for *government* related purposes [courts (8); compensation (3); municipality (5); project (2)]. Finally, *legal purposes* [projects (2); disputes (3); settlement (7); statutory valuation (1); selling and purchasing (9)] with some others [N/A (4); private (9); auction (2)]. Clearly, agents in the system are aware of the legal and administrative role of valuations. Hence, the assessed score for Q3 is between 'OK' and 'GOOD' or 3.5 on ascending Likert scale.

• Q4 [C-prof+admin+ & S-imp] *For each of the valuation purposes you just mentioned, could you please tell me something about the method(s) of valuation practised?*

The results above show that most respondents consider comparison and income cost as the mostly frequently used methods in valuation practice. The allocated Likert scale is 3.93 (see Q3 above).

• Q8 [IS-source & C-admin & S-dissem+S-imp] *In your work, how relevant are valuation standards?*

The allocated Likert scale is 2 (see Q4 above).

- Q13 [C-gov+admin] *Could you please explain how your department monitors and checks valuation accuracy?*

The allocated Likert scale score is 13 is 2.59 (see Q13 above).

Supportive and adaptive technologies (Q10)

- Q10 [IS-tech & C-tech] *In your work, to what extent is technology such as Geographic Information Systems or AVMs statistical mass appraisal software important for valuation?*

The allocated Likert scale score is 3.28 (see Q10 above).

Meta-cognition (Q12)

- Q12 [C-cog] *Talking about valuations conducted in your work, could you outline how you handle difficult or complex cases? [In term of seniority, use of AVM, review of cases]*

A robust valuation system needs to be able to handle complex valuation cases because valuation is both an art and science. Of some concern was the fact that 11 respondents thought the issue of complex valuation was not applicable at all; 5 respondents raised cases involving major projects; 2 respondents mentioned external and global review requirements; 2 respondents highlighted the lack of relevant information or overload; 2 respondents underlined the importance of site visits; 23 respondents admitted the uniqueness of each case; 7 respondents mentioned relevant department for valuation process; 3 respondents referred to top management for complex valuation; 2 respondents linked complex cases with infrastructure. Assuming that NA responses indicate unawareness of the complexity issue, this translates into an approximate Likert Scale score of 4.

P3 Conclusion

In conclusion, having systematically investigated the constituents for institutional capability via toolkit instrument questions [indicators] (Figure 13.3), the aggregate score for the Emirati RVS on Principal 3 (capabilities) is 2.94 which is just about 'OK' or reasonable. One issue that emerges is perhaps the need for more professionally trained Emiratis with internationally recognised qualifications such as MRICS.

P4 Trust

Principal 4: [Trust] RVS players trust system and its valuation outputs

P4 Indicators and questions

- Collaboration Q14
- Users trust valuations Q5, Q15
- Users trust system Q16.

P4 Results

Collaboration (Q14)

- Q14 [T-col] *How often does your department coordinate/meet regularly with other external institutions?(Such as Dubai Municipality, Taqyeem and other valuers)*

Assuming that responses indicate level of coordination, this translates into an approximate Likert Scale score of 2.96 – basically 'OK'.

Users trust valuations (Q5, Q15)

- Q5 [T-vals] *In general, how confident are you in residential valuation accuracy?*

Assuming that responses ascending strength of confidence, this translates into an approximate Likert Scale score for Q5 of 4.10 – basically 'GOOD'.

- Q15 [T-vals] *Thinking about recent meetings with various external real estate players, how confident are they in UAE valuations generally?*

Assuming that responses ascending strength of confidence, this translates into an approximate Likert Scale score for Q15 of 3.90 – moving towards 'GOOD'.

Users trust system (Q16)

- Q16 [T-syst] *To what extent do market participants trust the UAE RVS?*

Assuming that responses ascending strength of confidence, this translates into an approximate Likert Scale score for Q16 of 3.66 – moving towards 'GOOD'.

P4 Conclusion

The thesis investigated Emirati RVS Principal 4: [Trust] using the Framework indicators and questions, involving:

- Collaboration Q14
- Users trust valuations Q5, Q15
- Users trust system Q16.

The results in the Linked Analysis spreadsheet generate a P4 (T) score of 3.54 between 'OK' and 'GOOD.

P5 Salience

Principle 5: [Standards salience] valuation standards are widely disseminated, discussed and best practices implemented.

P5 Indicators and questions

- Valuation standards disseminated Q3, Q8
- Discussed Q9
- Implemented Q4, Q8.

P5 Results

Valuation standards disseminated (Q3, Q8)

- Q3 [C-admin & S-diss] *In your organisation, when are valuations required and for what purposes?*

The previously assessed Likert scale score is 3.5 (see Q3 above).

- Q8 [IS-source & C-admin & S-diss+imp] *In your work, how relevant are valuation standards?*

The previously assessed Likert scale score is 2 (see Q8 above).

Discussed (Q9)

- Q9 [S-discuss] *Thinking about your department, how frequently do you discuss issues related to valuation standards?*

The discussion of valuation standards among the respondents varies considerably. Two respondents claimed they are involved in daily discussions about standards but others were preoccupied with other matters; 6 discussed valuation standards weekly or monthly, whilst 7 only annually and 8 respondents never raised the issue of standards at all. These results suggest that the salience of valuation standards in the system is inadequate.

The Q9 results generated a score of 2.55, below 'OK'.

Implemented (Q4, Q8)

- Q4 [C-prof+admin & S-imp] *For each of the valuation purposes you just mentioned, could you please tell me something about the method(s) of valuation practised?*

The previously assessed Likert scale score is 3.93 (see Q4 above).

- Q8 [IS-source & C-admin & S-diss+imp] *In your work, how relevant are valuation standards?*

Reassuringly, most respondents (18/27 or 67 per cent) stated that valuation standards were at least 'quite important' but many of these considered them 'critical'. However, it is concerning that for a substantial minority of respondents, valuation standards are considered irrelevant for valuation.

The previously assessed Q8 Likert scale score is 2 (see Q8 above).

P5 Conclusion

Principle 5: [Standards salience] investigates if in the Emirati system, valuation standards are widely disseminated, discussed and best practices implemented. P5 Indicators and questions involved:

- Valuation standards disseminated Q3, Q8
- Discussed Q9
- Implemented Q4, Q8.

The results in the Linked Analysis spreadsheet generate a P5 (S) score of 2.76 or a 'FAIL' below 'OK'.

Conclusion of Operational Phase

Operational Phase interview evidence generates aggregated mean scores for four key aspects of UAE RVS performance:

- P2: Intelligence systems FAIL
- P3: Institutional capabilities FAIL
- P4: Trust OK
- P5: Standards salience FAIL.

Conclusion

To manage development pressures and improve resource allocation, smart cities need reliable information systems. The residential valuation system nests within the urban system but it vests in both public sector institutions and, laterally, in banks and commercial real estate firms. To determine whether the Emirati residential valuation system (RVS) is 'fit for purpose', the research developed an assessment toolkit. The literature and preliminary discussions suggested a framework with five key aspects:

- System output (reliability of valuations).
- Intelligence systems (data governance, source transparency, rich information field, systematic updates, analytics (statistics/forecasts/modelling).
- Institutional capabilities (governance, professionals, administrative competence, supportive technologies (adaptive), meta-cognition).
- Trust (collaboration, users trust valuations, users trust system).
- Standards salience (valuation standards disseminated, discussed, implemented).

The research evaluated the residential valuation system in Dubai using mixed methods, involving exploratory statistics, site visits, embedded observation, expert interviews and confirmatory discussions. The statistical analysis of property transactions and empirical site investigations found that the system generated reasonable valuations, notwithstanding market complexity and exposure to regional risks. In addition, most stakeholders trusted the residential valuation system. However, the research found indicative weakness in aspects of UAE residential valuation system, relating to information, capabilities and valuation standards.

Appendix 13.1 Linked Analysis

Principle	Concept	Indicator	Very poor 1	Weak 2	OK 3	Good 4	Excellent 5	Score		Check / (sRef)
P1	Valuations reasonable	Macro-examination E1			NA					
P1	Valuations reasonable	Qualitative examination E2			NA					
P1	Valuations reasonable	Statistical submarket examination E3.1			NA					
P1	Valuations reasonable	Statistical submarket examination E3.2			NA					
P1	Valuations reasonable	Statistical submarket examination E3.3			NA					
P1	Valuations reasonable	Statistical submarket examination E3.4			NA					
P1	Valuations reasonable	Output reasonable E4.1			3			3.00		
P1	Valuations reasonable	Output reasonable E4.2			3			3.00		
P1	**Valuations reasonable**	**Output reasonable**			**3**			**3.00**	**OK**	
P2	Intelligence systems (IS)	Data governance EB1_IS(g) (s6.3.4)	1					1.00		
P2	Intelligence systems (IS)	Data governance Q2		2				2.00		
P2 (g)	*Intelligence systems (IS)*	*Data governance AVERAGE*						*1.50*		
P2 (st)	*Intelligence systems (IS)*	*Source transparency Q8*		2				*2*		
P2	Intelligence systems (IS)	Rich information field Q1		2				2.00		
P2	Intelligence systems (IS)	Rich information field Q10	11	0	2	2	14	3.28		29
P2	Intelligence systems (IS)	Rich information field Q17a		2				2.00		
P2 (f)	*Intelligence systems (IS)*	*Rich information field AVERAGE*						*2.43*		
P2	Intelligence systems (IS)	Systematic updates Q1		2				2.00		
P2	Intelligence systems (IS)	Systematic updates Q17b	0	0	17	5	7	3.66		29
P2	Intelligence systems (IS)	Systematic updates Q17c	0	0	25	2	2	3.21		29
P2 (u)	*Intelligence systems (IS)*	*Systematic updates AVERAGE*						*2.95*		
P2	Intelligence systems (IS)	Technical analytics EB3_IS(t)		2				2.00		
P2	Intelligence systems (IS)	Technical analytics Q10	11	0	2	2	14	3.28		29
P2 (t)	*Intelligence systems (IS)*	*Technical analytics AVERAGE*						*2.64*		
P2	**Intelligence systems (IS)**	**Intelligence systems (IS)**						2.30	**FAIL**	**6.4.4.3 P3** **Conclusion**

Appendix 13.1 Continued

Principle	Concept	Indicator	Very poor 1	Weak 2	OK 3	Good 4	Excellent 5	Score	Check / (sRef)
P3	Institutional capabilities (C)	Governance EB1_C(g) (S6.3.4)	1					1.00	
P3	Institutional capabilities (C)	Governance Q2		2		5		2.00	
P3	Institutional capabilities (C)	Governance Q13	6	12	6	5	3	2.59	32
P3 (g)	*Institutional capabilities (C)*	*Governance AVERAGE*						*1.86*	
P3	Institutional capabilities (C)	Professionals EB2_C(p) (s6.3.4)		2				2.00	
P3	Institutional capabilities (C)	Professionals Q4						3.93	
P3	Institutional capabilities (C)	Professionals Q11	29	9	20	4	6	2.25	68
P3 (p)	*Institutional capabilities (C)*	*Professionals AVERAGE*						*2.73*	
P3	Institutional capabilities (C)	Administrative competence Q3						3.50	
P3	Institutional capabilities (C)	Administrative competence Q4						3.93	
P3	Institutional capabilities (C)	Administrative competence Q8						2.00	
P3	Institutional capabilities (C)	Administrative competence Q13						2.59	
P3 (c)	*Institutional capabilities (C)*	*Administrative competence AVERAGE*						*3.01*	
P3	Institutional capabilities (C)	Supportive technologies (adaptive) EB2_C(t) (s6.3.4)			3			3.00	
P3	Institutional capabilities (C)	Supportive technologies (adaptive) Q10						3.28	
P3 (t)	*Institutional capabilities (C)*	*Supportive technologies (adaptive) AVERAGE*						*3.14*	
P3 (mc)	*Institutional capabilities (C)*	*Meta-cognition Q12*						*3.96*	
P3	**Institutional capabilities (C)**	**Institutional capabilities (C) AVERAGE**						**2.94**	**FAIL**

Appendix 13.1 *Continued*

Principle	Concept	Indicator	Very poor 1	Weak 2	OK 3	Good 4	Excellent 5	Score		Check / (sRef)
P4	Trust (T)	Collaboration Q14	4	7	8	4	5	*2.96*		28
P4	Trust (T)	Users trust valuations Q5	0	0	1	25	4	4.10		30
P4	Trust (T)	Users trust valuations Q15	0	1	4	21	3	3.90		29
P4	Trust (T)	*Users trust valuations AVERAGE*						*4.00*		
P4	Trust (T)	Users trust system Q16	0	2	8	17	2	*3.66*		29
P4	**Trust (T)**	**Trust (T) AVERAGE**						**3.54**	**OK**	
P5	Standards salience (S)	Valuation standards disseminated Q3						3.50		
P5	Standards salience (S)	Valuation standards disseminated Q8						2.00		
P5	Standards salience (S)	*Valuation standards disseminated AVERAGE*						*2.75*		
P5	Standards salience (S)	Discussed Q9	8	7	6	6	2	*2.55*		29
P5	Standards salience (S)	Implemented Q4						3.93		
P5	Standards salience (S)	Implemented Q8						2.00		
P5	Standards salience (S)	*Implemented AVERAGE*						*2.96*		
P5	**Standards salience (S)**	**Standards salience (S) AVERAGE**						**2.76**	**FAIL**	

References

Creswell, J. W. (2009). *Research design: qualitative, quantitative, and mixed methods approaches.* Los Angeles, CA: Sage.

Creswell, J. W., Plano Clark, V. L., & Garrett, A. L. (2008). 'Methodological issues in conducting mixed methods research designs'. In M. Bergman (Ed.), *Advances in Mixed Methods Research* (pp. 66–83). Thousand Oaks, CA: SAGE.

Crosby, N. (2000). 'Valuation accuracy, variation and bias in the context of standards and expectations'. *Journal of Property Investment and Finance* 18(2): 130–161.

Crosby, N. and Hughes, C. (2012). 'The basis of valuations for secured commercial property lending in the UK'. *Journal of European Real Estate Research* 4(3): 225–242.

Davidson, C. (2008a). 'Dubai: the security dimensions of the region's premier free port'. *Middle East Policy* 15(2): 143–160.

Davidson, C. (2008b). *Dubai: The Vulnerability of Success.* New York: Columbia University Press.

Fama, E. (1980). 'Agency problems and theory of the firm'. *Journal of Political Economy* 88(2): 288–307.

Hepsen, A. and Vatansever, M. (2011). 'Forecasting future trends in Dubai housing market by using Box-Jenkins autoregressive integrated moving average'. *International Journal of Housing Markets and Analysis* 11(4): 210–223.

Kaplan (2012). Financial Knowledge Bank, available at: http://kfknowledgebank.kaplan.co.uk/KFKB/Wiki%20Pages/Agency%20theory.aspx (accessed 25 February 2017).

Lamberton, G. (2005). 'Sustainability accounting – a brief history and conceptual framework'. *Accounting Forum* 29(1): 7–26.

McGreal, S. and Taltavull, P. (2012). 'An analysis of factors influencing accuracy and the valuation of residential properties in Spain'. *Journal of Property Research* 29(1): 1–24.

McParland, C., Adair, A. and McGreal, S. (2002). 'Valuation standards: a comparison of four European countries'. *Journal of Property Investment & Finance* 20(2): 127–141, accessed at: www.emeraldinsight.com/doi/abs/10.1108/14635780210420025

RICS (2010). *Valuation Standards,* 6th Edition. Coventry: Royal Institution of Chartered Surveyors.

RICS (2014a). *Global valuation standards,* VPS 4 Bases of value, assumptions and special assumptions.

RICS (2014b). *The role of international and local valuation standards in influencing valuation practice in emerging and established markets.* London: RICS.

Streiner, D. L. (2006). 'Building a better model: an introduction to structural equation modelling'. *Canadian Journal of Psychiatry* 51(5): 317–324.

14

RETROFITS AND GREEN BUILDING REFURBISHMENT

Jonas Hahn[1]

Abstract

Smart development of urban built environments must consider modifications to the existing building stock to realize potentially tremendous energy savings and emission reductions. In fact, the energy-efficiency retrofitting of buildings is a crucial sustainability driver and a 'low-hanging fruit' for achieving climate and energy conservation goals on all regulatory levels. Besides its climatic benefits, several studies suggest that retrofitting is commercially viable. Lower operating expenses, higher willingness-to-pay by tenants and property value premiums were investigated and confirmed.

Several factors contribute to the success of urban retrofitting programs, including a high share of rental housing, high population density and major property investors. An energetic local authority to drive energy improvements is also helpful. In these circumstances, retrofits are very cost-efficient and 'win–win' solutions for all stakeholders. Curiously, despite its advantages, retrofit activity lags expectations. Information asymmetry explains the paradox because investors struggle to access adequate viability data.

This chapter outlines the factors affecting financial profitability of energy retrofits. It shows that building age and typology, energy prices and their forecasts, investor strategies and financing as well as the market reaction toward green retrofits all influence profitability. Also, a case study covering retrofitting measures from a real-life field project as well as the methodology for a simple, but profound profitability calculation are introduced and discussed.

Key terms

Urban built environment; energy efficiency; carbon footprint; green retrofits; profitability analysis; economic impact; financial model; investment incentives

Learning outcomes

- Energy-efficiency retrofits are not only politically desired and ecologically required. They also may make sound commercial sense and therefore generate value to financial investors.
- The property industry has a specific responsibility in terms of achieving the political goals, as it is a main causer of emissions and energy consumption.
- The main barriers to the implementation of retrofit measures are the user–investor dilemma, where one party pays the investment, but another party would gain the benefits; the high initial investment costs (typically co-financed by debt); as well as a lack of information and awareness about the financial profitability of such measures.
- The extent to which the energy-efficiency measures are economically beneficial depends on the interplay of certain parameters that can be summarized under typologies that describe the investor and their strategic approach and conditions, the market and its reaction towards retrofit measures, the property and the realizable energy savings as well as the country's legal/taxation regime.
- The variables within these typologies can act as drivers or as constraints of the commercial viability. The better these parameters complement each other, the higher the probability for commercial viability.
- Energy-efficiency retrofits should be analyzed by the Visualization of Financial Impact (VoFI) method. This method delivers a structured overview over time and an equity-oriented key figure, which is comprehensible to both private and professional investors. At the same time, it is thorough enough to comply with industry standards.

Introduction

On December 12, 2015, an agreement of historical dimension was made in Paris, France: for the first time in history, a global community of 195 sovereign nations unanimously signed a 'universal, legally binding global climate deal' (European Commission 2015). This is widely considered the – for the moment – greatest result of a long period of previous political negotiations, scientific analyses and engagement of non-governmental organizations, which altogether had anthropogenic climate change on their agendas on national and international levels. For instance, Europe postulated ambitious climate targets in 2008, which appeared as a tremendous self-burdening at the beginning, but which find themselves in good company with the latest Paris agreement, where the climate targets were drastically tightened on a global scale.

As buildings account for approximately 40 percent of the total energy consumption and more than one third of the carbon dioxide emissions in Europe, it is energy-efficiency retrofits in terms of space and water heating, insulation, appliances, cooling and ventilation, and lighting, which contain the greatest potential for progressing on emission reduction targets[2] (Paiho/Pinto-Seppä/Jimenez 2015: 2; Rapf/Groote 2015: 7).

However, modernization rates, that is the actual application of energy-efficiency retrofits within the total building stock, stay behind expectations. One of the reasons for this is identified as the lack of insight into commercial viability for investors. Therefore, this chapter aims to provide a methodology for the commercial evaluation of such retrofit measures. It embeds these measures into urban context and introduces them, before giving in-depth guidance on considerable drivers and constraints that impact the financial profitability and an easily applicable model for respective calculations.

Energy efficiency in the urban building stock

This chapter puts special emphasis on the role of energy-efficiency measures in the context of urban structures. While the rural area shows a generally higher volume of greenhouse gas emissions per household,[3] the urban area shows a higher leverage in terms of performing energy-efficiency investments and in terms of their financial impact. The next section outlines reasons for that, followed by an overview of typical retrofit measures in the urban context.

The urban context in energy efficiency

From the conceptual foundation for the worldwide energy shift, we know that the urban area plays a key role. Cities cause a total of 71–76 percent of the global carbon dioxide emissions and are responsible for 67 percent to 76 percent of the global energy use (Seto et al. 2014: 927). The identity of cities is largely driven by its historic and traditional buildings. These are not only different in terms of construction, but also regarding their technical equipment. In order to keep this building stock vital, it needs to be adjusted to current requirements and needs (Lewis/Hógáin/Borghi 2013: 6). This gains further urgency because of the international urbanization megatrend. In 2014, more than half of the world's population was settling in cities and forecasts for 2050 predict an increase to an average of 70 percent globally (United Nations 2014: 1; Eames et al. 2013: 504).

In the light of these insights, the urban lifestyle is both part of the problem as well as part of the solution. The high population share explains the high energy consumption in cities. On the other hand, cities provide a specifically high potential for energy savings, as in these high density settlement districts a more efficient implementation of energy-saving designs in living, transport and general service provision is possible (Lewis/Hógáin/Borghi 2013: 7). In terms of properties, the urban areas allow additional hypotheses for a different effectiveness in environment-friendly investments compared to the same investments in rural areas:

1   One investor manages (and can improve) a higher volume of cubature.
2   One investment can impact more than one housing unit, as multi-family houses and apartment blocks are not that common in the rural context.

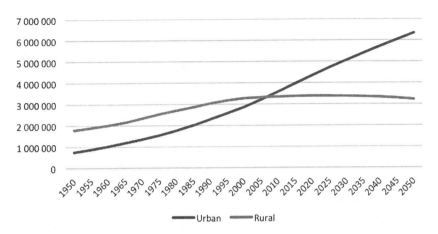

FIGURE 14.1 Urbanization worldwide between 1950 and 2050

Source: United Nations (2014: 7).

3   Domestic engineering is more efficient in operation, as maintenance costs per housing unit are lower than for areas with a high share of single-family houses. On the other hand, construction in cities may be subject to higher construction costs caused by lower supply available to satisfy high demand.

4   The investor in urban areas benefits from economies of scale regarding the investment costs and running costs of the measure.

5   The production of energy-friendly biomass energy as well as its distribution is not common in the urban area, while they support a 'green' energy mix in the countryside.

6   Solar energy production is typically significantly small scale compared to the possibilities of the broad expanse in rural areas.

7   The urban area has a higher share of rental housing. These are specific in terms of the existence of a user–investor dilemma: while the investor (owner) pays the energy-efficiency measure costs, the user (tenant) is the primary beneficiary of operating costs saved.

8   Depending on the legal situation, in rental housing both property-related advantages (in terms of energy savings and reduced operating costs) as well as investor-related advantages (for instance, rent increases after energy-efficiency retrofit) may be realizable.

For modern urban planning, the deciding challenge consists in developing profound concepts, how both obsolete building stock and urban infrastructure can be adapted to the changing framework conditions that arise from the progressing climate change and the scarcity of resources (Eames et al. 2013: 504).

In this context, designing and performing measures of energy-efficiency retrofits is a crucial instrument.

Overview and discussion of retrofit measures

Barriers and advantages

It was argued previously that energy-efficiency retrofits in the urban context are connected with several challenges. In addition to what was mentioned before, it is especially political will and political action that impacts urban planning and promotes – or hinders – the execution of retrofit measures (Amecke et al. 2013: 1). In addition to that, the following obstacles can also act as barriers:

Short-term costs versus sustainable long-term benefits
High initial investment costs are a barrier especially to private non-professional investors, as they may have difficulties to set up such financing in accordance with their private goals. But also for professional investors and housing companies, this may depict difficulties, as the savings and additional financial benefits are realized over time in a longer-term process. In this context, retrofit measures with higher amortization periods are neglected and only those measures may be considered that are very efficient in terms of money, but not necessarily very effective in terms of the energy savings lever. Public intervention can steer this by providing subsidized financing deals and by defining measures that can benefit from these offers.

The split incentive problem (or 'user–investor dilemma')
A very obvious barrier arises, if one party carries the cost for the retrofit measure, but another party would be the beneficiary – which is the case especially in urban rental housing. This leads to an incentive problem for investors or owners, who would provide utility to their tenants without any (immediate) benefits. This problem is best solved by the legal framework that should distribute the benefits out of the investment between both contractual parties and grant rights to increase the net rent up to neutrality in gross rent or per another reasonable rule.

Lack of information and awareness
Both parties – owners and tenants – regularly lack information on the energy savings potential from a retrofit and the financial benefit for both parties (Polly et al. 2011: 2). That is why these win–win situations are frequently not induced at all, although they might be beneficial. Also, the construction industry and their knowledge and awareness are drivers, but also barriers in case of non-existence, as these measures might be underestimated or avoided or finally performed inefficiently (Amecke et al. 2013: 5–6).

As soon as investors are aware of the barriers – or rather constraints – and consider them in their risk deliberations, they might be interested in what would be the financial benefits from these actions. A study by McKinsey & Company finds that the housing sector could save around 28 percent in energy consumption – and thereby energy costs – by 2020, if the scenario 'business as usual' would be replaced by an increased application of energy-efficiency retrofits. In fact, with a forecast of 560 TWh, existing buildings are estimated to have a five times higher leverage on avoiding consumption than new buildings and it is specifically measures in the context of the building envelope and the heating and cooling system that are considered high potential levers, emphasizing the importance of energy-efficiency retrofits in existing buildings (McKinsey 2009: 29–30).

In its *Business Case for Green Buildings*, the World Green Building Council (2013) finds sales premia of certified properties[4] of up to 20 percent, resulting from higher achievable rents, lower vacancy rates and lower operating costs (2013: 31, 37).

For the U.S., property owners reported double-digit increases in their Return on Investment (RoI) after energy-efficiency retrofits and confirmed that not only occupancy rates could be increased (at a median of 2.5 percent after retrofits), but also the actual rent charged. An average value increase of 6.8 percent after retrofit was observed (McGraw Hill Construction 2010: 10).

And not only practitioners, but also scientists have widely researched the operational and financial benefits of energy efficiency in the building context.[5] In their studies, they find evidence that not only have energy-efficient buildings lower operating costs, but they are therefore comparable to a hedge against increasing energy prices. They also show higher occupancy rates and higher average sales or rent prices. And furthermore, impacts on sociocultural level can be observed; besides an increase in the general level of comfort, specifically better working atmospheres and higher employee satisfaction were identified (World GBC 2013). The advantages of energy-efficiency retrofits should therefore be seen in a long-term context (Lewis/Hógáin/Borghi 2013: 5–6).

Specific measures and their impact

According to estimates of the U.S. Green Building Council, property owners in commercial real estate are estimated to invest almost one trillion USD until 2023 into sustainably retrofitting their existing buildings. Specific measures comprise mainly the improvement of energy efficiency in heating, air conditioning, ventilation as well as windows and lighting systems (U.S. GBC 2015: 1).

In a structured approach, retrofitting measures can be classified into the categories shown in Table 14.1.

For instance, replacing conventional bulbs with LED technology in residential real estate proves to be highly profitable financially, as approximately 160 EUR of net benefits can be generated per ton of CO_2 equivalents saved. However, the total emission savings potential is comparably low. As a synopsis, Table 14.2 contrasts emission saving potentials of the different property related category with the abatement cost effect, shown as the financial profitability behind the measures.

Assessing profitability

Various financial performance measures indicate project commercial viability, including a positive Net Present Value or Internal Rate of Returns (IRR) which

TABLE 14.1 Categories and exemplary measures of energy-efficiency retrofits

Category	Exemplary individual measures
BUILDING ENVELOPE	• Insulation of external walls and roofs • Replacement of windows and doors • Passive photovoltaic heating and cooling
TECHNICAL SYSTEMS AND EQUIPMENT	• Replacement of heating, cooling, air-conditioning and lighting instruments by installations with higher efficiency • Heat recovery • Insulation of pipes
RENEWABLE HEATING AND ELECTRICITY	• Groundwater heat pump • Photovoltaic and/or thermal solar collector • Wind and hydro energy
POWER TRANSMISSION AND DISTRIBUTION	• Access to the district heating network • Electronic appliances

Source: Based on Lohse/Zhivov (2014: 5).

TABLE 14.2 Profitability and potential of retrofit measures

Measures	Abatement profitability	Emission savings potential
LIGHTING	Very high	Rather low★
APPLIANCES	Very high	Moderate
HEATING, COOLING, AIR-CONDITIONING	High	Very high
INSULATION	High	High

Source: Based on McKinsey & Company (2010: 8).

Note ★ While the energy saving potential on the property level may be presumed high intuitively, lighting has only limited potential in terms of global emission savings.

exceeds a specific target rate. Here, for simplicity, we simply consider an energy retrofit investment profitable if it earns all capital costs. Capital costs consist of the payments due to debt financing (primarily interest payments) and the expected equity return. In practice, the question whether cost savings result from the energy-efficiency retrofit measure is quite complex. It depends on the highly individual situation and framework conditions of both investor and investment. The following section (1) introduces the perspective of typologies to understand the higher-level driving forces for the profitability of the investment; (2) outlines how these parameters can be driving or constraining the financial profitability by specific examples, before (3) suggesting a calculation methodology to determine the financial benefits arising.

The perspective of typologies

The profitability of an energy-efficiency investment depends on a whole spectrum of influencing factors. It makes sense to summarize all these variables under a higher-level category in order to understand which factors should be considered at an early stage.

Four critical domains affect the viability of a retrofit:

1   *The Investor.* Investor profile and investment vehicle structure influence retrofit profitability due to differences in risk appetite and in funding costs. For some investment vehicles, debt is the main source of finance and risk appetite is high when markets are rising. Less leveraged investment vehicles only seek to cover moderate equity yields. Another consideration is asset holding period which varies strongly between different investor types.

2   *The Property.* The characteristics of the property, the building age and condition, its envelope, its u-value improvements through past retrofits, and ultimately the building quality are the essential drivers for any energy saving to be achieved. Specifically, for very old buildings, energy-efficiency retrofits are financially more beneficial than for those constructed in the more recent

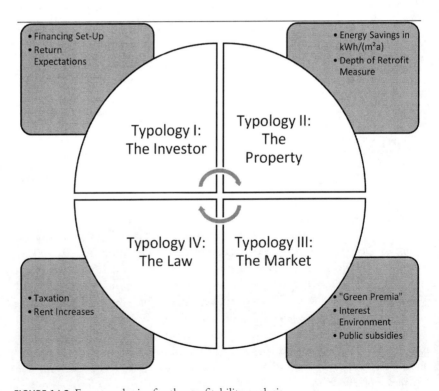

FIGURE 14.2 Four typologies for the profitability analysis

past (Paiho/Pinto-Seppä/Jimenez 2015: 2). Energy savings, however, must be distinguished from the financial benefits that arise to the investor, as the tenant may be the primary beneficiary of such achievements. Also, the property and the possibly realizable energy retrofit measures may be correlated or exclusive, depending on which particular measures are planned by the investor.

3   *The Market.* The market environment provides essential framework conditions in which the energy-efficiency investment is embedded. This comprises the interest level, the availability of public subsidies or the acceptance of higher energy standards in the form of a higher willingness to pay. Empirical evidence on this will be presented in the course of this chapter.

4   *The Law.* The legal framework is one of the underestimated forces of profitability. There are essentially two legal aspects that impact the profitability in both possible directions: first, the taxation framework (including tax rates, but also as it defines questions of deductibility of these measures) has direct influence on cash flows generated – or not generated – by the measure. As a second aspect, the possibilities of rent increases after energy-efficiency retrofits may be defined by legal standards.

While this section fundamentally introduced the driving typologies on a higher level, the following section will go into further details and explain how these typologies and the included aspects (variables) can influence the outcome of the profitability analysis.

Drivers and constraints of profitability

For each typology, this section outlines two examples on how in particular they and their contained variables have a positive or negative impact on the profitability assessment outcome.

Certainly, many additional functional chains in terms of how profitability is impacted by the typologies and their variables could be found. The following section will now illustrate an appropriate methodology for assessing the profitability of the measure.

The Visualization of Financial Impact (VoFI) methodology

The Visualization of Financial Impact – or simply, VoFI method – is an objective methodology for evaluating the profitability of investments and, while internationally available, is very common in Germany. The methodology contains the explicit *tabular reflection* of all the series of payments that are related to that investment (Grob 1993: 51). Contrary to the classical methods of investment appraisal, which are based on simplified formula from financial mathematics, this present concept offers the possibility to choose each underlying assumption unrestrictedly and to illustrate those realistically, transparent and nuanced over time.

TABLE 14.3 Impact of selected variables on the profitability assessment outcome

Typology	Variable	Impact
THE INVESTOR	Debt Portion	Adding debt to the retrofit project financing can increase the equity yield for the investor ('income gearing') by making use of the so-called *Financial Leverage Effect*, if the Return on Investment is higher than the borrowing interest rate for debt capital.
	Holding Period	The longer the holding period, the higher the forecasting uncertainties in terms of rent generation, vacancy rates or cost projections. As the validity of prognosis is limited, these risks can either increase or reduce the profitability.
THE PROPERTY	Building Age	Not every property can generate significant energy savings through energy-efficiency retrofits. Specifically, very old buildings without interim refurbishments have maximum potential in terms of saving energy costs through retrofits. With younger properties, especially those built in this century, the number of retrofit measures that can lead to profitable results, is very limited.
	Envelope Shape	The more complex the envelope shape, the higher the ex-ante energy consumption. However, also the ex-post consumption can be expected to be higher than in a comparable building of simple envelope shape. Also, the investment costs for retrofits are probably higher than in uncomplicated properties. Therefore, a simple ground plan is a good indication for lower investment costs at a given energy savings potential, therefore increasing the profitability of the measures.
THE MARKET	Interest Rate Level	A high interest rate for borrowing may over-compensate the achieved energy savings and therefore reduce profitability. The same is true for the Leverage Effect in the financial setup of the investment, which can be endangered; this would reduce the equity yield. In this context, investor should consider the availability of public subsidized financing, which may be rewarding if they are significantly lower than the market borrowing rate.
	Energy Price Level	It may appear paradoxical, but a high – and increasing – energy price level may be beneficial for the financial profitability of a retrofit investment. High energy efficiency is like a hedging instrument against these price tendencies as the energy costs saved increase in the same way like energy prices do, as energy unused does not need to be paid and avoided costs increase profitability.
THE LAW	Retrofit and Rent Increase Regulation	For the realization of financial benefits, also the legal framework plays a crucial role. Especially, if tenants would be allowed to reduce rent due to molestation from the retrofit construction work, this must be considered in the total calculation as it lowers profitability. On the contrary, a legal standard that defines the specific increasable amount of the net rent after retrofit, bursts the split incentive outlined in section Barriers and advantages and promotes the profitability of the measure.
	Tax Deductibility	If the investment costs for the retrofit are fully tax-deductible, their depreciation reduces the tax payment. Otherwise only the respective parts of the total volume can be deducted, reducing the overall profitability.

The considered payments do not refer, contrary to present value-oriented approaches, to the investment date, but to the planning horizon. Altogether, this methodology makes a highly comprehensible investment appraisal possible (Bone-Winkel et al. 2015: 604–605).

Structure and functionality

The VoFI calculation contains original data, that is payments that are directly relatable to the investment, as well as derivative data, such as payments for financing the investment as well as tax payments. These are usually not provided exogenously, but are based on underlying information and assumptions. These contain current tax rates, legal standards as well as relevant calculation commandments (Grob 2015: 105).

The initial payment comprises the investment costs for the energy-efficiency retrofit measure in $t = 0$. From the financing perspective, there is usually a certain equity portion as well as a certain volume in debt capital used, which are included in a separate financing plan. If such debt is borrowed in the form of an annuity loan, a periodical debt service must be paid, that is distinguished into amortization payment and interest payment (Grob 1993: 59). In this case, the annuity stays constant over time, while the interest portion is decreasing as it is coupled to the development of the residual debt. Alternative debt conditions may be installment loans (with constant amortization payments and decreasing interest payments over time) or bullet loans, with constant interest payments and one final amortization payment in the end.

Besides these liabilities, positive cash flows are considered in terms of (increased) rental payments or savings in energy costs that are due to the retrofit. These payments are considered available for a re-investment at the current interest rate, generating additional payments in the next period. Tax aspects are handled in a side calculation; its results are then transferred into the actual VoFI (Grob 1993: 59). An additional side calculation is the depreciation plan, which must determine the depreciation volume for each period as well as the residual book value of the retrofit over time. In its result, VoFI shows the absolute final equity value of the investor when performing the retrofit investment. If the final equity value is higher than in the scenario of not investing in the energy-efficiency retrofit, the measure is profitable and should be performed (Grob 1993: 61).

In order to increase the comparability of alternatives, the VoFI can also contrast the final equity value (K_N) with the original equity share of the total investment in $t = 0$ (EQ_0) to calculate a VoFI yield figure. This figure shows the mathematical average interest rate, at which the equity is invested periodically and can then be easily compared to market conditions for investments, such as government bonds (Bone-Winkel et al. 2015: 609). Also, comparisons with the expected equity return of the individual investor are possible in order to decide for the measure to be performed or not (Kofner 2010: 180).

The following formula illustrates the calculation of the VoFI Equity Yield r_{VOFI}:

$$r_{VOFI} = \sqrt[N]{\frac{K_N}{EK_0}} - 1$$

FORMULA 1: VOFI EQUITY YIELD – AVERAGE P.A.

Due to spatial limitations, this methodology is not extensively discussed at this point. Of course, there are aspects about it that could be discussed, as it is an economic model based on certain assumptions. From a positive point of view, however, this methodology is transparent enough and delivers results that are easy enough to understand even for non-professional investors. On the other hand, as it delivers an equity yield figure that is comparable to a Modified Internal Rate of Return (MIRR), it is complete and correct enough to meet industry standards.

(See Appendix 14.1 for an exemplary profitability analysis.)

CASE STUDY **AN EXTENSIVE ENERGY-EFFICIENCY RETROFIT IN THE CZECH REPUBLIC**

This study illustrates the case of a 2,500 m² apartment block in a city in the Eastern Czech Republic. The apartment block was undergoing a global retrofitting in 2010 and 2011, which included thermal insulation of the building envelope, insulation and repair of the roof, repair of the attic and lighting conductor, the replacement of wooden windows and balcony doors with plastic windows as well as the repair of entrance doors including lighting. The energy saving measures specifically comprised:

- a thermal Insulation Composite System (ETICS) with polystyrene EPS 70 F, 140 mm, on external walls,
- an extruded polystyrene XPS insulation of 100 mm thickness in perimeter and base area,
- the insulation of the basement ceiling with hydrophobic boards from mineral fibres (80 mm),
- the insulation of the roof with boards from fire-retardant foam polystyrene EPS 100 S Stabil (200 mm),
- the replacement of old wooden windows with new multi-chamber plastic windows, and of balcony doors with double glazing (frame U=1.3 W/m²K, window glass U=1.1W/m²K) as well as
- the replacement of old metal and wooden entrance doors with entrance doors made of aluminium profiles (U=1.5 W/m²K) with safety insulation double glass.

The investment costs of this 'deep renovation' were 590,000 EUR; however, parts of these costs are not directly linked with energy saving measures, but

FIGURE 14.3 Before the energy-efficiency retrofit

Source: Monika Kwapulinska.

FIGURE 14.4 Before the energy-efficiency retrofit

Source: Monika Kwapulinska.

additional repair measures performed at the same time. The energy-related investment cost amounted to 309,000 EUR net. Aside of an energy saving of 36.5 percent (206,111 kWh p.a. down from 327,725 kWh, annual average normalized with heating days), maintenance and running costs for the building went down by approximately 40 percent, with a value of 4,000 EUR p.a. before retrofit.

This financial profitability analysis shows the perspective of the equity investor, calculating the return on equity rather than the return on investment. As the tenant is the primary beneficiary of energy savings per Czech law, the investor mainly benefits from net rent increases,[6] reduced property running costs and a possible higher willingness-to-pay by other investors at the time of exit. These factors are reflected in the following:

Key building data

Construction Year:	1962
Stories and Housing Units:	5 / 34
Heated Rental Area:	1,700 m²
Total Floor Area:	2,500 m²
Occupancy:	100 percent (constant before and after retrofit)

Key investor characteristics (for illustration)

Debt Portion	216,300 EUR loan (70 percent)
Holding Period	Sale of the property intended at the end of 2015, five years after retrofit.
Payback of Debt	At the end of the holding period in one lump sum (bullet loan).
Green Value Assumption	It is assumed that the buyer in 2015 will be willing to pay 275,000 EUR additionally for the higher energetic standard.
Income Tax Rate	30 percent

The calculation delivers a 1.24 percent annual equity return. The relatively small return is due to the fact that the agreements on rent increase after retrofits are freely negotiable in the Czech Republic, which unfortunately prevents investors benefitting from the measures financially to a sufficient extent. Under the German framework, for instance, legal options are more regulated and would allow a rent increase of up to 1.67 per square meter a month – four times the amount that was achieved in the Czech retrofit example.

As the overall return of the measures is also lower than the interest cost of debt capital at that time, no use could have been made of the leverage effect from debt financing. To the contrary, the leverage effect was negative: if the measures would have been fully financed from equity, the return on equity would have amounted to 1.61 percent.

TABLE 14.4 Overview of full calculation pattern

Time Line	2010	2011	2012	2013	2014	2015
Original Payments						
Energy Retrofit Investment	-309000.00					
Rent Increase Income		9180.00	9180.00	9180.00	9639.00	9639.00
Development of Rent Levels compared to previous period			0.00%	0.00%	5.00%	0.00%
Benefits from reduced Running Costs		1600.00	1648	1697.44	1748.36	1800.81
Development of Running Costs compared to previous period			3.00%	3.00%	3.00%	3.00%
Additional Sales Price Payment						290000
Periodic Inflow Sum		10780.00	10828.00	10877.44	11387.36	301439.81
Derivative Payments						
Equity	92700.00					
Debt Sum	216300.00					
Credit Interest Payments 2,60 % fixed rate		-5623.80	-5623.80	-5623.80	-5623.80	-5623.80
Tax Payment or Tax Refund		1540.05	1505.56	1480.50	1345.66	-8400.68
Credit Amortization		0.00	0.00	0.00	0.00	-216300.00
Re-Investment Payback from t-1			6696.25	13472.97	20308.16	27458.00
Interest from Re-Investment t			66.96	101.05	40.62	27.46

TABLE 14.4 continued

Ex-Ante Assumptions on Interest Rate Development			1.00%	0.75%	0.20%	0.10%
Additional Re-Investment from Income t		-6696.25	-6709.76	-6734.14	-7109.22	-71115.33
Residual Debt		216300.00	216300.00	216300.00	216300.00	0.00
Running Equity Value		6696.25	13472.97	20208.16	27458.00	98600.79
Running Difference Between Residual Debt and Running Equity Value		-209603.75	-202827.03	-196091.84	-188842.00	98600.79
Final Equity Value						**98600.79**
Financing Plan						
Credit Borrowing	Bullet Loan					
	216300					
Credit Interest Payments 2,60 % fixed rate		5623.8	5623.8	5623.8	5623.8	5623.8
Credit Amortization		0	0	0	0	216300
Residual Debt		216300	216300	216300	216300	0
Depreciation Plan						
Depreciation Rate		3.33%	3.33%	3.33%	3.33%	3.33%
Depreciation Basis (Linear over 30 Years)		309000	309000	309000	309000	309000
Depreciation Amount		10289.7	10289.7	10289.7	10289.7	10289.7
Residual Book Value		298710.3	288420.6	278130.9	267841.2	257551.5

Taxation – Side Calculation

Regular Cash Flows	10780.00	10828.00	10877.44	11387.36	11439.81
Depreciation	-10289.7	-10289.7	-10289.7	-10289.7	-10289.7
Interest from Re-Investments	0.00	66.96	101.05	40.62	27.46
Loan – Interest Payments	-5623.80	-5623.80	-5623.80	-5623.80	-5623.80
Additional Sales Price Payment	0.00	0.00	0.00	0.00	290000.00
Residual Book Value	298710.3	288420.6	278130.9	267841.2	257551.5
Taxable Sales Profit					32448.50
Tax Rate	30%	30%	30%	30%	30%
Tax Basis	-5133.50	-5018.54	-4935.01	-4485.52	28002.27
Tax Payment	**-1540.05**	**-1505.56**	**-1480.50**	**-1345.66**	**8400.68**

VoFI Equity Yield   1.24%

FIGURE 14.5 After the energy-efficiency retrofit

Source: Monika Kwapulinska.

Notes

While all billing and rent payments were processed in Czech Koruna, the reporting currency for this investment calculation is Euros. Courtesy of STÚ-K, a.s. (Building Engineering Consultants, Prague).

Conclusion

The property industry has a specific responsibility in terms of achieving the political climate and energy goals, as it is a main cause of emissions and energy consumption. But energy-efficiency retrofits are not only politically desired and ecologically required. They can also provide a sound commercial profitability and therefore generate value to financial investors. Yet, modernization rates stay behind expectations on an international level. The main barriers to the implementation of retrofit measures are the user–investor dilemma, where one party pays the investment, but another party would gain the benefits; the high initial investment costs that must be typically co-financed by debt; as well as a lack of information and awareness about the financial profitability of such measures. The extent to which the energy-efficiency measures are economically beneficial depends on the interplay of certain parameters that can be summarized under typologies that describe the *investor* and their strategic approach and conditions, the *market* and its reaction towards retrofit measures, the *property* and the realizable energy savings as well as the country's *legal/taxation* regime.

This chapter has outlined which variables influence the outcome of the profitability assessment and their impact directions and functional chains. In addition, it introduced and explained the Visualization of Financial Impact (VoFI) method. This method delivers a structured overview over time and an equity-oriented key figure, which is comprehensible to both private and professional investors. At the same time, it is thorough enough to comply with industry standards. On that basis, the profound analysis of financial profitability of energy-efficiency retrofits is possible and the required transparency can be provided to capital investors.

Stakeholder implications

- Communities should promote energy-efficiency retrofits and make more and better use of their existing building stock – not only to conserve the spirit of their districts, but also to benefit from reduced energy consumption as well as reduced energy production and costs.
- Planners should align their actions with sustainable approaches by first being aware of the importance and utility of existing buildings and second by setting the pace and standards for a higher application rate in terms of retrofit measures. Also, planners should ensure the effectiveness of sustainable building use by informing politicians and public administrations of the possibilities and financial benefits arising in this context.
- Developers should conceive new business models for the systematic retrofitting and upgrading of existing buildings and should profile themselves as consultants that align their recommendations with economic/financial, but also social and ecological aspects.
- Investors should be aware of the increasing cost level for new buildings and the financial impact energy-efficiency retrofits can have. In consequence, they should generate a higher demand for energy-efficiency retrofits by providing capital for related measures, and see this investment class as an alternative (not only) in the low-interest rate environment.
- Professionals should aim for gathering know-how in terms of energy-efficiency retrofits and specific measures and implement this knowledge in their daily work.

Appendix 14.1

Exemplary profitability analysis

In this Appendix, I would like to illustrate the profitability analysis using the VoFI method through a simple fictional example. The investment has the following key data:

1   The investment requires an initial payment of 100,000 EUR.
2   Cash flows from the achievable rent increase (or alternatively energy cost savings, if the space is not rented or rented under gross rent) amount to 20,000 EUR per year.
3   The property is scheduled for sale after four years after performing the energy-efficiency retrofit.
4   The market will appreciate the retrofit at 85 percent of the investment volume after that period, that is at the time of exit, the next buyer will be willing to pay an additional 85,000 EUR for the retrofit and the resulting higher energy standard and its financial benefits.

Additionally, Table 14.5 outlines further market conditions and assumptions.
The calculation procedure contains mainly five steps:

1   Indicate the original payments and both equity and debt sum.
2   Compile the financing plan and transfer Credit Interest Payments into the main calculation.
3   Compile the depreciation plan and transfer the Depreciation Amount into the tax side calculation. Also, transfer Credit Interest Payments from step 2 and the other required data into the tax side calculation.
4   Calculate the tax payment for year 1 and transfer it into the main calculation.
5   Determine re-investable amount for $t = 1$ and investment this amount. In $t = 2$, this amount, its generated interest as well as the additional re-investable amount for $t = 2$ are then re-invested again. Continue this pattern for all the periods.

TABLE 14.5 Key assumptions for an exemplary VoFI

Key Assumptions	
Interest on Bank Deposits p.a. (Assumed constant over 4 years)	1.0%
Interest on Borrowing p.a. (Assumed constant over 4 years)	4.0%
Property Share of the Investment (tax-relevant portion)	100.0%
Depreciation Rate	5.0%
Equity Portion	40.0%
Debt Portion	60.0%
Income Tax Rate	30.0%
Debt Contract:	Bullet Loan

TABLE 14.6 Exemplary VoFI calculation

Time Line	$t = 0$	$t = 1$	$t = 2$	$t = 3$	$t = 4$
Original Payments					
Investment A_0	100,000				
Cashflows CF_t (= Rent Increases)		20,000	20,000	20,000	20,000
Additional Sales Price Payment R_n					85,000
Derivative Payments					
Equity	40,000				
Debt Sum	60,000				
Loan − Interest Payments (3.0%)		−2,400	−2,400	−2,400	−2,400
Tax Payment or Tax Refund		−3,780	−3,821	−3,863	−5,405
Credit Amortization		0	0	0	−60,000
Re-Investment Payback from t−1			13,820	27,737.20	41,751.57
Interest from Re-Investment t (1.0%)			138.20	277.37	417.52
Additional Re-Investment from Income t		13,820	−13,779	−13,737	(−37,195)
Payment Balance		0	0	0	0
Residual Debt	60,000	60,000	60,000	60,000	0
Running Equity Value	0	13,820	27,737.20	41,751.57	79,364.09
Running Difference Between Residual Debt and Running Equity Value	−60,000	−46,180	−32,262.80	−18,248.43	79,364.09
Final Equity Value					79,364.09
Financing Plan	Bullet Loan				
Credit Borrowing	60,000	600.000	456.584	308.865	156.715
Credit Interest Payments	0	2,400	2,400	2,400	2,400
Credit Amortization	0	0	0	0	60,000
Residual Debt	60,000	60,000	60,000	60,000	0
Depreciation Plan					
Depreciation Rate		5.0%	5.0%	5.0%	5.0%
Depreciation Basis (Linear)	100,000	100,000	100,000	100,000	100,000
Depreciation Amount		5,000	5,000	5,000	5,000
Residual Book Value	100,000	95,000	90,000	85,000	80,000
Taxation − Side Calculation					
Regular Cash Flows		20,000	20,000	20,000	20,000
Depreciation		−5,000	−5,000	−5,000	−5,000
Interest from Re-Investments		0	138,20	277,37	417,52
Loan − Interest Payments		−2,400	−2,400	−2,400	−2,400
Additional Sales Price Payment		0	0	0	85,000
Residual Book Value		95,000	90,000	85,000	80,000
Taxable Sales Profit					5,000
Tax Rate		30.0%	30.0%	30.0%	30.0%
Tax Basis (Positive cash flows minus depreciation minus negative cash flows)		12,600	12,461.80	12,877.37	18,017.52
Tax Payment		**−3,780**	**−3,821**	**−3,863**	**−5,405**
VoFI Equity Yield	**18.68 %**				

When calculating VoFI tables in the specific context of energy-efficiency retrofits, two fundamental aspects should be kept in mind: first, all originary cash flows should refer to a 'delta' consideration; that means that the calculation should incorporate *additional* ('delta') positive cash flows arising from the retrofit measure and *additional* negative ('delta') cash flows. However, this is not following a full-cost accounting approach on the whole property. This is important in order to actually evaluate the effect of the retrofit measure instead of looking at the building performance overall. In this context, investment costs must be distinguished from those that are actually energy-relevant and those that are arising in another context, e.g. regular maintenance or upgrading without energy effect. It is frequently observed that investors combine measures of energy-efficiency retrofits together with regular maintenance investments to reduce costs and burdens for tenants.

And second, the exemplary VoFI does not incorporate possible changes in maintenance costs for the retrofitted property. Typically, an additional line should be considered to reflect either reduced maintenance costs (due to more efficient equipment) or increased maintenance costs (due to higher complexity ex-post).

Notes

1   The author is thankful to Tomas Vimmr, Sven Bienert, Mira Bernard and other contributors from the RentalCal project for valuable comments and support on this chapter.
2   However, the scale of renovation within the European Union, for instance, would need to more than double to approximately 3 percent of the total building stock per year (Lewis/Hógáin/Borghi 2013: 9).
3   Glaeser and Kahn (2010), for instance, quantified that households in suburban Boston emit approximately 6 tons more carbon dioxide than households in the city center. This cannot completely be explained by higher household sizes.
4   It should be noted that energy efficiency is just one of the necessary features for a building to be called a 'certified' or 'sustainable' building. Additional characteristics depend on the certification label and may contain aspects such as the construction quality, the location site or planning process.
5   Amongst others, the following studies are of relevance in this research field: Leopoldsberger et al. (2011), Fuerst/van de Wetering/Wyatt (2013) and Cajias/Piazolo (2013).
6   Rent increases could be observed, amounting to an average of 0.45 EUR per square meter a month.

Bibliography

Amecke, H., Deason, J., Hobbs, A., Novikova, A., Xiu, Y., and Shengyuan, Z. (2013). *Buildings Energy Efficiency in China, Germany and the United States. CPI Report*, April. San Francisco.

Bone-Winkel, S., Schulte, K.-W., Sotelo, R., Allendorf, G., Ropeter-Ahlers, S.-E., and Orthmann, A. (2015). [Real Estate Investment], in: Schulte, K.-W., Bone-Winkel, S., and Schäfers, W. (Eds.), *[Real Estate Economics. Commercial Fundamentals]*, 5th edn, Berlin, pp. 581–648. (in German).

Cajias, M., and Piazolo, D. (2013). 'Green performs better: energy efficiency and financial return on buildings,' *Journal of Corporate Real Estate*, vol. 15, no. 1, pp. 53–72.

Eames, M., Dixon, T., May, T., and Hunt, M. (2013). 'City futures. Exploring urban retrofit and sustainable transitions,' *Building Research & Information*, vol. 41, no. 5, pp. 504–516.

European Commission (2015). *Paris Agreement,* [Online], Available: http://ec.europa.eu/clima/policies/international/negotiations/future/index_en.htm [3 January 2016].

Fuerst, F., van de Wetering, J., and Wyatt, P. (2013). 'Is intrinsic energy efficiency reflected in the pricing of office leases?' *Building Research and Information*, vol. 41, no. 4, pp. 1–11.

Glaeser, E., and Kahn, M. (2010). 'The greenness of cities: carbon dioxide emissions and urban development,' *Journal of Urban Economics*, vol. 67, no. 3, pp. 404–418.

Grob, H. (1993). *Capital Budgeting with Financial Plans. An Introduction.* Wiesbaden.

Grob, H. (Ed.) (2015). *[Introduction to Investment Appraisal. A Case Study History].* Munich. (in German).

Kofner, S. (2010). *[Investment Analysis for Real Estate].* Hamburg. (in German).

Leopoldsberger, G., Bienert, S., Brunauer, W., Bobsin, K., and Schützenhofer, C. (2011). 'Energising property valuation: putting a value on energy-efficient buildings,' *Appraisal Journal*, vol. 79, no. 2, pp. 115–125.

Lewis, J., Hógáin, S., and Borghi, A. (2013). *Building Enery Effiency in European Cities.* Saint-Denis, France.

Lohse, R., and Zhivov, A. (2014). *Investing into Energy Efficiency Projects: Why and How? Business and Technical Concepts for Deep Energy Retrofits of Public Buildings: Findings from IEA– EBC– Annex 61.* Brussels.

McGraw-Hill Construction (2010). *Green Outlook 2011. Green Trends Driving Growth.* New York City.

McKinsey & Company (2009). *Unlocking Energy Effiency in the U.S. Economy.* New York City.

McKinsey & Company (2010). *Impact of the Financial Crisis on Carbon Economics. Version 2.1 of the Global Greenhouse Gas Abatement Cost Curve.* New York City.

Paiho, S., Pinto-Seppä, I., and Jimenez, C. (2015). 'An energetic analysis of a multifunctional façade system for energy efficient retrofitting of residential buildings in cold climates of Finland and Russia,' *Sustainable Cities and Society*, vol. 15, July, pp. 75–85.

Polly, B., Gestwick, M., Bianchi, M., and Anderson, R. (2011). *A Method for Determining Optimal Residential Energy Efficiency Retrofit Packages.* Golden, CO.

Rapf, O., and Groote, M. de (2015). 'The active role of buildings in a transforming energy market.' BPIE Discussion Paper, Buildings Performance Institute Europe, pp. 4–20.

Seto, K., Dhakal, S., Bigio, A., Blanco, H., Delgado, G., Dewar, D., et al. (eds.) (2014). *Climate Change 2014: Mitigation of Climate Change. Contribution of Working Group III to the Fifth Assessment Report of the Intergovernmental Panel on Climate Change.* Cambridge.

United Nations (2014). *World Urbanization Prospects. The 2014 Revision Highlights.* New York.

U.S. GBC (2015). The Business Case for Green Building – Overall Trends, Available: www.usgbc.org/articles/business-case-green-building [3 January 2016].

World Green Building Council (2013). *The Business Case for Green Building. A Review of the Costs and Benefits for Developers, Investors and Occupants.* Washington, DC.

15

AFFORDABLE HOUSING AND PRIVATE RENTED SECTOR REFORM

Peter Smith

Abstract

Housing stress is not a feature of a smart city that should provide a range of decent and reasonably priced housing opportunities for all its citizens. In the UK, over the past few years, the role of the Private Rented Sector (PRS) has increased dramatically due to a range of factors. Chapter 15 investigates the affordability issues and dynamics in the sales and rental housing markets. The chapter reviews current UK policy and the legal framework as well as the limitations and challenges for reform, including the barriers for institutional investment. It investigates the factors behind the persistence of rents and capital value inflation, quality issues and tenure insecurity in the UK PRS compared to Germany where rent and other controls seems to have worked reasonably well. If implemented appropriately, a more tightly regulated market could help solve the UK housing crisis. However, the chapter calls for a multidisciplinary approach. On the demand side, it involves better management of PRS lettings and more active involvement of professional bodies to vet PRS quality and energy efficiency. On the supply side, responsible property owners and investors need tax and planning incentives. Making the PRS 'fit for purpose' involves informed policy tweaking, deliberation and collaboration to tackle multiple issues in an evolving system.

Key points

- The decline in social housing provision particularly followed the sale of council homes in the1980s.
- The increasing lack of affordability of house purchase exacerbated by the slow growth of incomes since the Global Financial Crisis (GFC).
- Institutional investors are only now beginning to enter the build-to-let market.

- Restrictions in the availability of mortgage finance, particularly for first-time buyers, necessitating higher levels of equity following the GFC.
- Competition between first-time buyers and individual buy-to-let investors following the availability of 'buy-to-let' mortgages has forced more households to rent, which in itself raises demand and rents and, consequently, reduces affordability.
- Break-up of families; rise in divorces resulting in more households seeking housing, and possibly renting.
- Increases in immigration.
- The PRS can play a part in the solution for sustainable UK housing provision.

Introduction

The Private Rented Sector (PRS) has virtually doubled in the past 15 years and this rate of growth is forecast to double again (Hardman and Coghill 2015).

Globally, as populations rise and lower-tier incomes are squeezed, chronic housing shortages and unaffordability affect many countries (RICS 2014). In India and China, population growth and concomitant urbanisation have increased pressures on housing provision. In China, the central government addressed the problem by sustained regulation, reversing previous policies of non-intervention (*ibid.*). The authorities there have set a target of 36 million new social homes to be built in the next five years, to head off potential unrest from the urban poor who have been priced out of the market following the sale of public housing (*ibid.*).

In the UK, also, the shortage of housing provision is chronic, not only in urban areas but also in rural areas. Capital values have risen well above inflation rates, while rents have also risen sharply in the private residential let sector, particularly since the global financial crisis in 2008, pricing many households out of the market. Endemic scarcity across all tenures has risen to the top of the political agenda in the UK. Wetzstein (2016) refers to the lack of housing affordability as a global phenomenon. In the 'anglophone world' (England, Ireland, the Commonwealth countries and the United States) housing affordability has declined substantially due to falling real household incomes combined with house price inflation (*ibid.*). However, because many governments have failed to invest in sufficient social housing due to high costs, the demand for private rented sector housing has sent rents rising to unaffordable levels (*ibid.*). Not only are households on low incomes finding house ownership unreachable, even middle income groups are being priced out of the market (*ibid.*).

In the UK, the housing crisis has been exacerbated by the failure to supply sufficient housing. The Barker Review in 2004 estimated that unsatisfied demand in the UK necessitated the production of around 240,000 new housing units per year. Successive governments have failed to deliver this number. The government's pledge (Housing and Planning Bill 2015) of one million new homes in England by 2020 remains merely an aspiration. Short-term, post-'Brexit' (EU referendum) uncertainty is unhelpful. It could dampen developer confidence and housing activity for some time to come (Smales 2016). In such a negative scenario, PRS pressures rise. Housing inflation shuts potential first-time buyers out of the owner-occupied market and back into the rental one. Inevitably, PRS rents rise but, if incomes fail to keep pace, the situation becomes unsustainable.

Strains on the PRS have been recognised by a number of professional bodies, including the Royal Institution of Chartered Surveyors (RICS) whose residential policy report expressed serious concerns that government policies are focused too much on increasing home ownership, and not concentrating on other forms of tenure, since not everyone is able to buy (Blackburn 2016). Surveyors expect sale prices and rents to rise by 25 per cent over the next five years from 2016 to 2021. To avoid rental escalation on such as scale, housing completions would need to increase by an unrealistic 80 per cent over the next 15 years (*ibid.*). One consequence of the construction shortfall is the surge in private sector renting. From being the smallest tenure until 2012, the PRS has become the second largest behind owner occupation with over 4 million households in private rented accommodation. Indeed, Savills predict a growth in the sector of more than 20 per cent over the period 2016–2021 (*ibid.*). In effect 'the die has been set'.

Although the GFC and the subsequent constraints on mortgage lending by the banks and mortgage lenders undeniably hastened the move away from house purchase into renting, the growth in the PRS has its roots elsewhere. The 'millennial generation' has exhibited a significant behavioural and cultural shift in contrast to previous generations, seen in lifestyle choices, how they spend their money and their financial aspirations for the long term. As a generation, they are more likely to delay marriage and starting a family (ONS 2013). Although home ownership is still an aspiration, the millennial generation is less prepared to defer spending habits at the expense of house purchase in the short term, and more likely to rent for longer (Mueller 2015). However, the Resolution Foundation (Clarke, Corlett and Judge 2016) recently reported that the 'millennial generation' or 'generation rent' may find it even more difficult to save sufficient amounts to make a down payment on a property because of rising housing costs where rents have risen in excess of incomes between 2003 and 2015.

In this period, the real average household incomes of private renters has risen by £8 per week (2 per cent) while real housing costs have risen by £19 per week (16 per cent) (*ibid.*). For Greater London, if all householders across all tenures are factored in, the figures reveal average household income has reduced by £29 per week (4 per cent) over the period, while real housing costs have risen by £36 (29 per cent) (*ibid.*).

In a post-GFC economy exhibiting low interest rates and sustained economic growth, for housing costs to continue to rise and take a larger share of the household budget, living standards will continue to be compromised (*ibid.*). In effect, the logic would suggest that the decision to rent is likely to be a permanent one, regardless of any aspirations to home ownership. Affordability, however, is not the only problem facing households in the PRS. The quality of dwellings is quite often poor with many let properties suffering from low energy efficiency (Energy Saving

TABLE 15.1 Trends in tenure

All households

| | Owner occupiers | | Social renters | |
	own outright	buying with mortgage	all	private renters
1981	4,313	5,546	**5,461**	**1,904**
1984	4,590	6,399	**5,034**	**1,920**
1988	4,834	7,414	**4,706**	**1,702**
1991	4,795	8,255	**4,435**	**1,824**
1992	4,815	8,255	**4,371**	**1,724**
1993	4,898	8,382	**4,317**	**1,833**
1994	5,008	8,421	**4,257**	**1,869**
1995	4,998	8,468	**4,245**	**1,939**
1996	5,115	8,407	**4,218**	**1,995**
1997	5,249	8,380	**4,170**	**2,078**
1998	5,404	8,413	**4,148**	**2,063**
1999	5,582	8,508	**4,072**	**2,000**
2000	5,764	8,575	**3,953**	**2,029**
2001	5,885	8,473	**3,983**	**2,062**
2002	6,019	8,540	**3,972**	**2,131**
2003	6,158	8,542	**3,804**	**2,234**
2004	6,288	8,389	**3,797**	**2,284**
2005	6,352	8,440	**3,696**	**2,445**
2006	6,425	8,365	**3,736**	**2,566**
2007	6,505	8,228	**3,755**	**2,691**
2008	6,653	7,975	**3,797**	**2,982**
2008–09	6,770	7,851	**3,842**	**3,067**
2009–10	6,828	7,697	**3,675**	**3,355**
2010–11	7,009	7,441	**3,826**	**3,617**
2011–12	6,996	7,392	**3,808**	**3,843**
2012–13	7,152	7,184	**3,684**	**3,956**
2013–14	7,386	6,933	**3,920**	**4,377**
2014–15	7,475	6,849	**3,912**	**4,278**

Sources:
1981 to 1991: DOE Labour Force Survey Housing Trailer;
1992 to 2008: ONS Labour Force Survey;
2008–09 onwards: English Housing Survey, full household sample

Trust 2016). The importance of the PRS in addressing the chronic shortage of housing is likely to grow in the years to come, and the critical issue of affordability and quality of accommodation needs to be addressed by governments with regulation of aspects of the PRS a possibility (Bate 2015). To understand the nature of policies required and whether regulation is indeed necessary, or appropriate, we first need to examine the key factors and influences that affect the PRS, the dynamics of the rental market and the criticisms of an unfettered rental market.

Key factors that affect the PRS and the current state of the rental market

The causes of the housing shortage in the PRS are due to secular trends, many of which are complex. On the demand side, increased household formation has undeniably been a major contributory factor with people living longer, and many living alone (Goddard 2016).

The break-up of the nuclear family has increased the demand for dwellings. The increase in divorces and, more recently, the significant increases in migrant workers (*ibid.*), together with the steady rise in the number of students in higher education have also played a part (Huston, Jadevicius and Minaei 2014). In order to deal with the large increases in migrant workers, a degree of regulation was introduced by the Coalition government with landlords having the responsibility to undertake 'right-to-rent' checks carried out as part of the Immigration Act 2014 (Goddard 2016).

In relation to student numbers, Huston *et al.* (2014) suggest that rapid university expansion without sufficient residential construction has meant 'rampant PRS growth' as well as a poor quality of dwellings. In addition, with a 'strong' landlord's market, it was noted the quality of management service was generally poor. Furthermore, with insufficient purpose-built accommodation for students in higher education, the student experience could be severely compromised, which could result in an erosion of long-term UK competitiveness. The authors conclude that the student's experience 'mirrors the general housing malaise around affordability, polarisation and the sustainable dwelling'. A clear link was made between the quality and affordability of rented accommodation for students in the PRS and the, perhaps less tangible, criteria of 'metropolitan productivity, sustainability and resilience' (*ibid.*). The authors postulate that high rents, low quality of accommodation and 'unprofessional management and service' could restrict the pool of talent (*ibid.*). Failing to provide decent and affordable rental housing restricts labour mobility and productivity. It has major social implications, as well, by increasing the division of income and the inequality of wealth, between those that rent and those who own their property, freehold or leasehold (Clark *et al.* 2016).

The PRS is dominated by small investor landlords whose portfolios are small in size (Crook and Kemp 2011). 'Private landlordism' is seen as a 'sideline activity' in England and most landlords have had 'no training or qualifications in property management or related experience to draw upon' (*ibid.*). Indeed, some are 'ignorant of landlord and tenant legislation' (Kemp and Kofner 2010). Kemp and Kofner identified a key theme: the lack of investment in the 'existing stock' of rented property and very little new construction, which will be covered later in this chapter. By comparison, larger landlords, defined as having more than ten properties, constitute only 1 per cent of landlords (DCLG 2012). However, the PRS has not significantly contributed to the supply of new housing according to the English Housing Surveys 2008–11 (Kemp and Kofner 2010).

It is interesting that Kemp and Kofner referred to the very high levels of 'tenant turnover' in the PRS. In 2007/08, 40 per cent of tenants in the sector had continuously occupied their accommodation for less than 12 months, while 70 per cent had occupied for less than three years (*ibid.*). The data do not mention the length of term that tenants were requesting from their landlords, and whether they were satisfied with the length of term offered to them. However, more recent data suggest that the periods occupied by tenants continuously in the same accommodation had lengthened, somewhat. A significant number of people are now looking to establish relatively long-term family homes in the PRS, and young professionals are looking to rent for longer periods while saving sufficient funds to purchase a home. As a result we are seeing 20 per cent of households in the PRS having been in their current accommodation for more than five years (DCLG 2012).

However, the underlying cause of the housing shortage, which has been identified by many reporters as a key weakness, has been the failure to sufficiently increase the supply of new homes of all tenure types, to meet the rising demand. Historically, statutory rent controls have discouraged the building of dedicated dwellings for private letting since the 1950s (*ibid.*). More individual landlords have entered the market after the deregulation of the PRS with the introduction of the Housing Acts 1988 and 1996, but the increase in the provision of rented accommodation had been piecemeal until assured shorthold tenancies became the default tenancy from 27 February 1997, the date the Housing Act 1996 became law (DCLG 2012). Unlike the Rent Act 1977, as amended, assured shorthold tenancies under the 1988 and the 1996 Housing Acts have no security of tenure and much reduced protection. The rents charged can be a market rent, or as agreed between the parties. Rent control and strong security of tenure under the Rent Act 1977, as amended, had been a major impediment to encouraging landlords to enter the market for fear of losing possession for up to three generations and receiving a rental income that was often insufficient to meet the landlord's repairing obligations. However, in more recent years, institutional investors have started to enter the 'build-to-rent' sector, which, potentially, can deliver a substantial amount of rented accommodation at no extra cost to the taxpayer. But because the build-to-rent sector is still very small, most PRS tenants will be heavily reliant on the small, relatively unregulated, private 'buy-to-let' investor landlords, or those who inherit their property, supplying the market (*ibid.*).

Since the late 1990s there has been a sharp decline in the number of dwellings being rented out by local authorities creating more pressures on the PRS. Much local authority housing stock has been sold off and this has not been followed by a compensatory increase in new development. There has been an increase in rented property provided by housing associations over the period, but this has certainly not made up the deficit (New Economy 2014).

As mentioned previously, there is a structural undersupply of housing in the UK. The chronic scarcity of housing in the market can be partially addressed by the development, on a large scale, of private sector rented accommodation, but this supply will need to be linked to the movement of the housing sales market. With limited availability of mortgage finance, the dynamics of the market will be influenced by the levels of equity and its subsequent movement, up or down (Gilmore 2014).

Affordability

The affordability of housing plays a key factor in the growth of the private rented sector. In the decade before the GFC in 2008, house prices rose well in excess of the growth in incomes. House price to earnings ratios rose steadily and, despite a temporary fall during the global financial crisis, has risen above the long-term average, in effect putting an end to many people's aspirations of home ownership (New Economy 2014).

The house price to earnings ratio, a direct measure of affordability for house purchasers, however, has an inverse relationship with the numbers seeking to rent. High and rising house price to earnings ratios will, *ceteris paribus*, be met by a rise in the number of people and households seeking to rent. The availability of mortgage finance to purchasers, which also influences affordability of housing, and indirectly measures the levels of equity purchasers have in their property purchases, also has a major impact on the PRS. Where potential purchasers are having difficulty in acquiring the necessary mortgage finance to buy freehold or long leasehold interests in residential property, there will be an increase, *ceteris paribus*, in the numbers of households looking to rent (Gilmore 2014).

The government has tried to address the shortage of equity through the 'Help-to-Buy' scheme, although as soon as interest rates start to rise again, mortgage finance will again prove difficult to acquire and, consequently, such initiatives are seen to be only temporary (New Economy 2014). When interest rates start to rise, demand for house purchase tends to fall, as finance becomes more expensive resulting in a *pari passu* increase in the demand for rented property (*ibid.*). The close relationship of the PRS to the home ownership market is again observed where house prices fall; the demand for rented accommodation, *ceteris paribus*, falls where households look to buy rather than rent as homes become more affordable to buyers. However, a steep and sustained fall in house prices would reduce residential development land values, resulting in new institutional PRS development schemes becoming more viable (Gilmore 2014).

The advantages of renting: a more flexible tenure

Renting is often seen to be a more flexible tenure allowing tenants to move around the country in search of work, and not chained to a location through difficulties in selling a house in, for instance, a poor market. Renting also allows people to be housed temporarily while searching for and starting employment (greater labour mobility), with many tenancy agreements being short term, commonly with terms of 12 months or less under the assured shorthold tenancy regime (Smith Institute 2014). The main reasons for the PRS offering greater flexibility are as follows:

- The speed in which tenancy agreements can be drawn up once a vacant dwelling is identified.
- The shorter nature of tenancy agreements in the PRS, typically only 12 months, or less.
- The greater legal, financial and marketing complexities involved in the buying and selling process which makes owner-occupied dwellings a relatively illiquid asset.

Tenants renting property also have the added advantage of avoiding housing debt, and a requirement to pay for structural repairs as required by s.11 Landlord and Tenant Act 1985. S.11 specifies that residential tenants cannot be required to pay for structural and external repairs in their accommodation provided their contractual term does not exceed seven years.

Drivers of growth of the PRS

Supply factors and buy-to-let mortgages

On the supply side, the steep rise in house prices in relation to earnings has been fuelled by the growth in the availability of buy-to-let mortgages for those with higher levels of equity. Specific buy-to-let property mortgages were introduced in 1996 and served to support an increase in the supply of rented accommodation in the PRS. It is estimated by the Council of Mortgage Lenders that there are currently over 1.5 million buy-to-let mortgages in existence (referred to as 'outstanding') by 2014 (Smith Institute 2014). In effect, though, these investors are competing with first-time-buyers, pushing prices even higher and forcing would-be first-time buyers into the rental market (*ibid.*). Incidentally, these buy-to-let investors, exhibiting high levels of equity, have been less affected by the GFC (Gilmore 2014). This is a conundrum that needs to be addressed in policy. The very buy-to-let suppliers of housing to rent in the PRS would appear to be competing with those households desperate to establish a foot on the first rung of the housing ladder, which serves to perpetuate renting by closing off alternatives, effectively polarising society.

Perhaps the entry of the larger build-to-rent investor increasing the supply of private rented accommodation might solve the problem of unaffordability and poor quality. However, there are structural impediments that might frustrate this. Niemietz (2015) suggests that constrictions on the supply caused by planning restrictions constitutes the main driver of high and rising housing costs (*ibid.*). These costs filter through to a decline in housing affordability, which appears to be affecting not only low income households but also, increasingly, middle income ones, as well. Interestingly, Niemietz in referring to ONS data, states that lower income groups are affected most. Middle income groups are able to adjust their occupational behaviour by living in smaller less expensive dwellings, living with parents for longer or delaying the purchase of the first home (ONS 2014). Before we see a large increase in the supply of buildings to rent, however, the structural impediments of the planning system will have to be addressed.

The increase in the supply of more build-to-rent dwellings with help from the government's 'Build-to Rent Fund' will help to increase supply (Bate 2016). If the scale of building is sufficiently high, it might ultimately create downward pressure on rents to more affordable levels, but it will, in the long term, improve the quality of accommodation on offer. By comparison, the small individual buy-to-let investor has shown an aversion to making energy efficiency improvements to their let dwellings (Hope and Booth 2014); an issue addressed later in this chapter. Lower more stable rents might eventually deter some potential second-home buy-to-let investors with short-term investment horizons from entering the market. By not having to compete with second-home PRS investors in the housing market, first-time buyers might see a fall in house price inflation, making house purchase a realisable dream for a section of PRS tenants. This relationship between the PRS and the housing market only emphasises the need for a good healthy balance between a wide range of coexisting tenures; a disequilibrium in one could easily destabilise others. In the interests of flexibility, care must be taken that any proposed regulation of the PRS does not create disharmony between the different tenures.

Energy use and efficiency: fuel poverty of tenants in let dwellings

As hinted previously, increasing PRS unaffordability is a worrying trend, but the quality of let property is also a serious problem. Fuel poverty impels tighter buildings heat insulation but ill-considered design and poor workmanship can compromise air quality with a pernicious increase in mould spores. Research carried out by Middlemiss and Gillard (2015) into fuel poverty of householders links energy vulnerability and fuel poverty gap. The 'fuel poverty gap' is defined as the average fuel shortfall faced by poor households in paying their fuel bills in the home (*ibid.*). A number of challenges to energy vulnerability of the poor were identified, including the quality of the buildings occupied, the costs of supplying energy to the dwellings, income stability of households in occupation, tenancy relations, that is relations between tenants and landlords, and ill-health. In respect of the latter, ill-health, often requiring increased levels of heating, raised energy vulnerability (*ibid.*).

In the PRS, the provision of structural improvements, such as insulation, will be the duty of the landlord; tenants are able to make only minimal improvements, most of these would fall, subject to contract, under the heading of 'decorative repairs' and 'keeping the property in a tenant-like manner'. S.11 of the Landlord and Tenant Act 1985, already mentioned, requires the lessor 'to be responsible for repairs to the exterior and structure of the dwelling', and for 'keeping in repair and proper working order all the installations required for the supply of the utilities – water, gas, electricity and for the provision of sanitation to the building'. Consequently, tenants have little power to influence landlords to carry out improvements that would enhance the structural integrity of the building by way of insulation. These improvements can enhance comfort of occupiers and reduce the costs of heating (Middlemiss and Gillard 2015). The legal requirement for the owner of the building to display at the dwelling an energy performance certificate, which describes the actual costs of space heating and provision of hot water and electricity at the dwelling, is not always apparent to the tenant when the tenancy agreement is signed, and is often overlooked. Future costs are difficult to predict with accuracy and, as a result, tenants may find it almost impossible to compare buildings for energy efficiency prior to signing a tenancy agreement; this, effectively, limits choice for tenants. Moving house in search of a better insulated building is, at best, an inconvenience, and possibly a substantial expense (*ibid.*).

For all of these reasons tenants, both in the public and private sectors, suffer significant energy vulnerability (Ambrose 2015). A summary of recent research carried out by the University of Exeter's Medical School European Centre for Environment and Human Health established clear links between energy efficiency, fuel poverty, indoor air quality and health. The findings suggest that greater levels of insulation in buildings undeniably reduce fuel poverty and cold-related illnesses. It also has the monetary benefit of reducing fuel bills through greater energy efficiency, whilst also reducing carbon dioxide emissions (Ambrose 2015). However, the report was not able to identify the impact of improved energy efficiency on the health of occupiers in homes that already experienced inadequate ventilation, condensation and mould growth.

The early indications are that damp in dwellings (either penetrating damp or damp from condensation) may trigger asthma and other lung diseases by increasing the incidence of mould and the production of mould spores throughout the dwelling (*ibid.*). The issue with all housing, but particularly PRS dwellings, is that improving energy efficiency achieved through reducing heat loss with improved insulation and reduced ventilation may also give rise to dampness and poor air quality. Tenants are particularly vulnerable to this as they have no powers under s.11 of the Landlord and Tenant Act 1985 to carry out the required works themselves, or enforce landlords to implement them; and, in any case, the short length of their tenancies would mean they might never reap the long-term benefits. Regulation is probably required to ensure let dwellings, as a starting point, are adequately ventilated as well as insulated, by granting tenants statutory powers to ensure their landlords bring the building up to the requisite standards for both energy efficiency and adequate ventilation.

The Energy Act 2011, has, admittedly provided primary legislation for the government to set minimum energy performance and carbon dioxide emission standards for both residential (the housing sector in general as well as the PRS) and commercial property. The Energy Efficiency Regulations 2015 set out these standards for England and Wales (Cook and Cross 2015). By April 2018, it will be unlawful for landlords to let dwellings below an Energy Performance Certificate (EPC) of E, the lowest energy efficiency rating in common with commercial property (*ibid.*). The data on energy efficiency provide sufficient evidence that tenants in the PRS are particularly vulnerable. In its report the British Property Federation (BPF) refer to government data that 13.5 per cent of PRS properties have an EPC rating of F or G, the poorest EPC rating (BPF 2013). In 2011, approximately 11 per cent of PRS properties (c.462,000) in England had an 'EPC rating of F or G (8 per cent F and 3 per cent G), compared to around 9 per cent (7 per cent F and 2 per cent G) in the owner-occupied sector, of which 65 per cent were built before 1919' (English Housing Survey 2011–12).

In addition to the measures mentioned above in relation to the PRS, the government specifies that landlords would not be expected to encounter upfront costs, which implies that the works required would have been funded and implemented through the Green Deal and the Energy Company Obligation (ECO) (Department of Energy and Climate Change 2011). Tenants might be tempted to say nothing to their landlords about energy efficiency and dampness in their homes for fear of not being allowed to renew their tenancies at the end of their contractual terms. Alternatively, they might apply to their local housing authority, which has been granted powers under the Housing Act 2004 to inspect properties using the Health and Safety Standard (HSS) to inspect homes where the conditions are perceived to be a health hazard. One such measure might be 'excess cold' which can be caused, in part, by a significant thermal inefficiency in a building (BPF 2013).

Attitudes of private sector landlords towards energy efficiency

Hope and Booth's (2014) research into the attitudes and behaviours of private sector landlords towards the energy efficiency of tenanted homes clearly identified the key problems facing tenants in the PRS. It focused on the 'attitudes and behaviours of private sector landlords towards energy efficiency' in dwellings occupied by their tenants. It is significant that there is a distinct lack of research relating to energy efficiency in the PRS, and, in particular, to landlords' attitudes towards making improvements. Most of the government initiatives, some of which have been referred to previously, have been aimed, specifically, at owner occupiers. Hope and Booth concluded that there needed to be an energy policy directed specifically at landlords in the PRS on the basis that the PRS is the fastest growing housing tenure and also the most energy inefficient (*ibid.*).

Hope and Booth (2014) drew attention to the fact that the 'disparate nature' of the PRS creates a barrier to the implementation of energy efficiency measures. In particular, they refer to the fact that the fragmentation of the sector, where most landlords own one property only, means there is a difficulty for government bodies to engage with the landlord body in general. It was suggested that all PRS landlords might be mandated to join the National Landlords Association (NLA), the UK's largest representative landlords' organisation. Policy makers would then have a unifying conduit through which they could focus their discussions. It would appear this might be one useful area of regulation on the sector (*ibid.*).

Hope and Booth's (2014) findings suggest that it is the large upfront costs of making a dwelling more energy efficient that discouraged landlords from implementing the relevant improvements. Most landlords consider that it is the tenant who benefits from the improvements as they pay the bills, rather than seeing the longer-term advantages of a dwelling that is easier to let in the future, and possibly at a higher rent. As mentioned earlier in this chapter, Hope and Booth conclude that tenants are much less likely to take the initiative of requesting energy efficiency improvements from their landlords for fear of losing possession after the contractual term ends (*ibid.*). Perhaps tenants should be given statutory powers to formally request these improvements, and landlords should be required to implement them; but this might be seen as the thin end of the wedge for more enhanced statutory regulation elsewhere in the PRS. However, it might be argued that regulation of this nature has been introduced in the commercial property sector, already, without compromising the free operation of an unfettered real estate investment market.

Efforts to solve the structural deficiencies of the PRS

The problems associated with the PRS that we have mentioned might, in part, be solved by institutional investors such as the pension funds and Real Estate Investment Trusts (REITs) entering the market with the building of homes to let. The DCLG's (2012) review of the barriers to institutional investment in private rented homes, the Montague Review, reported that the period of time residential tenants had remained in their homes uninterrupted had extended, indicating that many tenants might be prepared to take longer tenancies if they were offered them by their landlord. The indications are that longer terms might attract more institutional investors into the market. By comparison, commercial property investments are characterised by longer lease lengths, which have increased from 5.5 years in 2012 to 7.2 years by 2015 (MSCI, Strutt & Parker and the BPF 2015).

In the residential sector, longer tenancy lengths reduce the costs of voids over time, and the need to search for new tenants. The periods between tenancies are often used for deep cleaning and even redecorating, which can be expensive. These new market characteristics have started to attract the new breed of institutional investor. The Conservative government's Private Rented Task Force established the Build-to-Rent model (DCLG 2017b) as a consequence. The build-to-rent model is seen as the most effective way of increasing the supply of good quality affordable purpose-built housing in the PRS, increasing choice for tenants, and improving housing quality (*ibid.*).

Most tenancies in the PRS in the UK are Assured Shorthold Tenancies (ASTs) of 12 months' duration, but longer terms are beginning to appear. Increasingly, families are renting residential accommodation as high rising prices have effectively barred families from home ownership. The proportion of households in the PRS with dependent children has increased from 29 per cent in 2004 to 37 per cent by 2015 (English Housing Survey 2014–15). Having to move at the end of a short-term tenancy creates instability and uncertainty for families; in some cases, new schools have to be found for children. The short-term 12-month tenancy is effectively a symptom of a broken housing market for such families, and consequently longer term three-year family-friendly tenancies have been recommended for the build-to-rent market. The length of term could be enforced by a planning obligation (DCLG 2017a).

Greater involvement of institutional investors in the PRS through the build-to-rent initiative is recognised as being the most sustainable model for the PRS (*ibid.*). The sustainability of the PRS, dominated by the small buy-to-let investor, has been compromised as a viable tenure. It does not possess the capacity to increase the supply of good quality long-term housing for households. Huston *et al.* (2014) conclude that the inelasticity of supply prevalent in the PRS in general, and in the provision of student accommodation in particular, can only realistically be addressed by the institutional build-to-rent investors and this assertion is reinforced by the Housing White Paper (DCLG 2017b).

Entrance of institutional investors in build-to-rent

The Montague Review (DCLG 2012) registered a number of core impediments to attracting institutional investors, who are actively more involved in a number of EU countries and the USA. Many of the large institutional investors, such as insurance companies and pension funds, are not developers and seek income stability and longer term income growth. That is, they tend to be risk averse, choosing to avoid development of new homes (DCLG 2012).

Net yields in the PRS have tended to be low, which might deter institutional investors due to high costs of purchase. However, low yields have tended to reflect capital growth potential and lower risk, which, in itself, might be attractive. The demand for dwellings with vacant possession, particularly in the South of England, but also in many northern cities as well, has been very high. Indeed, the strength of the owner-occupied market has reduced the choice of rented accommodation available, forcing the institutional investors into considering the build-to-let market, or avoid involvement at all (*ibid.*).

A relative lack of familiarity by the development stakeholders (local authority planners, landowners and potential developers) with private sector residential development has resulted in a lack of confidence, generally. The need to bring together a wide range of disparate interests for the development to succeed, and the lack of tried and trusted models, also discourages supply (*ibid.*).

There is a general unease with the quality of management expertise available to institutional investors. This might partly be explained by the poor reputation that the private let sector has exhibited historically, resulting in a general lack of trust, by comparison with the commercial property sector (*ibid.*).

The potential of the PRS for institutional investors is beginning to be recognised. The rental market is subject to a structural and chronic undersupply and would constitute a strong defensive counter-cyclical investment where the basic fundamentals are sound. With housing demand outstripping supply in the PRS, any downturn in the economy coupled with any fall in sterling will see an increase in interest in the sector from overseas investors (*ibid.*). UK investors might also see a strengthening of the market as more people seek to rent rather than buy homes. Historically, the PRS demonstrated good defensive characteristics following the owner-occupied market downturns in the 1990s and following the global financial crisis in 2008 (Horti 2016b). Rents have generally kept pace with earnings growth over a long period and more recently have exceeded it. The sector offers diversification potential with other investments, and could provide a good inflation hedge. Net yields compare favourably with commercial property (DCLG 2012).

By 2016 as many as 30 large institutional investors were looking to invest in the sector (Watson 2015). Much of the growth in institutional investor involvement has been in build-to-rent mainly focused on London. Watson refers to the first, and one of the biggest, being the East Village, the former Olympic Athletes' Village (*ibid.*). Institutional investors will be more attracted to the larger scale purpose-built and professionally managed rental housing. However, the purpose-built private rented sector is still very small, around 1 per cent of the total rented market. Small landlords, however, still dominate the market, owning housing that was not specifically designed for PRS letting (*ibid.*).

As the housing market cooled in the short-term, after the EU referendum result and the following time of uncertainty, housebuilders have shown an eagerness to strike deals with investors in order to 'build out' sites from their land banks, in order to maintain sale rates (Horti 2016a). Horti argues that investors tend to avoid buying land outright and engaging in the planning process, as this increases the risk. The deals involve housebuilders providing the land and the investors providing 'forward funding' for the building of private rented sector blocks. The key advantage from the housebuilders' perspective is positive cash flow in a time of economic uncertainty. The window of opportunity for both investors and housebuilders might be short-lived while the prognosis for the economy is uncertain and housebuilders have built up land banks which they are keen to 'build out' in the absence of bank lending (*ibid.*).

In the longer term, as the larger institutional investor developers start to show interest in the sector, a number of professional bodies have recognised the need for greater professionalism in the management of residential properties (Souter and Kafkaris 2015). This demonstrates that there is an important role to be played by the Royal Institution of Chartered Surveyors (RICS) and also the British Property Federation (BFP) in the monitoring of standards in the PRS, and also potential training for its members. The RICS have produced a Private Rented Sector Code of Practice, 2015, which is aimed at raising professional standards of not just the buy-to-let landlords and the management of their properties, but also the institutional investors entering the market with build-to-rent (RICS 2015).

Attitudes of tenants to renting in the PRS and whether strong statutory regulation is required

In order to examine the possible need for strong regulation to address the problems of affordability and quality of accommodation in the PRS, it would be prudent to find out what tenants think of their accommodation and the relationship they have with their landlords. The attitudes of middle income tenants with families to renting can vary considerably according to location (Scanlon, Fernández and Whitehead 2014). In Greater London, the PRS has nearly doubled in size over the last 30 years. The sector has long seen an increase in the young, mobile and low income groups, but more recently there has been an increase in tenants from middle income households, including families. The latter group appears to have increased out of necessity rather than choice, in direct contrast to the cities of Berlin and New York (*ibid.*). Scanlon *et al.*'s work, which focused more on middle income families, shows a strong culture of renting in these cities over many years, encouraged by the legal and regulatory frameworks in the respective countries. Previous research suggests that where security of tenure and rent control exists, more families will seek to live in the PRS (*ibid.*). Germany has security of tenure for tenants in the PRS which offers tenants similar long-term benefits to owner occupation. The catalyst to tenants preferring to rent rather than owning their homes, therefore, might be seen as the existence of security of tenure, but in Holland where there is also strong security of tenure, there are few tenants in the PRS; a conundrum that requires further investigation (*ibid.*). The research attempts to identify the 'drivers behind the decision to rent, and what families liked and disliked about renting' in the four cities. Understanding these 'drivers' can help, from the tenant's perspective, to inform policy on regulation.

Most of the tenants interviewed in the study said they sought certainty of knowing they could stay in their home long term without having the responsibility for repairs, and not face landlords seeking to recover possession after short-term tenancies had expired (*ibid.*). Many were renting because they could not afford to buy, but the aspiration to buy was nevertheless present, even in Berlin where the PRS was more entrenched. Generally, in the PRS, poor maintenance was seen to be an issue, however (*ibid.*).

In conclusion, therefore, it would seem that more stringent regulation would be prudent in relation to the quality of housing provided, which has ramifications for public health, and the way the property is managed. The Deregulation Act 2015 was introduced to address some of these issues by providing some protection for assured shorthold tenants against retaliatory eviction where a tenant complains about poor and/or unsafe property conditions in the building. Landlords are prevented from recovering possession of the dwelling until appropriate improvements are carried out. Protection of a tenant's deposit has already been introduced in England and Wales with the introduction of the statutory Tenancy Deposit Protection Scheme (TDP) which commenced after 6 April 2007, with separate schemes operating in Scotland and Northern Ireland. Under the Deregulation Act if landlords do not provide protection of a tenant's deposit in a government authorised scheme, they will not be able to serve a s.21 'no fault' possession notice to bring the tenancy to an end, in addition to facing financial penalty.

On the subject of investor affordability, however, there is a danger that landlords, particularly the large institutional investors, might be discouraged from entering the market if there were to be a return to 'rent control' and 'security of tenure' for occupying tenants. It will be very difficult to overcome the prejudices of landlords who had experienced the difficult years of letting dwellings under the Rent Acts. That is not to say it is impossible, because strong security of tenure is present in Germany where the PRS works relatively well, but such countries have a different culture of renting, where there is no stigma attached to occupying on long-term leases.

Kemp and Kofner (2010) in their paper on the contrasting varieties of private renting in England and Germany provide an interesting comparison between the private rented sectors in Germany and England. In England, the deregulation of rents and the removal of security of tenure has encouraged growth in the PRS. This appears to be in direct contrast to Germany where the PRS is very large by comparison, and yet has security of tenure and regulation of rents, although the regulation of rent was less intense than 'fair rents' are currently, and have been under s.70 of the Rent Act in England. Kemp and Kofner draw attention to this contradiction. In Germany, regulation of rents and protection of tenants with security of tenure has increased stability in the PRS, and enabled the quality of housing to remain high, whereas in England rent regulation and security of tenure has both discouraged inward investment, and seen a resultant fall in the quality of let property. Indeed, the PRS had seen little inward investment in existing stock or the building of new dwellings (*ibid.*).

In Germany, private landlords (who are in the majority), tend to take a much longer-term view of residential property investment compared to residential landlords in England, and seek capital gains above rental gains (*ibid.*). The timing of investments is more important for English investors, which explains the short-term strategy often adopted by English landlords. English landlords seek a freedom to enter and exit the rental market in response to rises and falls in the capital (price) as opposed to the rental markets. With high levels of volatility in the English residential property market compared to Germany, high security of tenure reduces liquidity and the capacity to enter or sell investments quickly (*ibid.*). However, in Germany, rent regulation is softer than exists under the Rent Act 1977, or in the Housing Act 1988, as amended, with assured tenancies in England. Open-ended contracts cannot be cancelled arbitrarily. The landlord must have a 'legitimate interest' in the termination such as a breach of obligation, or that the landlord can prove hardship at an eviction trial (*ibid.*). In relation to rent regulation, landlords are prohibited from increasing rents to a level above the average being a 'reference rent' for the 'community' (local area) for that type of dwelling. In addition, landlords may not raise the rent of a dwelling by more than 20 per cent in any three-year period (*ibid.*).

In Germany, the capital sales market is less volatile and so landlords do not need to have the capacity to sell quickly to time purchases to tap into capital gains. On the demand side, tenants in England, particularly younger and professional/business class tenants, exhibit a short-term horizon as they seek to save to ultimately fund a down payment on a house purchase (*ibid.*). The timing of the purchase can be critical in a highly volatile housing market, and so short-term ASTs give landlords the necessary freedom and flexibility to time their purchases as vacant possession can be achieved relatively quickly.

In Germany, the market is less volatile and so having such flexibility is not so important to those tenants who are seeking to buy in the long term. In any case the numbers looking to buy their houses is much smaller in Germany. Thus, rent regulation and security of tenure in Germany is not seen as an impediment to the supply of rented accommodation, as it is in England. In England, landlords seeking to benefit from real house price gains will want to sell with vacant possession to benefit from the highest prices; houses sold with 'sitting tenants' sell at a substantial discount. Kemp and Kofner have identified this 'value gap' as the main reason why landlords in England are so opposed to security of tenure and rent regulation (*ibid.*).

In the UK, more recently, we have seen the introduction of some soft regulation of rents, vicariously through limiting housing benefit payments, although this is probably less to do with improving affordability of the PRS for tenants, and more to do with reducing government expenditure. The subject of housing benefits and local housing allowances is a separate subject in itself, and there are views that the implementation of housing subsidies reinforces social polarisation (Hills 2001). This view has been strengthened by attitudes taken by landlords and their insurers. Many PRS landlords choose not to let their properties to housing benefit recipients, which may be a strict requirement of their insurers, thus reducing housing quality and choice (Beveridge 1942, as cited by Hills 2001).

Conclusion

Ideally, a smart city should provide all its citizens with decent homes. Increasingly, UK cities struggle to fulfil this basic need. A combination of factors caused the UK housing crisis. Its catalyst was Thatcher's, ideologically inspired, rush to sell off council homes in the 1980s without proper consideration of the social ramifications. Since then, inexorable housing crisis drivers include chronic low levels of construction, impediments to house building caused by the planning system, persistent high net immigration and demographic shifts, reducing household sizes. A decline in the relative share of household incomes compared to capital returns, and the shortage of mortgage finance stimulated demand for rented accommodation, with the PRS almost doubling in size in the past 15 years. Rogue landlords thrived but, if the PRS is to play a more sustainable role in solving the UK housing crisis, it needs to lift its game. Quality and energy efficiency in the sector must improve. However, small buy-to-let investor landlords are unlikely to deliver substantive improvements as their risk appetite is for short-term, arm's-length investments. Tenants need the security of longer occupational terms, perhaps with the flexibility of 'tenant break' clauses built into their tenancy agreements. In the UK, the culture of rent regulation and more enhanced tenure security is unlikely to return. In Germany, the PRS is larger than the UK but the investment culture is longer term. Unlike in England where security of tenure is seen as an evil to be avoided, in Germany it enhances the sustainability of rental income. A sustainable PRS smart solution invokes industry collaboration, tighter but not overburdensome tenancy regulations, more institutional investment and informed tweaking of mortgage finance conditions. In short, a reformed PRS system can help provide the diversity of housing needed for a *smart city*. Making the PRS 'fit for purpose' involves informed policy tweaking, deliberation and collaboration to tackling multiple issues in an evolving system.

References

Ambrose, J. (2015) Studying the links: the relationship between fuel poverty, energy efficiency and health, *RICS Property Journal,* November, 34–36.

Barker, K. (2004) Review of housing supply, *Delivering Stability: Securing our Future Housing Needs.* London: HMSO.

Bate, A. (2015) Building the new private rented sector: issues and prospects (England). Briefing Paper, SN07094, 12 August.

Bate, A. (2016) Building the new private rented sector: issues and prospects (England). Briefing Paper, No. 07094, 12 December.

Beveridge, W.H. (1942) Social Insurance and Allied Services. Cmnd 6404. London: HMSO.

Blackburn, J. (2016) Crisis? What crisis? *RICS Residential Policy, RICS Property Journal,* July/August, 36–37, (online) available at rics.org/journals

BPF (2013) *A British Property Federation Guide to Energy Efficiency and the Private Rented Sector.* London: BPF.

Clarke, S., Corlett, A. and Judge, L. (2016) The housing headwind, the impact of rising housing costs on UK living standards, *Resolution Foundation Report,* June.

Cook, A. and Cross, S. (2015) Minimum energy efficiency standards for UK rented properties, *Out-Law,com, Legal news,* August, Pinsent and Masons, available at: www.out-law.com/en/topics/property/environment/minimum-energy-efficiency-standards-for-uk-rented-properties/ (accessed 7 August 2016).

Crook, A.D.H. and Kemp, P.A. (2011) *Transforming Private Landlords.* Oxford: Wiley-Blackwell.

Department for Communities and Local Government (DCLG) (2010) English housing survey. Household report, *National Statistics,* 2008–09.

Department for Communities and Local Government (DCLG) (2012) Review of the barriers to institutional investment in private rented homes, The Montague Review, available at: www.gov.uk/government/uploads/system/uploads/attachment_data/file/15547/montague_review.pdf (accessed 6 July 2016).

Department for Communities and Local Government (DCLG) (2017a) Planning and affordable housing for build-to-rent, Consultation Paper, February, available at: www.gov.uk/government/consultations/planning-and-affordable-housing-for-build-to-rent

Department for Communities and Local Government (DCLG) (2017b) Housing White Paper, Fixing our broken housing market, 7 February.

Department of Energy and Climate Change (2011) Local authorities and the Green Deal, *Information Note*, November. London: DECC.

Energy Saving Trust (2016) *Private Rented Sector*, National Archives, available at: http://web archive.nationalarchives.gov.uk/20160105160709/http://nomisweb.co.uk/query/select/getdatasetbytheme.asp?opt=3&theme=&subgrp= (accessed 22 August 2016).

English Housing Survey (2011–12) available at: www.gov.uk/government/publications/english-housing-survey-2011-to-2012-headline-report

English Housing Survey (2014–15) DCLG, available at: www.gov.uk/government/collections/english-housing-survey

Gilmore, G. (2014) The rental revolution: examining the private rented sector 2014, *Residential Research*. London: Knight Frank.

Goddard, J. (2016) Am I qualified for this? Right to rent is no solution to the public's immigration concerns, *RICS Property Journal*, 40–41. ISSN 2050-0106.

Hardman, J. and Coghill, J. (2015) In it for the long term: valuing build to rent, *RICS Property Journal*, May/June, 9–11.

Hills, J. (2001) Inclusion or exclusion? The role of housing subsidies and benefits, *Urban Studies*, 38(11), 1887–1902.

Hope, A.J. and Booth, A. (2014) Attitudes and behaviours of private sector landlords towards the energy efficiency of tenanted homes, *Energy Policy*, 75, 369–378.

Horti, S. (2016a) Can build-to-rent thrive in a Brexit-induced slowdown? *Property Week*, 29 July, 37–39.

Horti, S. (2016b) Housebuilders look to PRS investors as market cools, *Property Week* 5 August.

Huston, S., Jadevicius, A. and Minaei, N. (2015) Talent and student rented sector bottlenecks: a preliminary UK investigation, *Property Management*, 33(3), 287–302.

Kemp, P.A. and Kofner, S. (2010) Contrasting varieties of private renting: England and Germany, *International Journal of Housing Policy*, 10(4), 385.

Middlemiss, L. and Gillard, R. (2015) Fuel poverty from the bottom-up: characterising household energy vulnerability through the lived experience of the fuel poor, *Energy Research & Social Science*, 6, 146–154.

MSCI, Strutt & Parker, and the BPF (2015) *UK Lease Events Review*, November.

Mueller, K. (2015) Changing the way the UK rents, Residential Build to Rent, *RICS Property Journal*, July/August, 34.

New Economy (2014) Mapping the private rented sector for young professionals and mid incomes families in Greater Manchester, *New Economy*, available at: http://neweconomy manchester.com/publications/private-rented-sector-in-greater-manchester

Niemietz, K. (2015) *Reducing Poverty through Policies to Cut the Cost of Living*. York: Joseph Rowntree Foundation, 11–12, available at: www.jrf.org.uk/sites/default/files/jrf/reducing-poverty-cost-living-summary.pdf

ONS (2013) Households, families and people, *General Lifestyle Survey Overview* – a report on the 2011 General Lifestyle Survey, available from: www.ons.gov.uk/peoplepopulation andcommunity/personalandhouseholdfinances/incomeandwealth/compendium/general lifestylesurvey/2013-03-07/chapter3householdsfamilesandpeoplegenerallifestylesurvey overviewareportonthe2011generallifestylesurvey (accessed 20 August 2016).

ONS (2014) *Young Adults Living With Parents, 2013*. London: Office for National Statistics.

RICS (2014) Global Affordable Housing Report, *BRICS Plus Mortar*, St Andrews Centre for Housing Research.

RICS (2015) Private rented sector, *Code of Practice*, RICS, available at: www.rics.org/Global/ Private_Rented_Sector_code_PGguidance_amended_July_2015.pdf (accessed 7 August 2016).

Scanlon, K., Fernández, M. and Whitehead, C. (2014) *A Lifestyle Choice for Families? Private Renting in London, New York, Berlin and the Randstad*. London: Get Living London.

Smales, C. (2016) Housebuilders down tools until referendum storm has passed, Home Truths, Professional, *Property Week*, 19 August, 53.

Smith Institute (2014) The growth of private rented sector: what do local authorities think? available at: https://smithinstitutethinktank.files.wordpress.com/2014/09/the-growth-of- the-prs-what-do-local-authorities-think.pdf (accessed 27 February 2017).

Souter, J. and Kafkaris, A. (2015) The politics of private rent, *EGi, Legal, Estates Gazette*, 24 April, available at: www.egi.co.uk/legal/the-politics-of-private-rents/?keyword= private%20rented%20sector (accessed 28 August 2016).

Watson, S. (2015) Private Party, *RICS Modus*, 40–41.

Wetzstein, S. (2016) The global urban housing crisis and private rental in the Anglophone world: future-proofing a critical sector and tenure, *Housing Finance International*, Autumn, 31–34.

16

INTEGRATED AND SUSTAINABLE PRESERVATION OF ARCHAEOLOGICAL HERITAGE IN URBAN AREAS

Case study of Istanbul Sea Walls

Nisa Semiz

Abstract

Planning for the smart city requires understanding current urban patterns and broad agreement on social purpose to navigate contested land use conflicts. Understanding, preserving and conserving heritage cements place and cultural identity and, if wisely leveraged, can infuse development projects with an emotional authenticity that enriches quality of life for locals and attracts visitors. Drawing on the most iconic world city, Istanbul, Chapter 16 investigates the policy implications of smart urban development.

Introduction

Istanbul is an ancient trading hub that spans Europe and Asia. The historic core of today's metropolis sits on a peninsula at the south-western end of the Bosphorus, which links the Mediterranean with the Black Sea. The city's rich history has multiple layers of cultural heritage. Originally, historians thought that the Megarians established a colony with the name of Byzantion in the seventh century BC (Gilles 1988: 1–2). Yet recent archaeological excavations on the southern coast of the peninsula confirm that the city's foundations are prehistoric (Kızıltan 2008: 5–9). After coming under Roman rule in the second century AD, the designation of the city as the new centre of the Empire in the fourth century is the main turning point. Six years after he had inaugurated immense construction activities, Constantine consecrated Constantinople in 330 AD (Vasiliev 1952: 59). For more than a millennium, it remained the capital of the Eastern Roman or Byzantine Empire. After its conquest by Sultan Mehmed II in 1453, the city served as the capital of the Ottoman Empire until the twentieth century.

Archaeological and architectural monuments in the city manifest complex layers of Istanbul's diverse heritage. Since the 1960s, Istanbul heritage has been under continuous pressure from urban migration and consequent rampant development. In the historic quarters of Istanbul, seismic socio-demographic changes have weakened cultural bonds. Housing and development pressures also impact the Istanbul Sea Walls, which are among the primary components of archaeological heritage and engage with the historic urban landscape of the city.

The construction of the Istanbul Sea Walls runs parallel with the development of the historic city itself, beginning from its establishment as a Megara colony at the eastern apex of the peninsula until the last designation of the western borders of Constantinople with the reconstruction of the present Land Walls on the further west, in the fifth century. After the conversion of the city into the Ottoman capital in the mid-fifteenth century, the territory of the historic city enclosed by the Land and Sea Walls remained almost unchanged. The walled city located on the peninsula, which is called 'the Historic Peninsula' today, represented the city of Istanbul apart from its districts until the twentieth century. At present the Land and Sea Walls surrounding the historic city and the area enclosed by them is registered as a conservation zone and is under legal protection.

After losing their military function in the nineteenth century, the Land and Sea Walls were abandoned, and some parts were lost due to the modernization of the Ottoman capital. Then, the change and growth of the city in the Republic period caused more damage to and losses from the ancient walls. Today, only some parts of the Sea Walls are intact, and the remaining sections have been involved in the historic urban fabric of the city in different ways. Thus, the preservation of the Sea Walls and their immediate vicinity needs to be implemented considering all archaeological, architectural and urban assets.

Past and present Istanbul Sea Walls

Historical background

Istanbul Sea Walls have testified to all the development stages of the city beginning from its establishment as a Greek colony in the seventh century BC until reaching its final limits as the capital of the East Roman Empire in the fifth century AD (Figure 16.1). The historic city had always been open to hostile attacks by land and sea due to its position. The Greek settlement, namely 'Byzantion', was enclosed by walls (Yavuz 2014: 14). Although the city was restored and enlarged, with the building of new walls under Roman rule at the end of the second century (Van Millingen 1899: 9), the major change took place in the first half of the fourth century as the new centre of the Roman Empire. New fortifications were constructed by Emperor Constantine on the west, quadrupling the territory of the city (Gilles 1988: 10; Van Millingen 1899: 15; Bassett 2004: 22–23). After completion of the main urban arrangements, the city took the name of 'Constantinopolis', namely 'City of Constantine', with the consecration ceremonies in 330 (Vasiliev 1952: 59). Less than one century later, the city needed enlargement again due to an increase in its population because of its prestige and importance. Under the rule of Emperor Theodosius II, new Land Walls were constructed about 1.5 kilometres to the west of the old, including Blachernae which was a suburb surrounded with its own walls on the north-western corner of the city of Constantine (Bury 1958: 70). Then, the seaward fortifications were lengthened in order to fortify the open parts of the northern and southern coasts of the city except the coast of Blachernae (Schneider 1951: 82–83). Yet, after the Avar attacks in 626, northern walls were lengthened, with the Land and Sea Walls converging at the north-western corner of the peninsula (Dirimtekin 1948: 23–24). Thus, the defensive circuit reached the outline surrounding the city by land and sea continuously. As one of the earliest images of Constantinople, the engraving by Schedel, dated to the end of the fifteenth century, shows us the original characteristics of the walled city (Figure 16.2).

Compared to sophisticated design of the Land Walls which consist of a ternary system comprised of the main wall, the outer wall and the moat, the Sea Walls

FIGURE 16.1 Development of the city walls of Constantinople

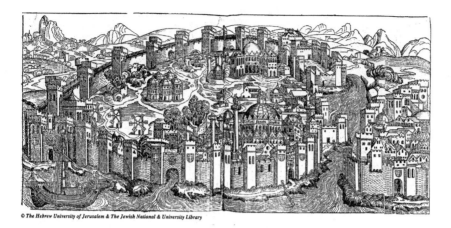

FIGURE 16.2 View of Constantinople prepared by H. Schedel, 1493

were comprised of one line of wall, studded with towers due to their safeguarding position along the coast. The Sea Walls were built either on the coastline or a short distance from it all along the shores of the Byzantine capital. The northern walls were aligned with space between the fortifications and the coastline to allow commercial activities and traffic along the Golden Horn which constitutes a deep and natural harbour on the north of the peninsula. On the other hand, the south coast is rocky and open to the rough waves, so the southern walls were erected on the coastline. There were also some important harbours enclosed by walls on the southern coast of the city. Furthermore, the gates on the Sea Walls led to the settlements, palace buildings or monasteries along the shores of the historic city from the sea.

In the Middle Ages, the Land and Sea Walls were maintained and repaired properly, and also some parts were rebuilt (Van Millingen 1899: 95, 180). After the conquest of the city by the Ottomans in 1453, the repairs of the old defence system were among the primary construction activities (Gibbon 2000: 730). However, due to the proliferation of firearms beginning from the sixteenth century, the city walls gradually lost their importance to the defence of the city. Housing was developed, especially in the vicinity of the Sea Walls, as is seen on the copy of the image of Constantinople by Vavassore dated to the sixteenth century (Figure 16.3).

Changes and losses

Although the city walls of Istanbul were not as militarily effective as before, in the Ottoman period the maintenance and repair of the defence system continued and the walls were protected until the nineteenth century (Sakaoğlu 2007: 32). With the military revolution in the first half of the nineteenth century, old defensive walls lost their significance entirely (*ibid.*: 35). Moreover, the effects of the

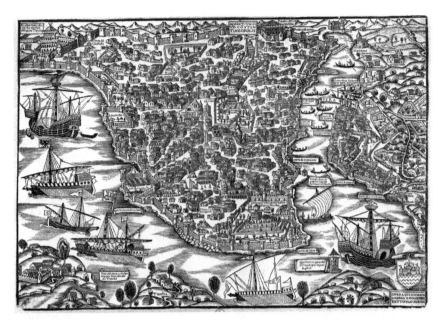

FIGURE 16.3 View of Constantinople. Engraving, c. 1530 according to a model around 1478/79–1490 prepared by Giovanni Andrea Vavassore (Buondelmonti 2005)

modernization of the Ottoman capital due to industrialization in this period endangered the continuity of the old defence system. These developments caused variable effects on the city walls; the Sea Walls especially were affected due to their position and structural strength. The construction of the Rumelia Railway between 1871 and 1872 caused the first losses on the walls. The southern walls, namely, Marmara Sea Walls, were interrupted at nine points, and also the eastern end of the northern walls, namely, the Golden Horn Sea Walls, were demolished for the construction of the railway. Beginning from the second half of the nineteenth century, the development of the capital towards to the north of the Golden Horn acquired a different extent with the removal of the Ottoman palace to Dolmabahçe (Çelik 1993: 39–42). The opposite banks of the Golden Horn were linked with two bridges in this period, consequently the continuity of the Golden Horn Sea Walls were affected by these developments adversely. Furthermore, the walls that had lost their function started to be given for rent in this period. The Ottoman documents dated to 1698 (BOA 1698) show the leased parts of the walls, namely, 'kule-i zemin' and their rental incomes. The Sea Walls, affected considerably by the developments in the nineteenth century, underwent functional transformation, and have become parts of the urban fabric in different ways.

In the Republic period, Istanbul lost its importance because the capital shifted to Ankara. Then a master plan of Istanbul was prepared and applied between 1936 and 1951. The planning decisions (Prost 1938) affected the old defence walls in different ways. The Land Walls were to be protected within a band, extending from the north end at the Golden Horn coast to the south end on the coast of the Sea of Marmara (*ibid.*: 6). However, parts of the Sea Walls, which were not accepted as important in terms of architecture and art history, were allowed to be demolished when the construction of new roads, or arteries were needed within the old city (*ibid.*). As a result of new political development in the country, Istanbul gained importance again beginning from the 1950s. At the end of the 1950s, as a consequence of a demolition and reconstruction process for the establishment of a modern transportation network within the city, the historic urban fabric was damaged severely (Tekeli 1994: 46–48). The building activities caused more losses on the old defensive walls, and also endangered the link between the city walls and their vicinity. In this period, the Golden Horn Sea Walls underwent important losses. Moreover, the increase in immigration from Anatolia to Istanbul caused changes in the social and demographic structure of the old quarters within the Historic Peninsula. This situation led to increase in damages on the Sea Walls.

On the other hand, the tendencies for the protection of historic urban fabric accelerated, with the result of new historic preservation legislation in 1973, and Istanbul Historic Peninsula surrounded by the ancient walls was declared a historic site (Ahunbay 2009: 21). Then, 'Historic Areas of Istanbul' which consist of four separately protected sites,[1] were accepted in the World Heritage List by UNESCO in 1985 (*ibid.*: 22) (Figure 16.4). Although the Land Walls are among them, only some parts of the Sea Walls are involved in two historic areas: the Archaeological Park and Land Walls.

Despite these developments encouraging the protection of the historic city, new urban projects carried out by the local authorities caused other damage to the historic and landscape values of the Sea Walls and their vicinity in the second half of the 1980s. Between 1984 and 1989, the industrial buildings located on both banks of the Golden Horn were demolished and the coastal areas were arranged as green areas by the Municipality of Istanbul (Semiz 2015: 65). These works were realized without any necessary preliminary survey of the historic sites, and caused damages and losses within the historic urban fabric. In addition, restoration work was started on some parts of the City Walls in 1986, yet the restorations took place ignoring the archaeological and landscape values of the Land and Sea Walls (Ahunbay & Ahunbay 1994: 79–80). These works were heavily criticized, and at the beginning of the 1990s some attempts were made for the restoration and conservation of Istanbul City Walls from a scientific perspective (Ahunbay & Ahunbay 2000). Then the restoration works were ceased in 1994 (Ahunbay & Ahunbay 1994: 80).

In the 2000s, the public works in the metropolitan area of Istanbul were accelerated due to the rampant increase in population (Figure 16.5). Recently developed projects for improving public transportation or urban regeneration have had detrimental impacts on the historic urban landscape of the city from various aspects. The projects have been realized regardless of the essential requirements for the protection of the historic urban fabric. Latest construction activities have also endangered the protection of the Sea Walls together with their cultural landscape values.

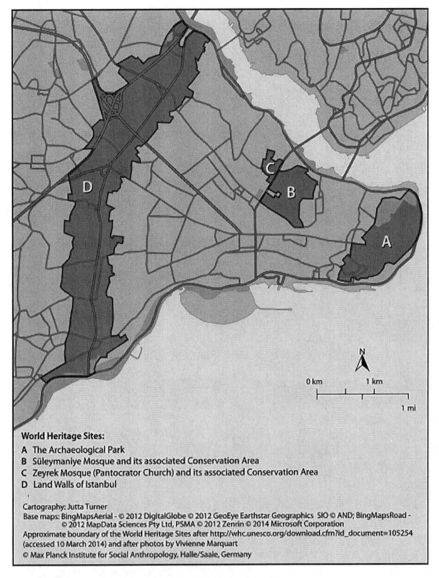

FIGURE 16.4 Historic Quarters of Istanbul in World Heritage List

Source: Marquart (2014).

Current state of preservation

Today, only some parts of the Sea Walls, which are separately known as the Golden Horn and Marmara Sea Walls, are intact (Figure 16.6). In some areas, the wall is preserved along with the towers, in others, only a tower or a part of the wall are still standing (Figure 16.7). The initial position of the Sea Walls has also changed

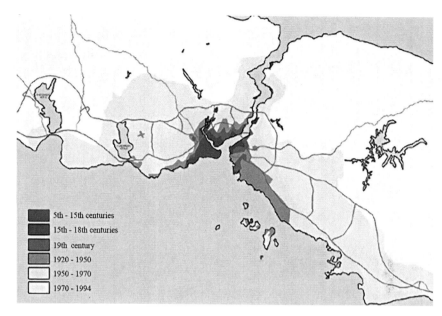

FIGURE 16.5 Development of settlement areas in Istanbul

Source: 1/100000 Ölçekli İstanbul Çevre Düzeni Planı Raporu (2009).

Legend:
- 5th - 15th centuries
- 15th - 18th centuries
- 19th century
- 1920 - 1950
- 1950 - 1970
- 1970 - 1994

due to extension of the shores. At present the surviving parts are located at a distance from the current coastline. For instance, the distance between the remaining parts and the coastline has reached 200 metres in Kumkapı and 1,000 metres in Yenikapı after the latest construction work along the southern coast of the Historic Peninsula.

The current state of preservation of each remaining part of the Sea Walls shows alterations according to the change in their original materials and construction techniques, and historic urban fabric surrounding them. The construction process of the Sea Walls encompasses a long time. They had been repaired and reconstructed due to the defensive needs during the Middle Ages. Then their maintenance and repair continued in the Ottoman time. Consequently, different construction techniques, which testify to different periods, can be seen on the remaining parts. Although the original construction techniques dated to the Byzantine and Ottoman periods have been protected well on some parts, the restorations in the 1980s and the 1990s caused some damage to the historic and archaeological values of the remaining parts.

Furthermore the vicinity of the Sea Walls has altered as a result of the urban changes in Byzantine and Ottoman times. So long as the defence system of the city maintained their military significance, housing was not allowed in the immediate vicinity of the walls. However, this situation changed after the considerable decrease in their military importance. On the image of Vavassore (Figure 16.3), the buildings stand close to the Sea Walls along the northern and southern coasts of the peninsula. Although housing in the immediate vicinity of the Sea Walls was

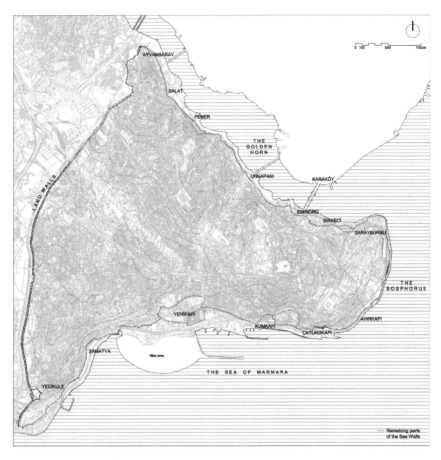

FIGURE 16.6 Remaining parts of the Sea Walls

FIGURE 16.7 Marmara Sea Walls at Sarayburnu

FIGURE 16.8 The remains of the Golden Horn Sea Walls and the vernacular architecture raised on them

restricted under the edicts of the Ottoman sultans at first, it could not be avoided (Ahmet Refik 1935: 58–59). Then, in the nineteenth century, the principal changes took place in the Ottoman capital. The urban fabric of the city underwent some changes as a consequence of industrialization and modernization (Çelik 1993: 49–81). Today, some examples of the civil architecture and also industrial buildings built in the nineteenth century are still standing on the remains of the Sea Walls (Figures 16.8 and 16.9). Therefore, the preservation of the Sea Walls does not mean only the surviving parts themselves, but also the protection of the historical and landscape characteristics of their immediate surroundings.

Assessment of conservation issues

Despite a history spanning over 1,600 years from their first construction, the ancient walls of Constantinople have survived substantially thanks to their firm structures and specific designs. However, the Sea Walls have been affected by the growth of the city and damaged much more in comparison to the Land Walls.

The surviving parts of the Sea Walls are in an advanced state of deterioration depending on physical and environmental conditions. Climate conditions have caused material decay to the walls. Some parts of the Sea Walls have structural problems due to the unstable ground they stand on. Moreover, the fortifications have suffered from earthquakes in their long history and the remaining parts are

FIGURE 16.9 Yedikule Gasworks and Railway Repair Shop raised on the western end of Marmara Sea Walls

Source: DAI-IST-R32771.

under seismic risk. According to historic sources, severe damage was caused by large-scale earthquakes, as a result of which the Sea Walls underwent repairs together with other parts of the city walls. In addition, the towers and adjoining walls have serious structural damage due to unstable ground. They are inclined vertically and, in some parts, the towers are separated from the adjoining walls. The structural problems are serious, and they cause risks not only to the heritage value of the walls, but also to the local communities living in the neighbouring areas.

On the other hand, the urban planning strategies and the public works from the Late Ottoman period to the present time have caused major losses on the Sea Walls. Moreover, the social and demographic structure of the old settlements along the coasts of the Historic Peninsula has changed considerably due to the rapid increase in the population of Istanbul beginning from the 1960s. Consequently, housing has densely risen in the vicinity of the walls. Then, in the 2000s, there was more building construction as a consequence of the growth of the city with the impact of neoliberal policies. Within this context, the urban problems affecting the conservation of the Sea Walls will be identified under the headings of planning issues, legal and administrative issues, and social and economic issues.

Planning issues

The urban planning projects in the Republic period caused many changes in old Istanbul. Some planning decisions led to detrimental effects on the historic urban fabric and historic landscape of the city. The Sea Walls were also considerably influenced by the urban changes, and lost many parts. At the end of the 1950s, ancient walls were destroyed due to the construction of the coastal road along the southern shore and at some parts of the northern shore of the Historic Peninsula. The clearing of the northern coastline in the 1980s not only caused more losses within the Sea Walls and their immediate vicinity, but also entailed decay of the walls due to the traffic.

Furthermore, in the 2000s, the increase in public works and building construction have affected the historic urban landscape of Istanbul dramatically. For instance, Yenikapı, located at the middle of the southern coast of the Historic Peninsula, was selected as the place for the main station of the Marmaray and Metro projects.[2] Before the construction, the Istanbul Archaeology Museums started a salvage excavation at the station area and an important harbour area, namely Theodosius Harbour dated to the fourth century, and also other archaeological remains dated from the Neolithic age to the Ottoman period were uncovered (Kızıltan 2008: 1–11). Some parts of the harbour walls are also among the recovered remains. Yet, today it is not possible to see the archaeological finds since almost all of the excavation area was covered and the main station building was constructed over it. Moreover, the coast of Yenikapı was enlarged to the south about 500 metres in order to create a meeting area for public demonstrations. Consequently, the last traces of historic urban fabric of the Yenikapı quarter were damaged considerably, and the natural historical topography of the Istanbul Historic Peninsula altered extensively. In addition, Avrasya Tunnel Project, developed as an underground passage from the European side to the Asian side for motor vehicles, led to irreversible impacts with the result of the rearrangement of the southern shore of the Historic Peninsula.

Unlike the Land Walls, it is hard to distinguish the remaining parts of the Sea Walls due to their inclusion in the urban fabric in most parts. So the relics of the Sea Walls must be identified properly. It is necessary to consider the archaeological remains when new urban planning decisions are improved, or rehabilitation and urban regeneration projects for the old quarters are developed. Consequently, an inventory of the Sea Walls was started as a part of a project called 'the Inventory of Cultural Heritage in Istanbul' in 2010 (Ahunbay 2011: 23). At present, this inventory has been revised and updated by the Directorate of the Cultural Heritage Preservation of the Metropolitan Municipality of Istanbul.

In addition, the restorations of some parts of the Istanbul City Walls from 1986 to 1994 have been criticized since they were realized without considering scientific approaches for archaeological restoration and protection of the historical landscape values. The need for interdisciplinary work was ignored. Moreover, reintegrations and reconstructions which are not proper to the original materials and construction techniques of the walls were applied at some parts. In addition, as important components of cultural heritage in Istanbul, the walls cannot be considered apart from their surroundings. So the restoration of the walls should have been integrated with the rehabilitation of their immediate vicinity.

At present, some regeneration projects of the old quarters in the Historic Peninsula are on the agenda of the Municipality of Fatih,[3] and they affect the vicinity of the Sea Walls at some parts. Ayvansaray Urban Regeneration Project, which comprises an old quarter surrounded by Land and Sea Walls at the north-western corner of historic city, has almost been completed. Yet, historic urban fabric dated to before the establishment of Constantinople was destroyed by changing the urban fabric which reflected the characteristics of the Ottoman neighbourhood, and constructing reinforced concrete structures (Esmer 2016: 47–49). As a consequence, the walls surrounding the quarter have been detached from their historic context.

Legal and administrative issues

The Land and Sea Walls surrounding the ancient city of Constantinople are registered as ancient monuments and the Metropolitan Municipality of Istanbul is the responsible body for their maintenance and protection. The ancient walls surrounding the historic city and the area enclosed by them is registered as a conservation zone and is under legal protection. The conservation zone is subject to national legislation, namely Legislation for the Conservation of Cultural and Natural Heritage (Law No. 2863, amended by Laws No. 3386 dated to 17 June 1987 and Laws No. 5226 dated to 27 July 2004). In 1985, the Historic Areas of Istanbul, which comprise the Archaeological Park, Süleymaniye, Zeyrek and Land Walls, were inscribed on the UNESCO World Heritage List. The Metropolitan Municipality of Istanbul has developed a conservation plan for the urban and archaeological assets of the walled city (Istanbul Historic Peninsula Site Management Plan 2011).

On the other hand, recent legislative regulations on protection of cultural properties have created legal confusions and gaps. In 2005, the law on the 'Preservation by Renovation and Utilization through Revitalizing Deteriorated Immovable Historical and Cultural Properties' (Law No. 5366) enables local authorities to prepare regeneration projects for degraded historical areas, namely 'renewal areas'. Despite the positive aspects with the result that local authorities take part in decision-making processes in more effective ways, the integrated preservation of the Historic Peninsula has been interrupted. At present, most of the remaining parts of the Sea Walls are involved in the renewal areas. This situation has caused contradictions in legal procedures for the preservation of the Sea Walls and their immediate vicinity in an integrated way. For instance, the Ayvansaray Urban Regeneration Project, which has been criticized due to irreversible damages in physical and social traits of the historic urban fabric (Marquart 2014: 44–46), was implemented after the declaration of the quarter as a renewal area under Law No. 5366.

The preservation of the Land and Sea Walls is a interdisciplinary work and needs equal participation of the related institutions. Dialogue between the Metropolitan Municipality of Istanbul, other local authorities, universities and non-governmental institutions, and also the participation of the public is necessary. The lack of dialogue and equal participation have disrupted the preservation of the walls, and consequently the maintenance of the walls has been neglected.

Social and economic issues

The population of Istanbul has risen dramatically due to immigration from rural areas to the city beginning from the 1960s as a result of industrialization (Tekeli 1994: 40). The population of Istanbul has increased from about 2.3 million in 1965 to about 14.7 million at present (URL-1). Consequently, the resulting unplanned urbanization has changed the social and demographic structure of old Istanbul considerably. Demographic change has disrupted the continuity of the relationship between locals and their environment. Moreover, maintenance of the relics and protection of the historic urban fabric have been neglected by the settlement of the people in the lower income group. In addition the lack of public awareness incapacitates every attempt for the preservation of the cultural heritage.

The protection of the Sea Walls and their surroundings is also related to the supply of economic resources. The maintenance and repair of the surviving parts of the walls need adequate funds. The recent public works in Istanbul indicate the availability of large funds. However, the highest priority has been given to the transportation projects due to rampant growth of the city.

Moreover, as a rapidly developing metropolis, Istanbul has become the financial and cultural centre of Turkey. The city has become an important centre for urban tourism at the same time (Uysal 2015: 9). Historical heritage in Istanbul is a leading factor for cultural tourism. Increases in tourism revenues promote the preservation and conservation of the historic fabric of the city. On the other hand, tourism demands and real-estate pressures have led to urban regeneration and renewal projects. This situation makes it difficult to sustain the historic fabric, and has also forced the changes in old settlements along the shores of the historic peninsula. It has caused environments to be disconnected from their original social and physical structure, as in the example of Ayvansaray Urban Regeneration Project. Consequently, the integrity of the remaining parts of the Sea Walls and their vicinity are open to more adverse effects of tourism and real-estate factors, unless these driving forces are used to provide sustainability.

Conclusions

Istanbul is an important city by means of natural, historical and cultural assets. With the effects of the Industrial Revolution, the city underwent a transformation in the second half of the nineteenth century. After the first urban plan of Istanbul at the end of 1930s, the social and physical traits of the city changed, and mainly as a result of political and economic decisions beginning from the 1950s. Consequently, the city has become a metropolis with a population of almost 15 million. These developments have affected the sustainability of the historic city adversely. The cultural heritage in Istanbul has been under some risks due to rapid growth of the city. In this study, Istanbul Sea Walls and their vicinity have been examined in terms of archaeological and urban perspectives as a tool for the sustainability of the city.

The surviving parts of the Sea Walls are significant components of the cultural and archaeological heritage in Istanbul. The study has identified the extant parts of the Sea Walls and documented the damages and deformations due to various factors (Semiz 2015).[4] It is clear that, if there is no immediate action, the planning, administrative, legal, social and economic issues can give rise to more damages, even losses. It is also essential for the sustainability of the historic urban fabric and the improvement of the social characteristics of their environment.

The establishment of a balanced, integrated and sustainable management process is the key factor for the conservation of a historic city (Bandarin & Van Oers 2012: 176). Holistic approaches are necessary for the preservation of the remaining parts of the city walls with their settings. Archaeological remains need multidisciplinary work and active and equal participation of all stakeholders for their protection and preservation. Establishment of a specific management system is fundamental. Constant watch over the Land and Sea Walls has to be adopted as a preventive measure. For their proper maintenance and repair, a conservation team comprised of specialists on masonry structures has to be created.

On the other hand, the preservation of the surviving parts is not only based on archaeological restoration, but also is related to proper management policy at the national level (ICOMOS 1990). Additionally, according to the Washington Charter 'the conservation of historic towns and other historic urban areas should be an integral part of coherent policies of economic and social development and of urban and regional planning at every level' (ICOMOS 1987). The remains of the Land and Sea Walls are integrated with their environments. So the preservation of historic urban fabric and the rehabilitation of social and demographic structure in the vicinity of the city walls is indispensable. Improving public awareness and providing participation of the locals constitute another essential phase for the protection of the city walls. In comparison to the Land Walls, the remaining sections of the Sea Walls are separate from each other and it is hard to distinguish them at some parts. Therefore, the presentation of the city walls also needs an integrated approach considering their heritage, archaeological and landscape values.

Providing funds for the preservation of the archaeological heritage is an important economic issue. Ideally, tourism revenues should be ring-fenced for culture-led urban regeneration projects (ICOMOS 1999). As Orbaşlı points out, 'Tourism is potentially an important catalyst for the safeguarding of historic fabric and the initiation of conservation on an urban scale' (2009: 42). However, the developing countries have distinctive issues facing conservation in building and urban scale, like urban growth in a much greater pace than in the West, limited urban resources which are under greater demand, inadequate economic supply for urban conservation, pressure of the desire to modernize urban environment (Orbaşlı 2009: 26). Istanbul is a developing metropolis and this situation has induced construction activities like new roads, bridges, transportation systems and so on.

Recent public works have caused degradation in natural, built and cultural environments in Istanbul. However, the main issue for a sustainable city is to establish a balanced relationship between preservation of cultural environment and support construction for today's needs (Rodwell 2007: 132). To strengthen sustainable urban development, Istanbul planning policies and practices need revision. Improvements should give greater weight to heritage protection, tighten public asset management, sharpen development application scrutiny and enforce planning regulations.

Notes

1 The Archaeological Park, Land Walls, Suleymaniye and Zeyrek.
2 Yenikapı Station is the junction point of the Marmaray and Metro lines. Marmaray enables the movement of trains from the European side to the Asian side by a channel under the Bosphorus. The main station at Yenikapı is connected to Sirkeci and the line continues under the Bosphorus, reaching Üsküdar on the Asian side of the city. On the other hand, two metro (subway) lines give access between Yenikapı Station and the western and northern parts of the city.
3 Fatih is the capital district that comprises Istanbul Historic Peninsula.
4 For further information, the PhD dissertation on the Golden Horn and Marmara Sea Walls and the proposals for their preservation can be viewed.

Bibliography

1/100000 Ölçekli Istanbul Çevre Düzeni Planı Raporu. (2009). Metropolitan Municipality of Istanbul, Department of Public Works and Urban Planning, Directorate of Urban Planning, Istanbul.

Ahmet Refik. (1935). *On Altıncı Asırda İstanbul Hayatı; 1553–1591: İstanbul'un Düşünsel, Sosyal, Ekonomik ve Tecimsel Ahval ile Evkaf, Uray, Beslev ve Gümrük İşlerine dair Türk Arşivinin Basılmamış Belgeleri*. Istanbul: Maarif Vekaleti.

Ahunbay, M., & Ahunbay, Z. (1994). Sur Onarımları. *İstanbul Ansiklopedisi* (Vol.7, pp.79–80). Istanbul: Tarih Vakfı Yayınları.

Ahunbay, M., & Ahunbay, Z. (2000). Recent Work on the Land Walls of Istanbul. *Dumbarton Oaks Papers, 54*, 227–239. Washington, DC: Dumbarton Oaks Publications.

Ahunbay, Z. (2009). Historic Areas of Istanbul. In G. Pulhan (Ed.), *World Heritage in Turkey* (1st ed., pp.15–123) (translated and edited by J. M. Ross). Istanbul: Republic of Turkey Ministry of Culture and Tourism Publications.

Ahunbay, Z. (2011). *İstanbul'da Kentsel Mimari*. Istanbul: Istanbul Bilgi Üniversitesi Yayınları.

Bandarin, F., & Van Oers, R. (2012). *The Historic Urban Landscape: Managing Heritage in an Urban Century*. Hoboken, NJ: Wiley Blackwell.

Bassett, S. (2004). *The Urban Image of Late Antique Constantinople*. Cambridge: Cambridge University Press.

BOA (Başbakanlık Osmanlı Arşivleri/Ottoman Archives) (1698). Maliyeden Müdevver Defterler (MAD.d), Gömlek no:2025 (03/Za/1109 H.).

Buondelmonti, C. (2005). Cristoforo Buondelmonti: Liber insularum archipelagi: Universitäts- und Landesbibliothek Düsseldorf Mp. G 13: Faksimile, herausgegeben von Irmgard Siebert und Max Plassmann; mit Beiträgen von Arne Effenberger, Max Plassmann und Fabian Rijkerp. Wiesbaden: Reichert.

Bury, J. B. (1958). *History of the Roman Empire from the Death of Theodosius I to the Death of Justinian*, Retrieved from http://archive.org/details/historyoflaterro01buryuoft

Çelik, Z. (1993). *The Remaking of Istanbul: Portrait of an Ottoman City in the Nineteenth Century.* Berkeley: University of California Press.

DAI-IST (Deutsches Archäologisches Institut-Istanbul) Photograph Archive

Dirimtekin, F. (1948). *Le siège de Byzance par les Turcs-Avars au vii. siècle.* Istanbul: Amis.

Esmer, M. (2016). Ayvansaray'daki Kentsel Yenileme ve Dönüşümün Değerlendirilmesi. *Mimarlık, 389*, 47–52. Ankara.

Gibbon, E. (2000 [1788]). *The History of the Decline and Fall of the Roman Empire.* Edited and abridged by David Womersley. London: Penguin.

Gilles, P. (1988) *The Antiquities of Constantinople.* 2nd edition, translated by John Ball & Roland G. Musto, based on the translation by John Ball. New York: Italica Press.

ICOMOS (1987). *Charter for the Conservation of Historic Towns and Urban Areas (The Washington Charter),* Retrieved from www.icomos.org/images/DOCUMENTS/ Charters/towns_e.pdf

ICOMOS (1990). *Charter for the Protection and Management of the Archaeological Heritage.* Retrieved from www.icomos.org/images/DOCUMENTS/Charters/arch_e.pdf

ICOMOS (1999). *International Cultural Tourism Charter – Managing Tourism at Places of Heritage Significance.* Retrieved from www.icomos.org/images/DOCUMENTS/Charters/ INTERNATIONAL_CULTURAL_TOURISM_CHARTER.pdf

Istanbul Historic Peninsula Site Management Plan. (2011). Retrieved 20 July 2016, from www.alanbaskanligi.gov.tr/files/Management_Plan_090312_TUM.pdf

Kızıltan, Z. (2008). Excavations at Yenikapı, Sirkeci and Üsküdar within Marmaray and Metro Projects, *Istanbul Archaeological Museums, Proceedings of the 1st Symposium on Marmaray-Metro Salvage Excavations* (pp.1–15) 5–6 May. Istanbul: Istanbul Archaeological Institute.

Law No. 2863 (1983, 23 July). Kültür ve Tabiat Varlıklarını Koruma Kanunu. *T. C. Resmi Gazete, 18113.* Retrieved from www.resmigazete.gov.tr/arsiv/18113.pdf

Law No. 5366 (2005, 5 July). Yıpranan Tarihi ve Kültürel Taşınmaz Varlıkların Yenilenerek Korunması ve Yaşatılarak Kullanılması Hakkında Kanun. *T. C. Resmi Gazete, 25866.* Retrieved from www.resmigazete.gov.tr/eskiler/2005/07/20050705.htm

Marquart, V. (2014). Insurmountal Tension? On the Relation of World Heritage and Rapid Urban Transformation in Istanbul. *European Journal of Turkish Studies.* Retrieved 30 September 2016, from: http://ejts.revues.org/5044

Orbaşlı, A. (2009). *Tourists in Historic Towns: Urban Conservation and Heritage Management.* London: Taylor & Francis.

Prost, H. (1938). *İstanbul'un Nâzım Planını İzah Eden Rapor 25.10.1937.* Istanbul: Istanbul Belediye Matbaası.

Rodwell, D. (2007). *Conservation and Sustainability in Historic Cities.* Oxford: Blackwell.

Sakaoğlu, N. (2007). Surlar Osmanlı Dönemi Onarımları. *Uluslararası Karasurlarının Korunması için Uygun Yaklaşım ve Yöntemler Sempozyumu,* 20–22 Ocak (pp.32–36). Istanbul: Istanbul Büyükşehir Belediyesi Yayınları.

Schedel, H. (1493). *Liber Chronicarum.* CXXX, Retrieved 20 July 2016, from: http://historic-cities.huji.ac.il/turkey/istanbul/maps/schedel_1493_CXXX.html

Schneider, A. M. (1951). *Die Blachernen.* Leiden: Brill.

Semiz, N. (2015) *Istanbul Haliç ve Marmara Surları: Belgeleme Çalışmaları, Tarihi ve Peyzaj Değerlerinin Korunmasına Yönelik Öneriler.* A Thesis Submitted in partial fulfilment of the Requirements of Istanbul Technical University for the Degree of Doctor of Philosophy. Istanbul: Istanbul Technical University.

Tekeli, İ. (1994). *The Development of Istanbul Metropolitan Area: Urban Administration and Planning.* Istanbul.

Uysal, Ü. E. (2015). *Urban Tourism in Istanbul: Urban Regeneration, Mega-Events and City Marketing and Branding.* A Thesis Submitted in partial fulfilment of the Requirements of

University of Helsinki for the Degree of Doctor of Philosophy. Helsinki: University of Helsinki. Retrieved from https://helda.helsinki.fi/bitstream/handle/10138/152740/UrbanTou.pdf?sequence=1

Van Millingen, A. (1899). *Byzantine Constantinople, the Walls of the City and Adjoining Historical Sites*. London: J. Murray.

Vasiliev, A. A. (1952). *History of the Byzantine Empire*, Volume 1: 324–1453. 2nd English Edition. Madison: University of Wisconsin Press. Retrieved from www.scribd.com/doc/127541886/Vasiliev-History-of-the-Byzantine-Empire-vol-I-upload-Salah-Zyada-pdf

Yavuz, M. F. (2014). Great Walls of Byzantium. *Annual of Istanbul Studies, 3*, 13–21. Istanbul: Istanbul Research Institute.

Websites

URL-1 <https://biruni.tuik.gov.tr/nufusmenuapp/menu.zul>, retrieved 17 January 2017.

17

CONCLUSION

Simon Huston

The 16 chapters in this volume provide an overview and eclectic insights into aspects of smart urban regeneration. The book's motivation is the crisis emerging in many rapidly growing cities, where air pollution, inadequate sanitation, broken housing markets, congestion and power shortages undermine human dignity, productivity and potential. The introduction noted complexity, evolution and conundrums. Smart urban regeneration is not a 'destination' but a 'mind-set'. Whilst it seeks to shape urban fabric toward elusive 'smart city' notions of technological, cultural or ecological hue, it is participatory and collaborative. Yet whatever their political constitution and urban form, all smart cities are resilient. Resilient systems can weather shocks and quickly adapt to new circumstances. Smart cities avoid fixation on short-term optimisation that can degrade system flexibility. Rather, resilient systems incorporate redundancy to cope with unexpected disaster, enable alternate responses or facilitate innovation.

The first chapter, on smart urban planning, found that the undisputed necessary conditions for smart urban development include a constitutional and legislative framework to guarantee freedom, privacy and due process in a thriving economy with universal access to essential services. Beyond human rights, dignity and empowerment for human flourishing, paradox vests in three irreconcilable tensions. First is *dirigisme* or evolution in institutional networks. Second is freedom or community cohesion. Third is strategic fragmentation. Productivity, low carbon or humanist imperatives pull in different directions (resilience vs. short-term optimality, hedonic vs. eudemonic well-being). Transformative constraints include land designation and environmental quality thresholds, local identity, culture and funding. Smart urban regeneration often involves measured projects rather than mega urban regeneration ones. It blends judicious and measured processual strategic interventions and learning in collaborative institutional networks. As well as the vertical political system in different spatial tiers, the network encompasses horizontal commercial, non-government institutions and bottom-up engagement from local bodies or citizens, empowered by digital technology and land information.

The second chapter on green infrastructure focused on linkages and connectivity for multifunctional benefits. The chapter noted the limitations of attempts to monetise GI's positive spin-offs, and acknowledged the challenges in cost–benefit analyses of its correlated ecosystem services. The chapter suggests a pragmatic approach to the design and implementation of GI focusing on the more immediate and localised functional benefits that are simpler to justify in financial terms, for example the delivery of ancillary water-related or public access services, and on its connectivity roles. Smart city plans that incorporate GI can bring a host of incidental and intangible benefits. GI spin-offs such as biodiversity protection or cleaner air are difficult to isolate and problematic to monetise but are vital.

The third chapter explored the nexus between primary food production, urban food and waste and outlined strategies to produce a more holistic food system. To reduce the physical and mental distance between urban and rural communities, smart planning should consider food supply and foodscapes. Most urban areas have lost local and regional food sovereignty. Indeed, cities are clusters of sustenance dependents, technically, socially and spatially separated from food producers. The Foodscapes concept provides an opportunity for urban communities to re-connect with their food systems in a more agro-ecologically resilient way; it also provides a vehicle for community cohesion and partnerships. To tackle food waste and urban rainwater problems and strengthen food security resilience, smart urban development must involve foodscapes.

The fourth chapter investigated the notion of 'urban consciousness' and its relationship with 'smart cities', and quality of life. Smart city residents are relatively tolerant and enlightened and benefit from a 'conscious atmosphere'. A range of concrete measures help stimulate urban consciousness including augmenting digital infrastructure, improving public transport, footpaths or cycleways and preserving historical buildings to protect cultural identity.

The fifth chapter stressed the importance of effective local partnerships and urban innovation incubators or hubs for creative ecosystems. Important innovation facilitators include education; global connectivity to create a local buzz, informed by global trends and producer pipelines. In propitious policy environments with positive collaboration, heritage precincts provide the ideal milieu to foster a local buzz to spread tacit cultural knowledge for fashion innovation.

The sixth chapter investigated issues around urban megaprojects in smart development. It found megaprojects are unique and, almost invariably, contentious. Some megaprojects radically transform but others backfire and waste resources or damage the environment. Concerns are vested interests, expulsion, governance, stakeholder management, forecasting errors, value capture and funding models. Practical difficulties are local opposition, macroeconomic turbulence, policy flux and project complexity in diverse settings. Given political contentions, mega-project judgements must consider ecological, social and commercial aspects. Success depends on planning engagement, institutional structure, design, project management and sustainable funding. London, with its global status, provides a rich milieu to investigate megaprojects. The research focused on three projects: King's Cross, Olympic Park and Nine Elms. The research interviewed over 30 experts involved in these projects to find that a clear project vision and public realm enhancement help project success.

The seventh chapter investigated smart urban development in the Far East and found multiple helpful practices, including brownfield infill, revitalisation of decayed or historic precincts, pedestrianisation, mixed uses, higher densities, public transit systems, local identity, public safety, traditional streets. In their rush to develop, major conurbations, like Singapore and Kuala Lumpur, thoughtlessly tore down historical precincts contrary to sound urban design principles. Developers simply erased old historic precincts, wiping out identity and collective memory. Kuala Lumpur became the 'Los Angeles of Southeast Asia' because its ten different municipalities instigated a spate of uncoordinated and ill-considered developments. A belated massive highway system failed to integrate the fragmented metropolis. Both cities have learnt from their mistakes. Now, traditional buildings are restored, river corridors revitalised and tropical public spaces created. Intensive tree planting in Singapore has fostered a 'tropical garden city' ambiance. In KL, administrative fragmentation undermines efforts to shift development from a predatory piecemeal approach to a smart one, underpinned by sound urban design rooted in unique culture.

Chapter 8 examined the role of construction firms in smart development. Government cannot build smart cities without them. The research found that the keys to construction firms' competitiveness were a robust corporate strategy, organisational capabilities and a sound financial system. However, despite the importance of individual performance, smart urban development extends to the resilience of private and public sector partnerships in regulated but competitive markets. Market regulations include stringent construction industry guidelines. Smart city pilot projects strengthen policy learning and improve the resilience of construction project partnerships.

Chapter 9 noted that in many cities, public transport investment lags rather than leads growth. One way to fund it is to capture the uplift in real estate prices. Hedonic modelling provides a useful tool to articulate the impact of transport facility upgrades on property prices and help fund smart city infrastructure-led urban regeneration. Jaipur transport system investment improved accessibility, inflated property values and revitalised brownfield sites. The Jaipur transit project demonstrates that the taxation of adjacent property price uplifts can fund smart city infrastructure. Funding aside, other infrastructure challenges include political shenanigans, vested interests, liquidity constraints and construction-induced congestion.

The indicative link between smart cities and university ranks, and extensive econometric evidence suggest that universities are central to smart cities. Chapter 10 investigated how universities can catalyse smart development via expansion, research, promulgation of smart disciplines and skills training.

Smart development balances local human and ecological needs. Chapter 11 investigated critical issues associated with green corridors to provide anthropological and ecological services. For wildlife, this does not mean green spaces should necessarily be contiguous as species can diffuse hierarchically via a chain of habitats. Conservation efforts though should focus on target species, preferably with local relevance. Anthropologically, corridors provide environmental services, particularly by including trees, shrubs and hedges. Ideally, green corridors should improve intra-urban walkable links but a relatively small corridor is sufficient to meet the human need for green space. To prevent neglect and cut crime risk, corridor maintenance is essential via regular direct works or, indirectly, via environmental education to help foster and sustain local community interest.

The twelfth chapter noted China's rapid transformation is unprecedented in human history. It has unsettled traditional society and radically transformed landscapes, via urbanisation and intensification. By 2030, one billion people will live in Chinese cities. To cope with the challenge, the Chinese government has designated 285 pilot smart cities and 41 special pilot projects. However, China's smart notion is techno-centric. Its 'digital' smart cities programme involves deploying devices, strengthening networks, providing platforms and tweaking applications. Whether such measures are distractions or real solutions to many environmental and other challenges remains an open question. Notwithstanding the prognosis, some tentative policy suggestions emerge from reviewing the approach:

•   Establish an appropriate government actor to coordinate smart *city* planning
•   Support reform initiatives such as smart city task forces
•   Continuously engage with relevant stakeholders.

To manage development pressures and improve resource allocation, smart cities need reliable information systems. Chapter 13 considered the residential valuation system which nests within the urban system but vests in both public sector institutions and, laterally, in banks and commercial real estate firms. To determine whether the Emirati residential valuation system (RVS) is 'fit for purpose', the research developed an assessment toolkit with five criteria: valuation output reasonableness, stakeholder information transparency, administrative capability, end-user trust and valuation standards salience. The research collected evidence, using embedded observations, transactions analysis, expert interviews and discussions. It found indicative weakness in aspects of the UAE residential valuation system, relating to information, capabilities and valuation standards.

Smart development of urban built environments must consider modifications to the existing building stock to realise potentially tremendous energy savings and emission reductions, which is examined in Chapter 14. In fact, the energy efficiency retrofitting of buildings is a crucial sustainability driver and a 'low-hanging fruit' for achieving climate and energy conservation goals on all regulatory levels. Apart from solvent, committed and savvy property investors, market, spatial and building factors contribute to the success of urban retrofitting programmes. At the market level, forecast energy prices and market reaction toward green retrofits influence scheme profitability. Fruitful milieus are densely populated and have a high share of rental housing. An energetic local authority also helps. At the building scale, age and typology influence retrofit outcomes. Curiously, despite its advantages, retrofit activity lags expectations. Information asymmetry or legacy legal/taxation peculiarities explain the paradox.

Smart cities provide all their citizens with decent homes. Increasingly, UK cities struggle to fulfil this basic need, which is examined in Chapter 15. Explanations vary but include vocal NIMBY activism, a maladapted planning system, high net immigration, demographic shifts and deteriorating working conditions. Notwithstanding the social and economic cost of the broken housing system, unscrupulous 'rogue' landlords thrived. If the private rented sector is to play a more sustainable role in solving the UK housing crisis, it needs to deliver balanced performance. Rental sector quality and energy efficiency standards must rise. However, although tenants need more security, regulated tenancies could exacerbate the housing supply situation. A smart solution invokes industry collaboration, tighter but not overburdensome tenancy regulations, more institutional investment, mortgage finance tweaking, tighter PRS lettings management, more active involvement of professional bodies and energy and tax incentives.

The smart city must preserve and conserve its heritage to cement place and cultural identity, and infuses development projects with an emotional authenticity that attracts visitors and enriches quality of life for locals. Drawing on the most iconic world city, Chapter 16 uses Istanbul to draw out some smart urban development policy implications. Due to rapid growth, the city's cultural heritage and particularly the Istanbul Sea Walls are at risk. To preserve the remaining wall artefacts calls for holistic and multidisciplinary management, vigilance and continuous maintenance and repair work overseen by a conservation team, comprised of masonry and building specialists. Helpful at a national scale are coherent policies and robust regional and urban planning systems. Istanbul's planning policies and practices should give greater weight to heritage protection; tighten public asset management, sharpen development application scrutiny and enforce planning regulations. Other aspects of heritage-informed (smart) development include public awareness and local participation. Tourism can help fund culture-led urban regeneration projects.

While by no means exhaustive, each chapter illustrates the multiple angles involved in smart regeneration. It cannot resolve political or other conundrums, because it involves ongoing dialogue, policy learning and 'muddling through'. It blends astute processual strategy, intelligently regulated markets, judicious catalytic interventions and evolution in collaborative networks. Its realisation involves practical wisdom and partnerships between robust political, commercial and social institutions to select, administer and deliver substantial low-carbon investments or tackle multiple issues at different spatial scales in complex fragmented societies. To conserve or appropriately transform built environments, system constituents need foresight, ambition, expertise, deliberation, adaptable systems, trust and standards, nurtured by sound policy and continuous learning. As well as the vertical political system with its administrative tiers, the network encompasses horizontal commercial, non-government institutions and bottom-up feedback from local bodies or citizens, empowered by digital technology and land information embedded in local knowledge with a deep respect for place, cultural identity and historical roots.

INDEX